Computer-Aided Structural Analysis

Computer-Aided Structural Analysis offers a novel and comprehensive approach to teaching advanced structural analysis using computer programming, specifically through MATLAB®. By focusing on developing fundamental programming skills, this book encourages learners to move beyond the "black box" mentality of commercial software. The inclusion of real-world examples, clear explanations, and practical tutorials ensures students gain not only theoretical knowledge but also the confidence to design their own programs for specialized applications. Additionally, the book promotes enhanced accuracy, speed, and problem-solving ability, making it an invaluable resource for both students and practicing engineers. It equips readers with the critical skills needed to excel in the modern era of AI-driven infrastructure projects and complex engineering challenges. It serves as a textbook for courses in engineering disciplines such as Ocean Engineering, Civil Engineering, Structural Engineering, Applied Mechanics, and Aerospace Engineering, and provides students with a deep understanding of the essential principles behind structural analysis and the application of computer-aided tools.

- Offers clear explanations, real-world examples, tutorials, and MATLAB® code in each chapter, enabling students to apply theoretical concepts to practical scenarios.
- Includes an Instructor's Guide and customized MATLAB® files for adopting professors.
- Bridges the gap between academic learning and real-world applications, helping students develop skills that are directly applicable to current industry demands and infrastructure projects.

Designed cover image: Shutterstock

First edition published 2026
by CRC Press
2385 NW Executive Center Drive, Suite 320, Boca Raton FL 33431

and by CRC Press
4 Park Square, Milton Park, Abingdon, Oxon, OX14 4RN

CRC Press is an imprint of Taylor & Francis Group, LLC

ISBN: 978-1-041-07159-4 (hbk)
ISBN: 978-1-041-07160-0 (pbk)
ISBN: 978-1-003-63915-2 (ebk)

DOI: 10.1201/9781003639152

Typeset in Times
by KnowledgeWorks Global Ltd.

Access the Instructor and Student Resources/Support Material: routledge.com/9781041071594

Computer-Aided Structural Analysis

Integration Techniques and Case Studies

Srinivasan Chandrasekaran

CRC Press
Taylor & Francis Group
Boca Raton London New York

CRC Press is an imprint of the
Taylor & Francis Group, an **informa** business

Contents

Author Biography vii
Preface ix

Chapter 1 Structural Form and Numerical Simulation 1

Chapter 2 Planar Orthogonal Structures 30

Chapter 3 Planar Non-Orthogonal Structures 85

Chapter 4 Planar Truss Structures 129

Chapter 5 Three-Dimensional Analysis of Space Frames 160

Chapter 6 Unsymmetric Bending 207

Chapter 7 Stability of Structural Systems 242

References 313

Index 321

Author Biography

Srinivasan Chandrasekaran is currently a Professor with Higher Academic Grade in the Department of Ocean Engineering, Indian Institute of Technology Madras. He is also a Visiting Professor in the Department of Structural Engineering, University of Naples Federico II, Italy, and Visiting Professor in the Faculty of International Maritime Studies, Kasetsart University, Sriracha Campus, Thailand. He holds the prestigious Fellowship conferred by the Ministry of Italian University Research (MiUR Fellow). He commands a rich experience in teaching, research, and industrial consultancy of about 31 years. He has supervised many sponsored research projects and offshore consultancy assignments, both in India and abroad. His active research areas include dynamic analysis and design of offshore structures, structural health monitoring of ocean structures, risk and reliability, fire-resistant design of structures, use of functionally graded materials (FGM) in marine risers, and health, safety, and environmental (HSE) management in process industries. He has authored about 210 research papers in peer-reviewed international journals and refereed conferences organized by professional societies worldwide. He has authored 23 textbooks published by various publishers of international repute. He is an active member of several professional bodies and societies, both in India and abroad. He has also conducted about 35 distance-education programs consisting of about 60 video lectures in each of them on various engineering subjects for the National Program on Technology-Enhanced Learning (NPTEL), Government of India. Through his impressive academic and research records with a Google Scholar h-index of 28 and i-10 index of 80, he is placed in the top 2% of world-leading scientists. Chandrasekaran has been recognized for his excellence, receiving the Best Reviewer Award from the Korean Society of Steel Structures and the Best Teacher Award from IIT Madras in 2022. He is a vibrant speaker and has delivered many keynote addresses at international conferences, workshops, and seminars organized in India and abroad. He holds a life membership of many national and international professional societies.

Preface

The book titled *Computer-Aided Structural Analysis* is a comprehensive tool for understanding classical structural analysis using high-end computational tools. The book's contents are tailored to aid the analysis using the MATLAB program, which is developed exclusively to solve the numeric problems described in the book. Civil engineering structures are unique in the field of engineering, as they pose challenges in the development and conceptualization of the design. In particular, there are exceptional designs whose geometric forms pose challenges to the designers. As innovative structural forms are envisaged, detailed analysis using computer tools is inevitable. Chapters in the book are organized such that the reader gets an overall idea of computer-aided analysis of various types of structural forms.

Chapter 1 describes various structural forms and equivalent numerical models to represent their support conditions, loading plane, and deformation behavior. Bio-inspired forms and the approach to simulate their numeric models help learners understand the process of drawing board to design table. To facilitate the learners, the fundamentals of the MATLAB program are also included. Chapters 2 and 3 present computer methods of analysis of orthogonal frames, beams, and non-orthogonal plane frames in detail. Numeric examples are solved using MATLAB codes, while preparing input to run the code is explained step-by-step. While Planar truss analysis is dealt with in detail in Chapter 4, 3D space frame analysis is discussed in Chapter 5. The unique approach used to solve 3D space frames is very useful, and the learners will find it quite interesting and simple. The objective is to make teaching this course very simple and comfortable without indulging in complex equations and iterative solutions. Unsymmetric bending, presented in Chapter 6, is quite comprehensive and useful for designers to handle curved beams and crane hooks analysis. The stability of structures using the stiffness approach through the MATLAB program is a unique addition to this book, which is quite novel and simple. High-end computations for stability equations are handled more simply through design charts and tables, enabling quick and correct solutions to such problems using MATLAB program codes.

Detailed descriptions of the fundamentals are explained in a classroom style, which will make the readers more user-friendly and comfortable to understand. Basic concepts are emphasized through simple illustrative examples and exercises. MATLAB program shall enrich the readers and improve their confidence level in computer-aided analysis. Contents and order of presentation are the outcome of the rich teaching experience the author commands in the subject domain. The book is also recommended for practicing and consulting structural engineers who require revisiting the basic concepts in a simplified form.

The author thanks the Centre for Continuing Education, Indian Institute of Technology Madras, India, for the administrative support in publishing this book. Further, the author also thanks MathWorks, Inc., USA, for enabling the use of MATLAB program throughout this book as a tool for problem-solving.

Srinivasan Chandrasekaran
10 March 2025

1 Structural Form and Numerical Simulation

1.1 STRUCTURAL FORMS AND FUNCTIONS

Structures are constructed to meet various functional requirements (Aslam, 2009). They are not constructed in solitude but cater to demands raised by social, economic, and political frameworks. Buildings must be able to overcome limits in natural and constructed surroundings while capitalizing on the development potential. As a result, when designing and erecting a structure, carefully considering the contextual factors on the site is critical. Beyond environmental factors, regulatory frameworks such as zoning rules influence how and where buildings are built. These restrictions consider existing land-use practices, specify permissible uses for the site, and set limits on building size, shape, and placement. Site circumstances influence building design, construction material, construction type and method, and maintenance, which places demands on transportation networks, utilities, and other infrastructure services. Calculating how much development a place can support without overloading these systems, overconsuming energy, or creating environmental damage is a significant difficulty. In this context, sustainability becomes a prominent element, necessitating a careful balance between development and the environment, ensuring that structures suit human requirements while reducing their environmental impact. Under special loads such as fire, blast, and explosions, the performance of buildings needs to be assessed critically (Chandrasekaran, 2024; Chandrasekaran and Pradeepa, 2023; Chandrasekaran and Nagavinothini, 2018, 2020a, 2020b; Chandrasekaran and Srivastava, 2017, 2022; Chandrasekaran and Suja, 2024; Chandrasekaran and Uddin, 2020; Chandrasekaran et al., 2021). Furthermore, it is also vital to formulate the problem for analysis from the fundamental principles, as the existing software tools may not suffice for advanced problem formulation (Chandrasekaran, 2013, 2014, 2015, 2016, 2017, 2018, 2019a, 2019b).

1.1.1 Geometric Forms

The concept of geometric forms conceived by architects delves into architecture's creative and functional expressions, where form and structure intertwine to bring buildings to life. A building's geometric form is not merely an aesthetic choice but a response to numerous factors such as functionality, material properties, environmental context, and technological advances. Over the centuries, architects have developed and conceived various structural forms that have transformed how we interact with spaces, addressing the needs of their time while pushing the boundaries of design and engineering. In counteracting complicated excitation loads (and their combinations), geometric form plays an important role. It is on this basis that offshore and large floating structures

DOI: 10.1201/9781003639152-1

are conceived. Instead of design focused on the material strength (or capacity), static redundancy in the geometric form helps build mode capacity (Chandrasekaran et al., 2023; Chandrasekaran and Lognath, 2015). It shifts the design focus from a strength-based concept to a form-dominant. For example, floating offshore platforms are seen as futuristic investments for space applications to launch rockets (Ashish et al., 2024; Chandrasekaran et al., 2023). While various researchers are examining conventional energy harvest designs for wave energy, conceived from a form-dominant approach, Basanagouda et al. (2024a, 2024b), Chandrasekaran and Sricharan (2018, 2019, 2020, 2021), Chandrasekaran and Raghavi (2015), Chandrasekaran et al. (2014), Chandrasekaran and Deepak (2013), Chandrasekaran et al. (2014), Chandrasekaran et al. (2022; Chandrasekaran et al., 2012; Chandrasekaran et al., 2013; Chandrasekaran and Harinder, 2010, 2011, 2012) showed novel design of wave energy harvest using bio-inspired floating devices. As the intuition to conceive geometric forms is growing limitless, a strong ideology to model them numerically is also required. While dynamic analysis becomes very vital for form-dominant designed structures, matrix methods of analysis become the most preferred method (Amin Gohli, 2017; Chandrasekaran, 2014, 2015, 2015, 2017, 2018, 2024; Chandrasekaran et al., 2022a, 2022b; Chandrasekaran and Jamshed, 2017; Chandrasekaran and Shivam, 2007; Chandrasekaran and Varun, 2007a; 2007b; Chandrasekaran et al., 2006; Chandrasekaran et al., 2024). Smart technologies for sustainable development follow innovative geometric forms that are chosen to disburse the applied loads without a strong dependence on the strength of materials (Chandrasekaran et al., 2013, 2014, 2020; Chandrasekaran et al., 2024; Chandrasekaran, Hari et al., 2019; Chandrasekaran, Jain, 2013; Chandraekaran and Madhavi, 2021; Chandrasekaran, Serino et al., 2024; Chandrasekaran et al., 2024; Chapman, 2015; Kassundi, 2009; Lopez, 2019; Pradeepa and Chandrasekaran, 2024; Siauw and Bayen, 2014).

In India, the rich diversity of landscapes—encompassing mountains, plains, coastal areas, deserts, and even tropical forests—demands various architectural approaches, structural systems, and foundation techniques. This geographic diversity, climatic variations, cultural influences, and traditional knowledge have shaped a unique architectural heritage across the country. Each region's architectural styles and construction techniques are inherently linked to the landscape, addressing local environmental challenges while utilizing region-specific materials.

1.1.2 Modern and Contemporary Forms

The evolution of structural forms in architecture reflects advancements in materials and construction techniques and changes in cultural, environmental, and functional needs (Chandrasekaran et al., 2024; Chandrasekaran and Thennavan, 2024). From ancient pyramids to modern parametric designs, architects continue to innovate, balancing the aesthetics of form with structure demands. Each era has its defining structural forms, shaped by the technology and ideas of the time, and the future of architecture will likely see even more complex and sustainable forms emerge through the use of digital design tools and new materials. The design of tall structures has seen significant advancements over the years, driven by the need to create efficient and resilient buildings in the face of increasing urbanization and environmental challenges. Modern engineering has introduced innovative structural systems, allowing taller buildings and improving their stability, performance, and sustainability. Critical structural systems such as tube-in-tube, outrigger, and bio-inspired structures have emerged as leading solutions for constructing skyscrapers and high-rise buildings. Quite often, these structural forms are also victims of vertical and horizontal asymmetricity, which makes the analysis and design more intensive (Chandrasekaran, 2014, 2017; Chandrasekaran and Utpal, 2004; Chandrasekaran et al., 2005, 2006, 2007, 2021).

Tube-in-tube structures enhance lateral load resistance by combining an inner tube, typically formed by the core of the building, with an outer tube made up of the exterior framework (Chandrasekaran et al., 2008, 2011). This dual-system approach provides stability against wind and seismic forces while optimizing material usage. On the other hand, Outrigger structures utilize horizontal beams that link the central core to the exterior columns, distributing the lateral forces and reducing the building's sway, making it possible to construct taller, slender buildings. Additionally, the field of architecture has increasingly turned to bio-inspired structures, which draw inspiration from natural forms and systems to achieve both efficiency and adaptability. These designs often mimic the load distribution and flexibility seen in nature, resulting in buildings that are more energy-efficient, sustainable, and resilient to environmental stresses.

Together, these structural systems reflect the evolution of engineering and architecture in response to the growing demands for taller, more efficient, and resilient buildings in modern cities, blending form, function, and sustainability in ground-breaking ways.

1.1.2.1 Tube-in-Tube Structures

Tube-in-tube structures are an efficient solution for tall buildings (Figure 1.1), where the exterior and interior are primary structural elements to resist lateral forces like

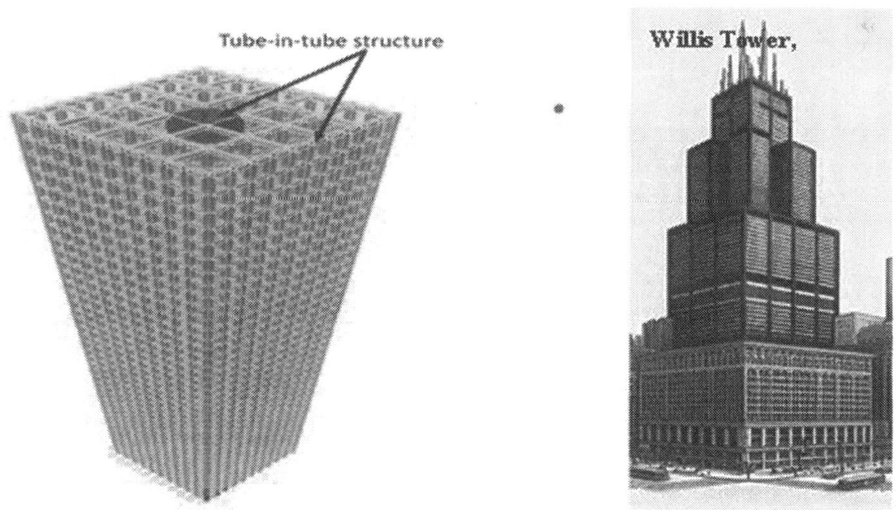

FIGURE 1.1 Tube-in-tube tall structure.

wind and seismic activities. This design maximizes the building's strength while minimizing material usage, making it highly efficient for skyscrapers.

Design Overview: The tube-in-tube structure consists of an **outer tube** formed by closely spaced exterior columns and beams and an **inner tube** created by the elevator core or shear walls. Both tubes work together to resist lateral loads and distribute them evenly throughout the building.

Lateral Load Resistance: The exterior tube takes on most of the wind forces, while the inner tube supports the building against additional lateral loads such as seismic activity. The dual system allows for a more robust, stiffer structure.

Efficiency and Cost-Effectiveness: Because the load is distributed between two systems, less material is needed than a conventional solid core structure, making it a cost-effective solution for tall buildings.

Applications: Tube-in-tube designs are widely used in high-rise office buildings and residential towers, including iconic structures like the **Sears Tower (now Willis Tower)** in Chicago.

1.1.2.2 Outrigger Structures

Outrigger structures represent another advancement in the design of tall buildings (Figure 1.2), where horizontal beams or trusses—called outriggers—are connected between the central core and the exterior columns. This system enhances the building's resistance to wind and seismic forces by extending the load-bearing capacity outward.

- **Design Overview**: In an outrigger structure, outriggers are horizontal structural elements that extend from the central core of the building to the perimeter columns, essentially linking the core with the outer frame. These outriggers act like rigid arms that prevent the core from bending under lateral loads.
- **Lateral Stiffness**: The outriggers significantly increase the lateral stiffness of the building by engaging the perimeter columns, which absorb a portion

FIGURE 1.2 View of outrigger structures.

of the wind or seismic forces. This reduces the bending and swaying of the central core, allowing for taller, slender structures.

- **Reduction of Drift**: One of the primary advantages of the outrigger system is that it reduces lateral drift—the movement of the building under wind loads—by distributing forces throughout the structure.
- **Applications**: Outrigger systems are standard in super-tall buildings, such as the Burj Khalifa in Dubai (Figure 1.2) and the Shanghai Tower in China, where the building height requires additional stabilization.

1.2 BIO-INSPIRED GEOMETRIC FORMS AND SIMULATION

Bio-inspired structures take inspiration from nature, applying biological principles and organic forms to architecture and engineering. These structures mimic natural systems' efficiency, adaptability, and resilience, leading to innovative and sustainable design solutions, especially for tall buildings (Jayashree et al., 2024). In particular, offshore structures that are designed for large displacements resist lateral loads mainly from their geometric form and member arrangement, not by the material strength (Chandrasekaran and Jamshed, 2017; Chandrasekaran and Kiran, 2018; Chandrasekaran and Lognath, 2015; Chandrasekaran and Muthuselvakumar, 2023; Chandrasekaran and Pachaiappan, 2023; Chandrasekaran and Rao, 2024; Chandrasekaran and Shyba, 2024). In fact, member cross-section plays a vital role in load disbursement (Chandrasekaran and Kiran, 2018; Chandrasekaran and Nagavinothini, 2020a; Chandrasekaran and Vinothini, 2020b, 2020c). This helps in designing the system with additional resistance offered by the geometric shape of the structure and the cross-section shape of the member (Chandrasekaran, 2020, 2024). In case of structures with large displacements, though the lateral loads are resisted by in-plane motion of the structure, their performance becomes critical under earthquake loads and unstable conditions (Chandrasekaran and Madhuri, 2012, 2013; Chandrasekaran et al., 2012; Chandrasekaran et al., 2019). However, novel materials are also used in the building construction to enhance the performance of the structure; such materials are functional-dominant and termed as functionally graded materials (FGM) (Chandrasekaran and Jain, 2016; Chandrasekaran and Thennavan, 2024; Chandrasekaran et al., 2013). Force reduction is also achieved by altering the structural form; one such example is by providing perforated cover around an offshore member (Chandrasekaran and Madhavi, 2014a, 2014b, 2014c, 2014d, 2015a, 2015b, 2015c, 2015d, 2015e, 2015f, 2016, Chandrasekaran et al., 2014, 2015). Highly complex forms that support offshore wind turbines are a classical group of structures under bio-inspired forms (Chandrasekaran and Purushotham, 2024).

- **Design Overview**: Bio-inspired structures often draw inspiration from natural forms, such as the exoskeletons of insects, the skeletons of trees, or the cellular structures of plants and animals. These natural systems demonstrate efficient load distribution, flexibility, and adaptability, which can be applied to modern architecture. Plastic analysis methods are employed to carry out fracture analysis of complicated structures (Chandrasekaran and Ajesh, 2017, 2019).

- **Energy Efficiency and Sustainability**: Bio-inspired structures are often designed to be more energy-efficient by incorporating passive ventilation, heat regulation, and lightweight yet strong materials. Nature-inspired forms also reduce material waste, making the construction process more sustainable. Offshore compliant platforms, for example, are classical referents for form-dominant, bio-inspired designs. Triceratops, inspired by the first-generation dinosaurs, are proven to be more efficient design for offshore platforms (Chandrasekaran and Madhuri, 2015; Chandrasekaran and Senger, 2017; Chandrasekaran and Yuvraj, 2013; Chandrasekaran et al., 2013, 2014, 2023). Semisubmersibles are equal representatives of bio-inspired design (Chandrasekaran and Syed, 2021).
- **Load Distribution and Resilience**: Like in nature, where structures evolve to carry loads and adapt to environmental stresses efficiently, bio-inspired designs are often used to enhance the building's ability to withstand lateral loads, such as wind or seismic activity. The natural forms, such as branching or cellular networks, allow for better load distribution throughout the structure.

A few examples of bio-inspired forms are shown in Figure 1.3. The Eden Project in the United Kingdom, which mimics the structure of soap bubbles for its geodesic dome; the Beijing National Stadium (Bird's Nest), whose lattice structure is inspired by a bird's nest; and Cybertecture Egg, Mumbai, India, are a few examples of how bioinspiration can be applied to architecture. In skyscrapers, bio-inspired designs may resemble trees with branching supports or the sponge-like skeletons of sea organisms that can handle tension and compression, allowing the structure to flex without breaking. Bio-inspired skyscrapers, like the Taipei 101 in Taiwan employ earthquake-resistant designs inspired by natural structures' flexibility and

FIGURE 1.3 Bio-inspired structural forms.

resilience. Other examples include towers inspired by termite mounds promoting natural ventilation and cooling.

1.3 STRUCTURAL ANALYSIS AND DESIGN

The most primitive step in the structural analysis is the problem formulation using an appropriate mathematical model. There are two models widely used in classical structural analysis, namely, (1) statically determinate model and (2) statically indeterminate model. Statically determinate models are relatively easier as they use only the basic equations of static equilibrium to solve the problem. Hence, one should look for ways to solve problems related to statically indeterminate models through computer methods (Beaufait et al., 1970; Bhatt, 1986; Chandrasekaran, 2024; Fleming, 1989; Forsythe and Moler, 1969; Hibbeler, 1984; Holzer, 1985; Igor and Olga, 2010; Jack McCormick, 2007). To solve the problems, it is important to formulate a standard procedure, which should be generic and not problem-specific. Thus, it is important to note that the models should be restrained to be solved only by using the standard equation of statics. This can be done by grouping the formation. Grouping is done by two methods, namely, (i) flexibility method and (2) stiffness method.

It is critical to monitor the performance and functional safety of the bio-inspired and form-dominant structures. Structural Health Monitoring (SHM), combined with artificial intelligence (AI) tools, makes the SHM process not only efficient but cost-effective as well (Chandrasekaran and Thailammai, 2016, 2018; Chandrasekaran et al., 2016). Both methods are equally powerful and useful; there is no supremacy of one method over the other. Any method can be used for grouping based on the user's convenience. However, the easily programmable method should be preferred, as the main objective of the book is to make the problems solvable by computer methods. There is a significant difference in identifying the unknowns for formulating the problem. In the flexibility method, the unknowns are actions such as shear force, axial force, and bending moment. On the contrary, the unknowns in the stiffness method are displacements such as translational and rotational displacements. The basic assumptions applicable to both methods are as follows:

1. Linear relationship exists between applied load and the resulting displacement of the structure. This makes the principle of superposition valid through the formulation.
2. The material of the structure must obey Hooke's Law, which says that the material must not be stressed beyond its elastic limit.
3. Equations of static equilibrium shall be developed using the geometry of an un-deflected model. The change in geometry caused by the imposed loads is negligible when compared to the original geometry.

1.3.1 Importance of Structural Analysis in Engineering Design

Structural analysis is the backbone of engineering design, ensuring structures can safely withstand applied loads, environmental forces, and the test of time. Whether the structure is a small residential building, a sprawling bridge, a towering skyscraper, or a dynamic offshore platform, accurate analysis is indispensable for its stability, safety, and functionality. Detailed analysis is essentially required to understand the structure's response under accidental loads such as fire and earthquakes (Chandrasekaran and Chandak, 2004; Chandrasekaran and Roy, 2006; Pradeepa and Chandrasekaran, 2024a, 2024b).

1.3.1.1 Buildings and Bridges

For buildings, structural analysis ensures that the load-bearing elements, such as beams, columns, and slabs, can support dead loads (self-weight), live loads (occupants, furniture), wind, and seismic forces. The analysis is critical for high-rise buildings, where lateral loads caused by wind and earthquakes are significant. Similarly, in bridges—whether suspension, cable-stayed, or arch types—structural analysis determines how forces are transferred through the members, ensuring durability while minimizing material usage. For example, load distribution in trusses and frames is meticulously analyzed to balance strength and economy.

1.3.1.2 Ships and Offshore Structures

Due to dynamic loading from waves, currents, and wind, structural analysis becomes even more intricate in maritime and offshore engineering. Ships must maintain their integrity while navigating through turbulent seas, and offshore platforms (e.g., oil rigs) are subject to constant hydrodynamic forces, requiring precise modeling to ensure resilience. Space frames and 3D beam systems are pivotal in achieving stability for these complex structures.

By connecting fundamental structural analysis principles with practical engineering marvels like skyscrapers, bridges, and offshore systems, learners can better appreciate the real-world significance of the subject. Environmental loads acting on ships and offshore structures belong to a special classification as they induce critical responses (Chandrasekaran et al., 2011; Chandrasekaran and Srivastava, 2017, 2022). Offshore wind turbines play a novel role in energy harvesting and the blue energy revolution. Structures supporting such harvesting systems are highly intelligent and self-resilient (Chandrasekaran et al., 2009, 2024).

1.3.2 Fixed Beams: Idealization Across Structures

The **fixed beam**, a fundamental structural element, is an idealized model in numerous applications, abstracted from various real-world structures. Fixed beams are characterized by their ends being restrained against rotation, resulting in zero slope at the supports. This property allows them to carry higher moments than supported beams, making them a critical element in many engineering applications.

1. **Buildings**: In multistory buildings (Figure 1.4a), continuous beams are fixed to columns to resist bending moments and lateral forces. For example, reinforced concrete beams in high-rise structures are often modeled as fixed beams (Figure 1.4b and c) to minimize deflection and improve load distribution.
2. **Bridges**: Fixed beams form integral parts of deck systems in short-span bridges (Figure 1.5c). When analyzed, their rigidity reduces vertical displacements

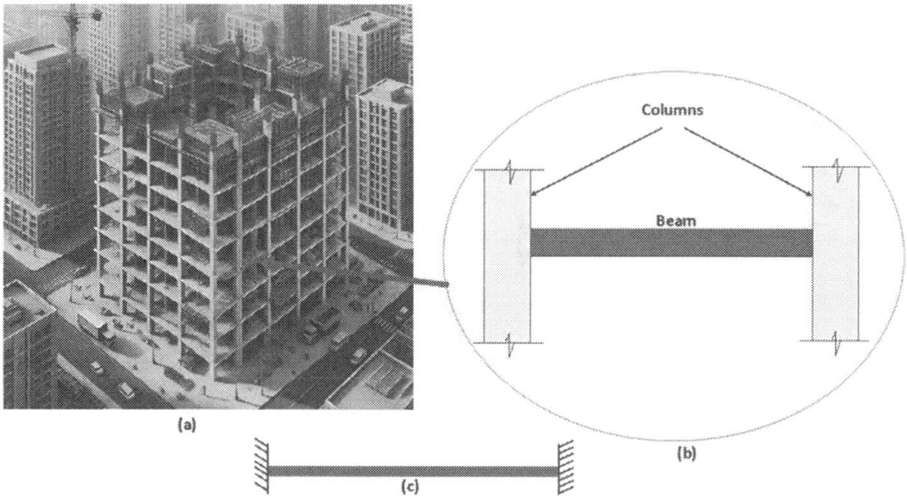

FIGURE 1.4 (a) Reinforced concrete building with various structural elements; (b) enlarged view of a selected beam column arrangement; and (c) idealized fixed beam.

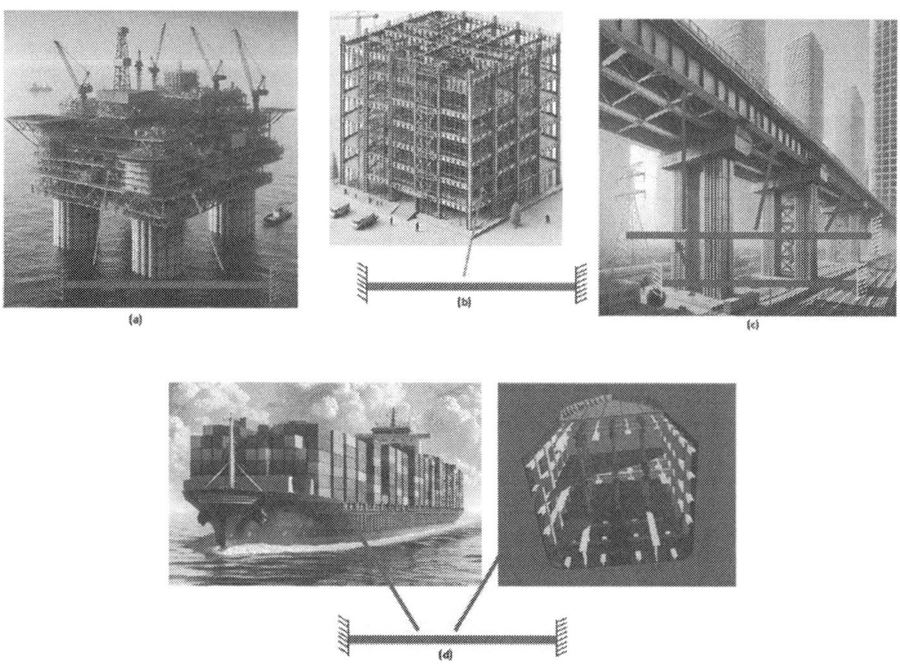

FIGURE 1.5 (a) Offshore platform; (b) steel structure; (c) bridge; and (d) cargo vessel with idealized fixed beams.

and enhances the structure's capacity to carry vehicular loads. Fixed boundary conditions help designers minimize mid-span bending moments.

3. **Ships and Offshore Structures**: Structural members like girders and stiffeners are modeled as fixed beams to withstand bending caused by wave forces. Offshore platforms and ships incorporate fixed beams (Figure 1.5a and d) in their jacket or topside structures to provide resistance against environmental loading.
4. **Frames and 3D Space Structures**: In advanced systems such as space frames (Figure 1.5b), fixed beams are idealized to connect nodes rigidly, ensuring that forces and moments are transferred seamlessly across the structure. Such systems are widely used in bridges, aircraft hangars, and long-span roofs.

Readers can better understand their significance in designing efficient and resilient structures by examining how fixed beams are idealized from such diverse structural systems. This linkage reinforces the practicality of structural analysis, bridging theoretical concepts with applications in real-world engineering.

1.4 INDETERMINACY

Both flexibility and stiffness methods circumscribe the problem formulations around the term indeterminacy. It is important to understand indeterminacy in terms of problem formulation by either flexibility or stiffness method. There are two types of indeterminacy, namely, static and kinematic indeterminacy.

1.4.1 Static Indeterminacy

Static indeterminacy is related to flexibility approach and defined as the number of actions that must be released to transform the structural system into a stable statically determinate system. For example, the actions that should be released are shear force, axial force, and bending moment, which can be either external or internal. Thus, the objective is to convert the known structural system into a statically determinate and stable system, for which the number of actions has to be identified. **Degree of static indeterminacy** is defined as the number of released actions, which specifies the number of special independent equations that must be developed in terms of the released actions to analyze the system. So, the approach used in the formulation and solution is *flexibility approach.*

1.4.2 Kinematic Indeterminacy

Kinematic indeterminacy is related to stiffness approach and referred to as the number of independent components of joint displacements, required to describe the response of the system under any arbitrary load, which is required to describe response of the system under any arbitrary load. It could be either internal or external. It, therefore, deals only with unrestrained displacements at the supports and joints of a structural system. Kinematic indeterminacy is useful for stiffness method of problem solver as it converges to a generic solution. It is important to note that the structure must be restrained to convert or transform the system into kinematically determinate structure. Structure with all joint displacements restrained is the formulation. *Degree of kinematic indeterminacy* is defined as the number of unrestrained components of the joint displacements (both rotational and translational). It is highly important to know that the degree of kinematic indeterminacy specifies the number of independent equations that must be written in term of unrestrained displacements, if the system is to be analyzed using stiffness approach.

1.4.3 Degrees of Freedom

There are two degrees of freedom (DOF), namely, static and kinematic. They are explained as follows.

1.4.3.1 Static Degrees of Freedom

While structural systems can be derived from any geometric form, the basic structural system is referred to as a statically determinate model. Structural members such as beams can be supported at one or more sections. In structural analysis, understanding static DOF is essential for determining the stability and equilibrium of a structure. A beam with only one end supported is referred to as a cantilever (examples are sunshades and chajjas). A beam with two supports is referred to as a simple supported beam and more than two supports are termed as a continuous beam. Static DOF is the number of independent reactions offered by the supports. There are three types of supports, namely, roller support, hinged support, and fixed support (Figure 1.6).

As seen in the figure, while roller support offers a reaction only in one direction (vertical), hinged support offers a reaction in two directions (both horizontal and vertical). A fixed support offers reactions in two directions as same as a hinged support; in addition, it also offers a moment constraint. Hence, if one has to classify the above mathematically, the following conditions hold good:

No. of reactions (unknowns) of a roller support = 1 (V_y)
No. of reactions (unknowns) of hinged support = 2 (H_x, V_y)
No. of reactions (unknowns) of a fixed support = 3 (H_x, V_y, M_z)

As it is understood that a structure can possess more than one support, the total number of static DOF is the sum of the unknowns offered by each support in a

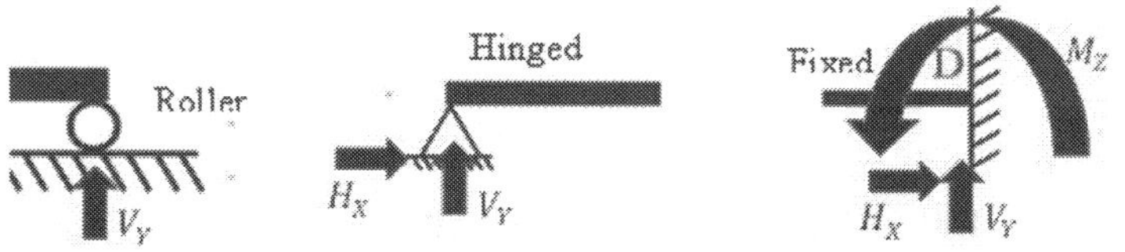

FIGURE 1.6 Types of supports and reactions.

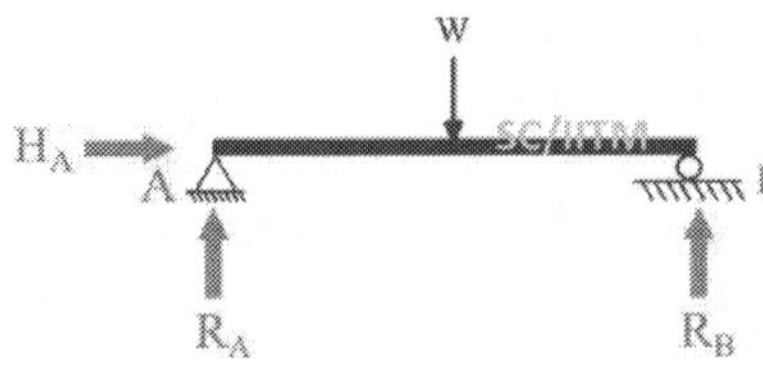

FIGURE 1.7 Simply supported beam.

structure. For a 2D analysis, the number of equations of static equilibrium are as follows:

- The algebraic sum of all forces acting along the X-axis is set to ZERO: $\sum H_X = 0$
- The algebraic sum of all forces acting along the Y-axis is set to ZERO: $\sum V_Y = 0$
- Algebraic sum of moments of forces about any section, about Z-axis is set to ZERO: $\sum M_Z = 0$

The static degree of indeterminacy is the difference between the static DOF and the number of static equations of equilibrium. For example, let us consider a simply supported beam shown in Figure 1.7. The total number of reactions offered by two supports (A, B) is three, as one end is hinged and the other is on roller. As the total number of static equations of equilibrium is also three, the static degree of indeterminacy is ZERO. Thus, a simply supported beam is said to be statically determinate. In case the number of reactions is more than the number of equations of static equilibrium, then the member is said to be statically indeterminate. The degree of static indeterminacy can vary from one to infinity.

Let us consider an example of a continuous beam as shown in Figure 1.8. The total number of supports is four (A, B, C, D). While the support is marked as a hinged support, the rest are on rollers. As explained earlier, roller support has one reaction marked as R_B, R_C, R_D. The hinged support has two reactions, but only one is marked in the figure (R_A). It is important to note the assumption made in this

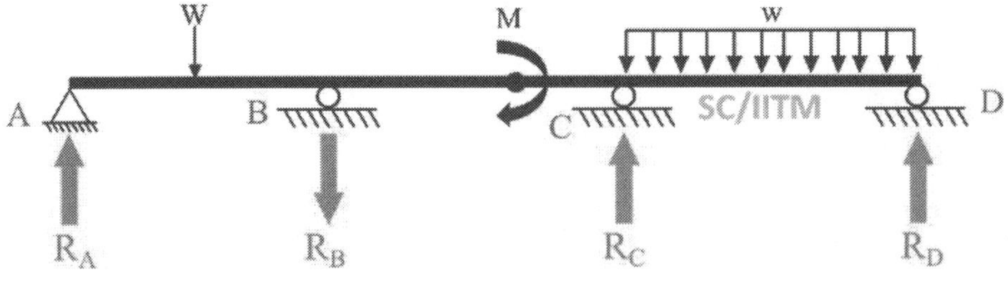

FIGURE 1.8 Continuous beam.

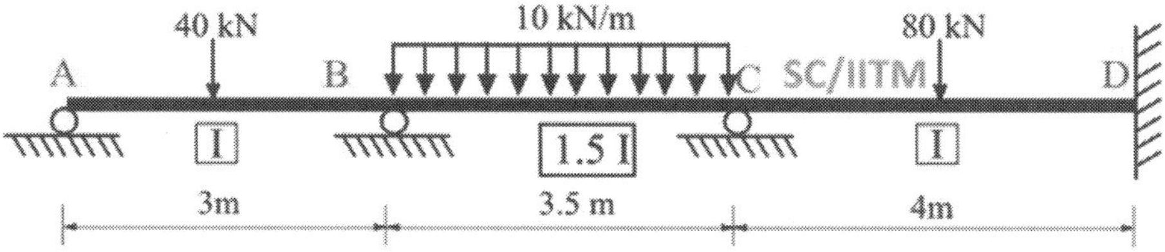

FIGURE 1.9 Continuous beam with a fixed support.

case: axial deformations are neglected. Hence, no horizontal reaction at support is marked. It is also wise to correlate the valid number of equations of static equilibrium. Instead of three, it is only two as $\sum H_X = 0$ is not applicable in this case. Hence, the static degree of freedom of the continuous beam is four (as there are only four reactions, offered by the supports). The static degree of indeterminacy is two (4 – 2). The beam is statically indeterminate to the order of TWO.

Let us now look at one more example, shown in Figure 1.9, which is also a continuous beam. The total number of supports is five (one at each support A, B, C, and two at D). Please note the assumption made in the analysis as axial deformation was neglected and hence the support D has only two reactions (R_D, M_D). For the equations of static equilibrium, the support is only two since $\sum H_X = 0$ is not applicable in this case, the static degree of freedom of the continuous beam is five. The static degree of indeterminacy is three (5 – 2). The beam is statically indeterminate to the order of THREE.

1.4.3.2 Kinematic Degrees of Freedom

Kinematic DOF are related to the displacement capabilities of a structure without considering the forces. Kinematic analysis helps in understanding the possible displacements of a structure under the applied loads. It is important to note that displacements are also measured only at the supports as reactions. There are two types of displacements possible in a joint/support. They are rotational and translational. Further, translational displacements can be along the X-axis and Y-axis, while rotational displacements are about Z-axis only.

Similar to the static degree of freedom, it is crucial to estimate the kinematic DOF of a structural member. It helps assess the ability of a structure to identify the potential mechanisms or modes of failure. For instance, a planar truss can be analyzed for its kinematic DOF to ensure that it remains stable and does not collapse under load. The kinematic analysis involves checking the geometric configuration and connections of the structure to ensure that it is rigid and stable. Similar to the static DOF at each support, one can also associate possible kinematic DOF with each support. Figure 1.10 shows the types of support and the displacements associated with each of them. While identifying the displacements, please note that rotations are about Z-axis while translations can be along X-axis and Y-axis, in a 2D analysis.

Table 1.1 helps understand the kinematic DOF for each support. In the figure, unrestrained (free to move) degrees are marked in green while restrained degrees

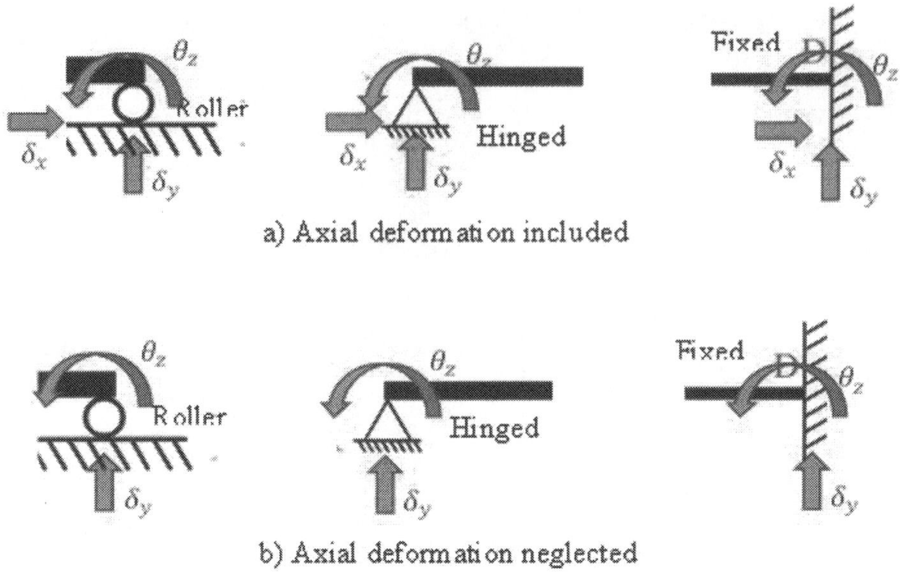

FIGURE 1.10 Types of supports and displacements.

are marked in red, for clarity. All sign conventions marked for the displacements and rotation are considered positive.

Let us consider an example of a continuous beam shown in Figure 1.11. There are four supports (A, B, C, D). While support D is a fixed support, rest are on rollers. With reference to Table 1.1, following observations can be made:

TABLE 1.1
Kinematic Degrees of Freedom at Each Support

	Degree of Freedom, if Axial Deformation Included		Degree of Freedom, if Axial Deformation Neglected	
Support	**Restrained**	**Unrestrained**	**Restrained**	**Unrestrained**
Roller	Translation δ_y	Rotation θ_z Translation δ_x	Translation δ_y	Rotation θ_z
Hinged	Translation δ_x Translation δ_y	Rotation θ_z	Translation δ_y	Rotation θ_z
Fixed	Translation δ_x Translation δ_y Rotation θ_z	Nil	Translation δ_y Rotation θ_z	Nil

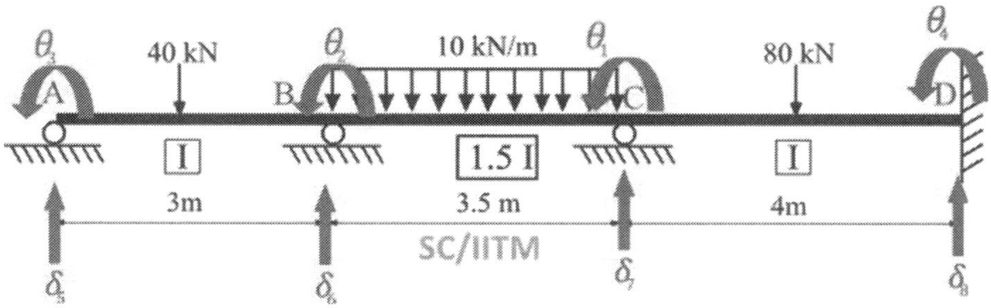

FIGURE 1.11 Continuous beam marked with kinematic degrees of freedom.

	Degree of Freedom, if Axial Deformation Neglected	
Support	**Restrained**	**Unrestrained**
A	Translation δ_5	Rotation θ_3
B	Translation δ_6	Rotation θ_2
C	Translation δ_7	Rotation θ_1
D	Translation δ_8 Rotation θ_4	Nil

The degree of kinematic indeterminacy is the total number of unrestrained kinematic DOF. As seen in the above case, kinematic degree of indeterminacy is three.

Example 1.1:

For the two span continuous beam shown below in Figure 1.12, estimate the static and kinematic degree of indeterminacy. Neglect axial deformation in the beam.

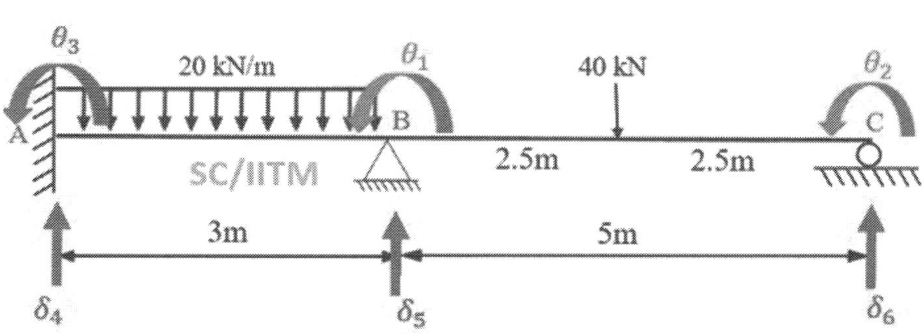

FIGURE 1.12 Two-span continuous beam.

Support	Static Degree of Freedom	Kinematic Degree of Freedom Axial Deformation Neglected	
		Restrained	Unrestrained
A	2 (VA, MA)	Rotation θ_3 Translation δ_4	nil
B	1 (VB)	Translation δ_5	Rotation θ_1
C	1 (VC)	Translation δ_6	Rotation θ_2
Degree of indeterminacy	Static: 4 – 2 = 2	Kinematic: 2 (only the sum of unrestrained degrees of freedom)	

Example 1.2:

For the frame in Figure 1.13, estimate the kinematic degree of indeterminacy.

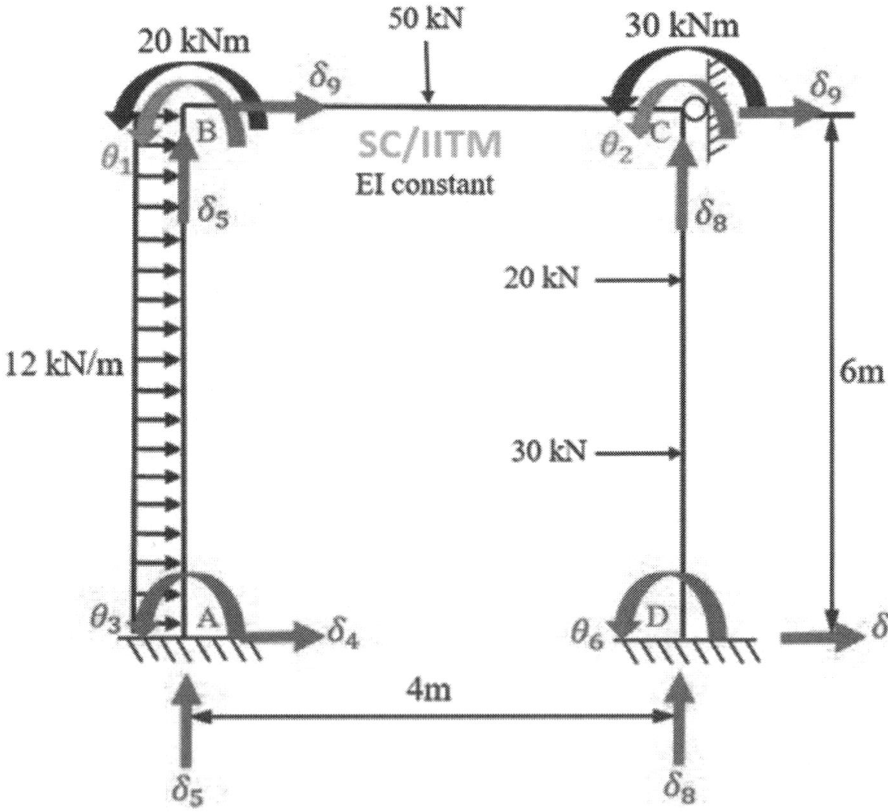

FIGURE 1.13 Frame.

	Kinematic Degree of Freedom Axial Deformation Neglected	
Support	**Restrained**	**Unrestrained**
A	Rotation θ_3 Translation δ_4 Translation δ_5	nil
B	Translation δ_5 Translation δ_9	Rotation θ_1
C	Translation δ_8 Translation δ_9	Rotation θ_2
D	Rotation θ_6 Translation δ_7 Translation δ_8	Nil
Degree of indeterminacy	Kinematic: 2 (only the sum of unrestrained degrees of freedom)	

Example 1.3:

For the stepped frame shown below in Figure 1.14, determine the kinematic degree of indeterminacy.

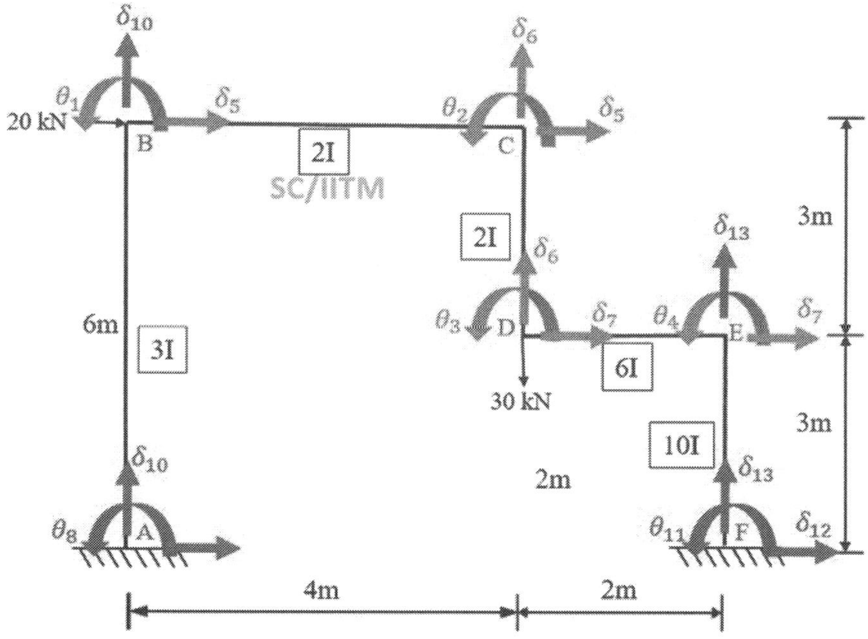

FIGURE 1.14 Stepped frame.

	Kinematic Degree of Freedom Axial Deformation Neglected	
Support	**Restrained**	**Unrestrained**
A	Rotation θ_8 Translation δ_9 Translation δ_{10}	nil
B	Translation δ_{10}	Rotation θ_1 Translation δ_5
C	Nil	Rotation θ_2 Translation δ_5 Translation δ_6
D	Nil	Rotation θ_3 Translation δ_7 Translation δ_6
E	Translation δ_{13}	Rotation θ_4 Translation δ_7
F	Rotation θ_{11} Translation δ_{12} Translation δ_{13}	Nil
Degree of indeterminacy	Kinematic: 10 (only the sum of unrestrained degrees of freedom)	

In structural analysis, it is essential to use consistent units of measurement. Common units include length (in meters or millimeters), force (in newtons (N), kN, MN, etc.), and stress (in Pascals (Pa), kPa, MPa).

Static DOF is related to release of actions like shear force, bending moment, axial force, etc. It is associated to flexibility approach. Kinematic DOF is related to displacements. It is associated to stiffness approach. The DOF mentioned in the definition of stiffness is related to displacement and hence kinematic DOF. In the above, there are four DOF (two rotations and two translations). One should give unit displacement in each DOF to find the forces at the respective degrees by keeping the remaining DOF restrained. k_{ij} is also defined as moment in ith DOF by imposing unit rotation at jth DOF by keeping all other DOF restrained. Imposing unit displacement represents $\delta_r = 1$ or $\delta_s = 1$ and unit rotation implies $\theta_p = 1$ or $\theta_q = 1$.

1.5 METHODS OF ANALYSIS

Structural analysis determines the internal forces (moments, shear, and axial forces), stresses, and deformations in structures under applied loads. Several methods exist to perform structural analysis, primarily categorized into **classical** and **matrix methods**. Two important techniques are the **stiffness and flexibility methods**, which are matrix-based approaches widely used in modern structural analysis (Chandrasekaran and Kiran, 2018; Chandrasekaran and Lognath, 2017; Chandrasekaran and Madhuri, 2015; Chandrasekaran and Sricharan, 2021; Chandrasekaran and Thailammai, 2018; Chandrasekaran and Thennavan, 2023; Chandrasekaran and Vinothini, 2020).

Both stiffness and flexibility methods form the foundation for solving indeterminate structures. They rely on different principles:

- **Stiffness Method**: Directly relates displacements to applied forces.
- **Flexibility Method**: Relates applied forces to resulting displacements.

These methods differ in computational approach and are suited to specific problems depending on the structure and complexity.

1.5.1 Stiffness Method

The stiffness method, also called as **displacement method**, is based on the principle of **force-displacement relationships**. It focuses on calculating displacements (translations and rotations) at specific points in the structure (usually nodes). It is ideal for structures with a higher degree of indeterminacy and is used in modern finite element analysis (FEA) software.

1. **Stiffness Matrix**:
    - Each structural element is represented by a stiffness matrix, which relates forces at the nodes to the displacements.
    - The global stiffness matrix is assembled by combining the stiffness matrices of all elements in the structure.
2. **Equation of Equilibrium**: $K{\cdot}D = F$
   where K is the global stiffness matrix, D is the displacement vector, and F is the force vector.
3. **Procedure**:
    - Formulate the global stiffness matrix.
    - Apply boundary conditions.
    - Solve for unknown displacements.
    - Back-calculate forces and moments using displacements.

1.5.2 Flexibility Method

The flexibility method, also called as **Force method**, is based on the principle of **compatibility of deformations**. It is focused on determining unknown forces (or moments) in the structure and calculating displacement. It is suitable for systems with fewer redundancies compared to unknown displacements.

1. **Flexibility Matrix**:
    - Each structural element is characterized by a flexibility matrix, which relates forces to displacements.
    - The global flexibility matrix combines the flexibility properties of all structural elements.
2. **Compatibility Equations**: $F{\cdot}Q = \Delta$
   where F is the flexibility matrix, Q is the unknown force vector, and Δ is the known displacements due to applied loads

3. **Procedure**:
   - Identify redundant forces or moments.
   - Formulate compatibility equations.
   - Solve for redundancy using the flexibility matrix.
   - Compute final reactions and internal forces.

1.5.3 Comparison and Applications

The differences between flexibility and stiffness methods are summarized as follows:

Flexibility Approach	Stiffness Approach
This deals with static indeterminacy.	This deals with kinematic indeterminacy.
The unknowns are actions such as shear force, axial force, bending moments, etc.	The unknowns are joint displacements such as rotational and translational displacements.
The problem formulation converts the structural system into a statically determinate structure.	The problem formulation converts the structural system into the kinematically determinate structure.

Thus, static and kinematic indeterminacies of any structural system are indicators of the amount or extent of computational efforts required to analyze the structural system either using the flexibility approach or stiffness approach. The structural analysis can be carried out by any of the two methods, which are equally useful and powerful numerically. The unknowns are released to convert the structural system into a statically or kinematically determinate system so that the standard equations can be used to solve the system under applied loads. If the problem formulation reduces the number of unknowns, then it is the best formulation attempted by a mathematician or an engineer. The number of unknowns in the system of equations purely depends on the choice of the method demanded. For the computer method of structural analysis, one should keep in mind that the method recommended should be more or less generic and not problem-specific.

Degree of static and kinematic indeterminacies of standard problems are given below:

1. *Continuous beam:*

   Consider a continuous beam of three spans with one hinged joint and three roller joints with reactions R_1, R_2, R_3, and R_4, as shown in Figure 1.15. The

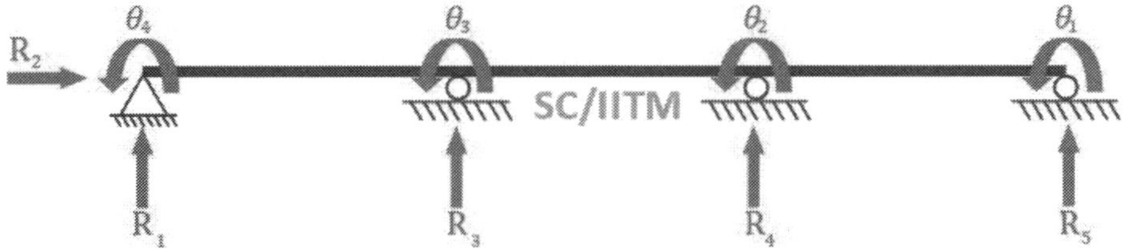

FIGURE 1.15 Continuous beam.

displacement unknowns, neglecting axial deformations are θ_1, θ_2, θ_3, and θ_4. We all know that there are three systems of standard equations available to solve the problem. The degree of static indeterminacy and the degree of kinematic indeterminacy are as follows:

Degree of static indeterminacy
= Number of unknown reactions – system of standard equations

$$= 5 - 3 = 2.$$

Degree of kinematic indeterminacy = Number of displacements

$$= 4$$

2. *Fixed beam:*

   Let us now consider a fixed beam with reactions R_1, R_2, and R_3 at the support A. Similarly, the reactions at support B are R_4, R_5, and R_6 as shown in Figure 1.16. The rotational displacements that are free to move are zero. Thus,

   Degree of static indeterminacy = 6 – 3 = 3
   Degree of kinematic indeterminacy = 0

3. *Simply supported beam:*

   Let us consider a simply supported beam with reaction components R_1, R_2, and R_3. The displacements are θ_1 and θ_2 as shown in Figure 1.17. Thus,

   Degree of static indeterminacy = 3 – 3 = 0
   Degree of kinematic indeterminacy = 2

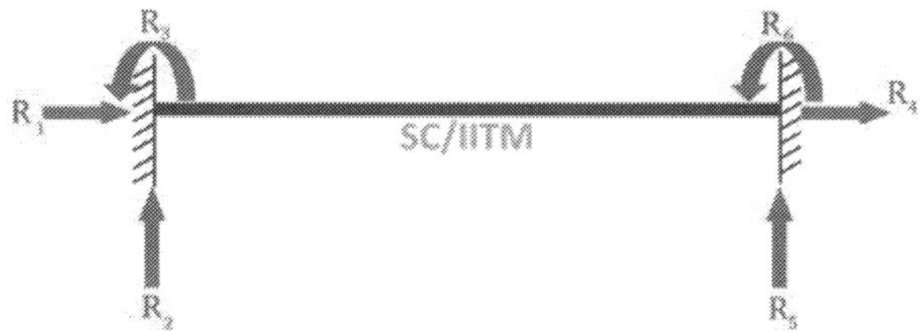

FIGURE 1.16 Fixed beam.

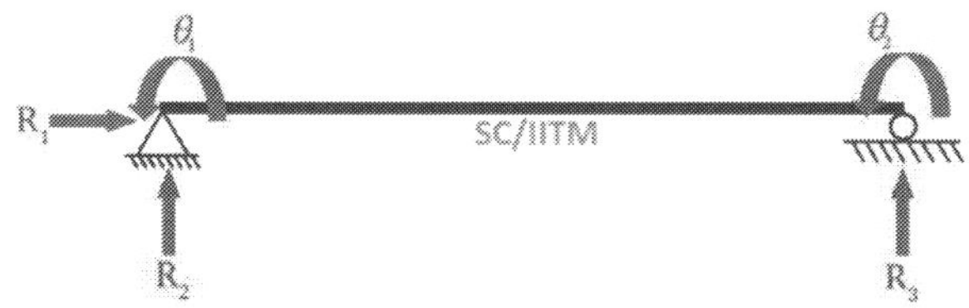

FIGURE 1.17 Simply supported beam.

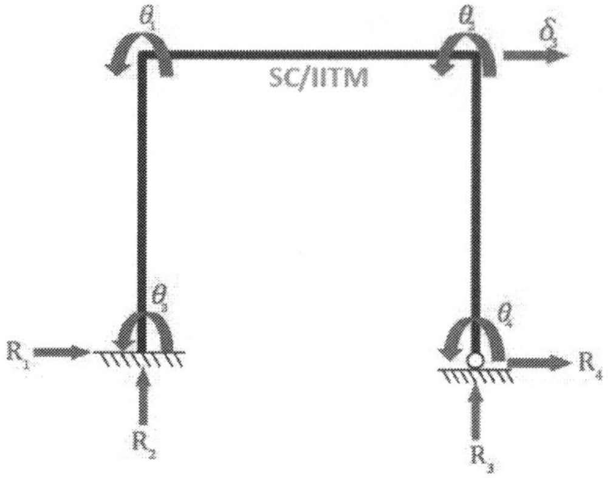

FIGURE 1.18 Single story single bay frame.

4. *Frame:*

    Let us consider a single-story bay frame with one end fixed and the other end on roller support as shown in Figure 1.18. The unknown reactions are R_1, R_2, R_3 and R_4. The independent displacements are θ_1, θ_2, θ_3, and θ_4 by neglecting axial deformation.

    Degree of static indeterminacy = 5 – 3 = 2
    Degree of kinematic indeterminacy = 4

From the above examples, it is seen that the number of unknowns is different depending upon the choice of the method. To summarize the choice of method of analysis that governs the system of equations, the following should be noteworthy:

1. There are essentially two methods to solve the statically indeterminate structures: flexibility method and stiffness method. So the choice of the method depends on the computational convenience.
2. For the flexibility method, there are several alternatives for redundant or unknowns. Thus, the choice of redundant has a significant effect on the computational effort.
3. On the other hand, there is no choice of unknown quantities in stiffness method because there is only one possible restrained structure. Therefore, this has a set of standard procedure.
4. Based on computer methods of structural analysis, one can say that the choice of the method should not be geometry specific. It should be more generic and repetitive in nature.

Thus, stiffness method is a better choice fulfilling all the above requirements (Chu Kia Wang, 1953; Edward Wilson, 2002; Kardestuncer, 1974; Kenneth et al., 2002, 2013, 2022; Livesley, 2013; Mau, 2012; McGuire and Gallagher, 1979; Meek, 1971; Meyers, 1983; Przemieniecki, 1985; Ramon, 2007; Reddy, 2011, Ronald Sack, 1994; Rubinstein, 1970; Siegfried Holzer, 1985; Smith David, 2010; Vanderbilt, 1974; Wang, 1970; William McKenzie, 2006; William and James, 2012; William et al., 2000). In this book, the stiffness method is used for elaborating the application procedures. After identifying the variables in a given system such as rotational or translational displacements for every joint, it will result in the formation of a set of linear equations.

1.6 MATRIX ALGEBRA: BASICS

1.6.1 Matix Operations

Let us express the matrix A as a system of equations,

$$x_1 + 5x_2 + 2x_3 = 2$$

$$4x_2 + x_3 = 5$$

$$2x_2 + x_3 = 4$$

In matrix form, these equations can be written as follows:

$$\begin{bmatrix} 1 & 5 & 2 \\ 0 & 4 & 1 \\ 0 & 2 & 1 \end{bmatrix} \begin{Bmatrix} x_1 \\ x_2 \\ x_3 \end{Bmatrix} = \begin{Bmatrix} 2 \\ 5 \\ 4 \end{Bmatrix}$$

$$\begin{Bmatrix} x_1 \\ x_2 \\ x_3 \end{Bmatrix} = \left[A\right]^{-1} \begin{Bmatrix} 2 \\ 5 \\ 4 \end{Bmatrix} = \begin{bmatrix} 1 & 1/2 & -3/2 \\ 0 & 1/2 & -1/2 \\ 0 & -1 & 2 \end{bmatrix} \begin{Bmatrix} 2 \\ 5 \\ 4 \end{Bmatrix} = \begin{Bmatrix} -6.5 \\ 0.5 \\ 3 \end{Bmatrix}$$

If one can generate system of equations with unknowns as variables, then these set of equations can be solved using matrix inversion.

$$\{x\} = \left[A\right]^{-1} \{B\}$$

This is true only when $[A]^{-1}$ exists. It should also be noted that $\{x\}$ purely depends on $\{B\}$ and $[A]^{-1}$ does not change to get the value of $\{x\}$. Assume matrix A is a stiffness matrix of a given system, B is a load vector, and x is a displacement vector. By this comparison, it can be seen that the value of the displacement vector for the changed load vector can be found without changing the inverse of the stiffness matrix. In case, if $\{B\}$ is zero and when $[A]^{-1}$ exists, then the solution possible is said to be a trivial solution, that is, $x = 0$. In case, $[A]^{-1}$ does not exist, then the above set of equations will lead to a non-trivial solution (Alston, 2013; Amin et al., 2017).

1.6.2 Linear Equations

A system of 'm' linear equations with 'n' unknowns is expressed as follows:

$$\begin{aligned} a_{11}x_1 + a_{12}x_2 + \cdots + a_{1n}x_n &= b_1 \\ a_{21}x_1 + a_{22}x_2 + \cdots + a_{2n}x_n &= b_2 \\ \cdots\cdots & \\ a_{m1}x_1 + a_{m2}x_2 + \cdots + a_{mn}x_n &= b_m \end{aligned} \tag{1.1}$$

The above set of equations can be written in matrix form as follows:

$$\begin{bmatrix} a_{11} & a_{12} & a_{1n} \\ a_{21} & a_{22} & a_{2n} \\ a_{m1} & a_{m2} & a_{mn} \end{bmatrix} \begin{Bmatrix} x_1 \\ x_2 \\ x_n \end{Bmatrix} = \begin{Bmatrix} b_1 \\ b_2 \\ b_m \end{Bmatrix}$$

$$[A]\{x\} = \{B\} \tag{1.2}$$

Pre-multiply the above equation with A^{-1} on both sides,

$$[A]^{-1}[A]\{x\} = [A]^{-1}\{B\}$$

$$[I]\{x\} = [A]^{-1}\{B\}$$

$$\{x\} = [A]^{-1}\{B\} \tag{1.3}$$

Thus, the above equation gives the unknown 'x' by multiplying the inverse of matrix A and B vector. Now, the problem is to compute the inverse of a matrix.

1.6.3 Inverse Matrix

The inverse of matrix A is given by,

$$[A]^{-1} = \frac{adj\ A}{|A|} \tag{1.4}$$

In a given square matrix, replace each element a_{ij} of the matrix $[A]$ by its cofactor α_{ij}. Transform the cofactor matrix to obtain an adjoint matrix. The following simple example will give the procedure to find the inverse of a matrix.

Consider a matrix, $A = \begin{bmatrix} 1 & 5 & 2 \\ 0 & 4 & 1 \\ 0 & 2 & 1 \end{bmatrix}$. Find $[A]^{-1}$ by the adjoint method.

$$|A| = 1\{(4\times 1)-(2\times 1)\} = 2$$

Cofactors are given by,

$$\alpha_{11} = (-1)^{1+1}\begin{vmatrix} 4 & 1 \\ 2 & 1 \end{vmatrix} = 2$$

$$\alpha_{12} = (-1)^{1+2}\begin{vmatrix} 0 & 1 \\ 0 & 1 \end{vmatrix} = 0$$

$$\alpha_{13} = (-1)^{1+3} \begin{vmatrix} 0 & 4 \\ 0 & 2 \end{vmatrix} = 0$$

$$\alpha_{21} = (-1)^{1+3} \begin{vmatrix} 5 & 2 \\ 2 & 1 \end{vmatrix} = -1$$

$$\alpha_{22} = (-1)^{2+2} \begin{vmatrix} 1 & 2 \\ 0 & 1 \end{vmatrix} = 1$$

$$\alpha_{23} = (-1)^{2+3} \begin{vmatrix} 1 & 5 \\ 0 & 2 \end{vmatrix} = -2$$

$$\alpha_{31} = (-1)^{3+1} \begin{vmatrix} 5 & 2 \\ 4 & 1 \end{vmatrix} = -3$$

$$\alpha_{32} = (-1)^{3+2} \begin{vmatrix} 1 & 2 \\ 0 & 1 \end{vmatrix} = -1$$

$$\alpha_{33} = (-1)^{3+3} \begin{vmatrix} 1 & 5 \\ 0 & 4 \end{vmatrix} = 4$$

The cofactor matrix is written as:

$$\alpha_{ij} = \begin{bmatrix} 2 & 0 & 0 \\ -1 & 1 & -2 \\ -3 & -1 & 4 \end{bmatrix}$$

$$\text{Adj A} = \left[\alpha_{ij}\right]^T = \begin{bmatrix} 2 & -4 & -3 \\ 0 & 1 & -1 \\ 0 & -2 & 4 \end{bmatrix}$$

Thus,

$$[A]^{-1} = \frac{adj\,A}{|A|} = \frac{1}{2}\begin{bmatrix} 2 & -4 & -3 \\ 0 & 1 & -1 \\ 0 & -2 & 4 \end{bmatrix} = \begin{bmatrix} 1 & 1/2 & -3/2 \\ 0 & 1/2 & -1/2 \\ 0 & -1 & 2 \end{bmatrix}$$

To check:

$$[A]^{-1}[A] = \begin{bmatrix} 1 & 1/2 & -3/2 \\ 0 & 1/2 & -1/2 \\ 0 & -1 & 2 \end{bmatrix} \times \begin{bmatrix} 1 & 5 & 2 \\ 0 & 4 & 1 \\ 0 & 2 & 1 \end{bmatrix} = [I] = \begin{bmatrix} 1 & 0 & 0 \\ 0 & 1 & 0 \\ 0 & 0 & 1 \end{bmatrix}$$

1.6.4 Solution of Linear Equations

Let us express the matrix A as a system of equations,

$$x_1 + 5x_2 + 2x_3 = 2$$

$$4x_2 + x_3 = 5$$

$$2x_2 + x_3 = 4$$

Now, this set of equations has to be solved to get the variables $(x_1,\ x_2,\ x_3)$. In matrix form, the equations can be written as follows:

$$\begin{bmatrix} 1 & 5 & 2 \\ 0 & 4 & 1 \\ 0 & 2 & 1 \end{bmatrix} \begin{Bmatrix} x_1 \\ x_2 \\ x_3 \end{Bmatrix} = \begin{Bmatrix} 2 \\ 5 \\ 4 \end{Bmatrix}$$

$$\begin{Bmatrix} x_1 \\ x_2 \\ x_3 \end{Bmatrix} = \left[A\right]^{-1} \begin{Bmatrix} 2 \\ 5 \\ 4 \end{Bmatrix} = \begin{bmatrix} 1 & 1/2 & -3/2 \\ 0 & 1/2 & -1/2 \\ 0 & -1 & 2 \end{bmatrix} \begin{Bmatrix} 2 \\ 5 \\ 4 \end{Bmatrix} = \begin{Bmatrix} -6.5 \\ 0.5 \\ 3 \end{Bmatrix}$$

If one can generate system of equations with unknowns as variables, then these set of equations can be solved using matrix inversion. This is an easy method to solve for the variable as given by Eq. (1.3).

$$\{x\} = \left[A\right]^{-1} \{B\}$$

This is true only when $\left[A\right]^{-1}$ exists. It should also be noted that $\{x\}$ purely depends on $\{B\}$ and $\left[A\right]^{-1}$ does not change to get the value of $\{x\}$. Assume matrix A is a stiffness matrix of a given system, B is a load vector, and x is a displacement vector. By this comparison, it can be seen that the value of the displacement vector for the changed load vector can be found without changing the inverse of the stiffness matrix. In case, if the is $\{B\}$ zero and when $\left[A\right]^{-1}$ exists, then the solution possible is said to be a trivial solution, that is, $x = 0$. In case, $\left[A\right]^{-1}$ does not exist, then the above set of equations will lead to non-trivial solution. MATLAB program is very useful and simple to solve numerical problems using matrix approach (César, 2014).

1.7 FUNDAMENTALS OF MATLAB PROGRAM

MATLAB - Matrix Laboratory

- MATLAB is a high-level language and interactive environment
- It allows you to analyze data, develop algorithms, and create models and applications.

- It is widely used in academia and industry for its powerful capabilities and ease of use.

Capabilities of MATLAB

- Mathematical computations and simulations
- Data analysis and visualization
- Algorithm development
- Modeling and prototyping
- Application development, including graphical user interface building

Simple Actions that MATLAB can perform

- Performing basic arithmetic calculations
- Creating and manipulating matrices and arrays
- Plotting data in 2D and 3D
- Solving equations and systems of equations
- Importing and exporting data

Introduction to MATLAB

MATLAB (Matrix Laboratory) is a high-performance language used for technical computing. Its strength lies in matrix manipulations, plotting of data and functions, and implementation of algorithms, making it invaluable in structural analysis for tasks like matrix stiffness method, eigenvalue problems, and optimization (Stephen and Chapman, 2015; Timmy and Alexandre, 2014).

Getting Started

- **Workspace**: Stores variables during a session.
- **Command Window**: For direct command execution.
- **Script Files**: Written in. m files for reusable code.

Basic Syntax and Operations

```
Defining Variables:
a = 5; % Scalar
b = [1, 2, 3]; % Row vector
c = [1; 2; 3]; % Column vector
```

Arithmetic Operations:

```
sum = a + 2; % Addition
prod = a * b (1); % Multiplication
```

Relational and Logical Operators: These are used in conditional checks like if statements.

Key Commands for Structural Analysis
Help and Documentation:

```
help eig % Displays help for eigenvalue problems
```

Clearing Workspace:

```
clear; % Clears all variables
clc; % Clears the command window
```

MATLAB Functions

Built-In Functions

MATLAB comes with numerous built-in functions essential for structural analysis:

Linear Algebra:

- inv(A): Computes the inverse of matrix AAA.
- det(A): Calculates the determinant of AAA.

Structural Analysis Specific:

- eig(A): Finds eigenvalues and eigenvectors (used in vibration analysis).
- svd(A): Singular value decomposition.

Data Visualization:

- plot(x, y): Creates 2D plots for stress-strain or load-displacement graphs.

User-Defined Functions

Custom functions can be created to solve structural problems.

Example: **Function to Calculate Member Stiffness**:

```
function k = member stiffness(E, A, L)
% E: Young's Modulus, A: Cross-sectional Area, L: Length
k = (E * A) / L;
end
```

Anonymous Functions

Short, inline functions for quick computations:

```
stress = @(F, A) F / A; % Calculates stress as Force/Area
```

Matrix Operations in MATLAB

Matrices are fundamental in structural analysis for methods like the **stiffness method**.

Creating Matrices

Row and column matrix:

```
A = [1, 2, 3; 4, 5, 6; 7, 8, 9]; % 3x3 Matrix
B = zeros(3); % 3x3 Zero matrix
```

Sparse matrices for large systems:

```
S = sparse(10, 10); % 10x10 Sparse matrix
```

Basic Matrix Operations

Addition and Subtraction:

```
C = A + B;
```

Multiplication:

```
F = A * B; % Matrix multiplication
G = A . * B; % Element-wise multiplication
```

Transpose:

```
T = A';
```

Structural Applications

Global Stiffness Matrix Assembly: Use loops to assemble a global stiffness matrix from member stiffness matrices.

Solving Systems of Equations:

```
K = [6, -2, -1; -2, 5, -3; -1, -3, 7];
F = [10; 20; 30];
```

```
D = K \ F; % Solves K * D = F
```

Running Program Codes and Debugging

Running MATLAB Codes

Save the script with .m extension (e.g., analysis.m).

Run using:

```
analysis; % Executes the script
```

Input data can be prompted using input:

```
E = input ('Enter Young's Modulus: ');
```

Debugging

Debugging ensures error-free execution of structural analysis programs.

Setting Breakpoints: Click beside the line number in the editor to pause execution.

Common Commands:

- dbstop: Stops execution at a specified line.
- dbstep: Steps through code line by line.
- dbquit: Exits debug mode.

Common Errors in Structural Analysis Scripts

Matrix Dimension Errors:

- Ensure matrix sizes align for operations like multiplication.

Incorrect Indexing:

- MATLAB indexing starts at 1, not 0.

Uninitialized Variables:

- Always initialize variables before using them.

Example Applications

Stress Analysis

```
F = [100; 200; 300]; % Forces in N
A = [0.01; 0.02; 0.03]; % Areas in m²
Stress = F ./ A; % Stress calculation
```

Eigenvalue Problem for Natural Frequencies

```
M = [2, 0; 0, 3]; % Mass matrix
K = [6, -2; -2, 4]; % Stiffness matrix
[Phi, Lambda] = eig(K, M); % Phi = mode shapes, Lambda =
eigenvalues
Natural Frequencies = sqrt(diag(Lambda)); % Convert
eigenvalues to frequencies
```

2 Planar Orthogonal Structures

2.1 STANDARD BEAM ELEMENT

A beam element is one of the basic elements to be used in the structural analysis of planar orthogonal frames. There are a few sign conventions that need to be followed before deriving the stiffness matrix of the beam element. In addition to the anticlockwise end moment, joint rotation and joint moments are considered positive; reaction (or force) and displacement of the joint toward the positive y-axis are also considered positive. Reaction force or axial displacement at the joint toward the positive x-axis is considered positive. Let us limit the discussion to planar orthogonal structures in this chapter, which have members either horizontal or vertical. It is important to note that all members of the frame are mutually orthogonal and coplanar. The X-Y plane is under consideration, and hence, bending takes place about the Z-axis.

The fixed beam is considered a basic model. Consider a fixed beam undergoing deformation due to bending, neglecting the axial deformation. The standard fixed beam is shown in Figure 2.1. The two joints of the beam element are 'j' and 'k,' and the length of the member is L_i. In terms of mathematical conditions, the considered beam element is fixed at nodes 'j' and 'k' with a constant flexural rigidity (EI) over its entire length. The jth node is taken as the origin, and the positive x-axis is marked such that the whole beam lies on the positive *x-y* axis system. The member is designated as ith member; (x_m, y_m) are local axes of the member, as marked in the figure. It is very important to note the axis system. With its origin at the jth end, x_m is directed toward the kth end; y_m is counterclockwise 90° to x_m axis. Therefore, the (x_m, y_m) plane defines the plane of bending the beam element. For axial deformation neglected, one should identify the possible translational and rotational displacements (kinematic degrees of freedom) at each end of the beam, as shown in Figure 2.2.

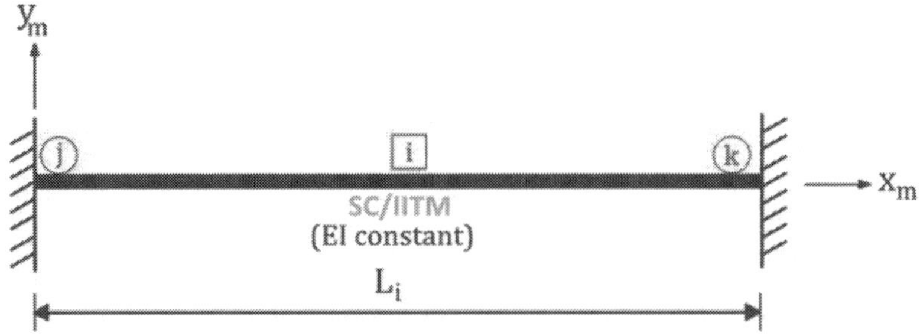

FIGURE 2.1 Standard beam element.

DOI: 10.1201/9781003639152-2

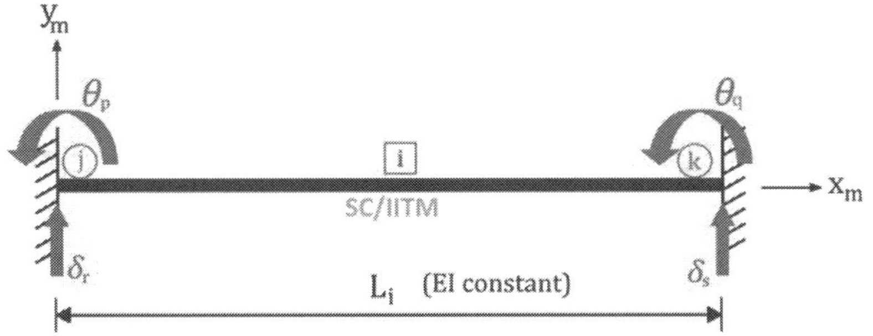

FIGURE 2.2 Rotational and translational displacements of the beam element.

Suitable subscripts are used to denote the rotational and translational moments. Note the order by which the moments are marked, which can make the readers understand the computer programs easily. The displacements at the jth end and kth end are (θ_p,δ_r) and (θ_q,δ_s), respectively.

Now, let us derive the stiffness coefficient, k_{ij}. By classical definition, k_{ij} is the force in the ith degree-of-freedom by imposing unit displacement (either translational or rotational) in the jth degree-of-freedom while restraining all other degrees-of-freedom. Let us give unit rotation at the jth end, keeping all other degrees of freedom restrained, as shown in Figure 2.3. This will invoke members with end forces, k^i_{pp}, k^i_{qp}, k^i_{rp}, k^i_{sp}. k^i_{pp} is the force in pth degree-of-freedom by giving unit displacement in pth degree-of-freedom in the ith member. Similarly, k^i_{qp} is the force in the degree-of-freedom by giving unit displacement in pth degree-of-freedom in the ith member.

The second subscript in all the notations is common, which is ‘p’. It confirms that the unit displacement is imposed at a pth degree of freedom, enabling the generation of the elements of the stiffness matrix column-wise. Thus, the obtained stiffness coefficients correspond to the first column of the stiffness matrix. Similarly, apply unit rotation at the end k of the member, as shown in Figure 2.4. The stiffness coefficients, in this case, will be $k^i_{pq}, k^i_{qq}, k^i_{rq}, k^i_{sq}$. The stiffness coefficients are again obtained by applying unit displacements at the jth end and the kth end. As shown in

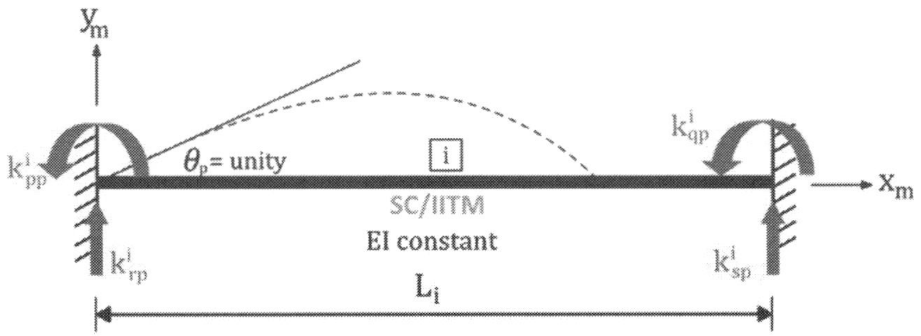

FIGURE 2.3 Fixed beam under unit rotation at jth end.

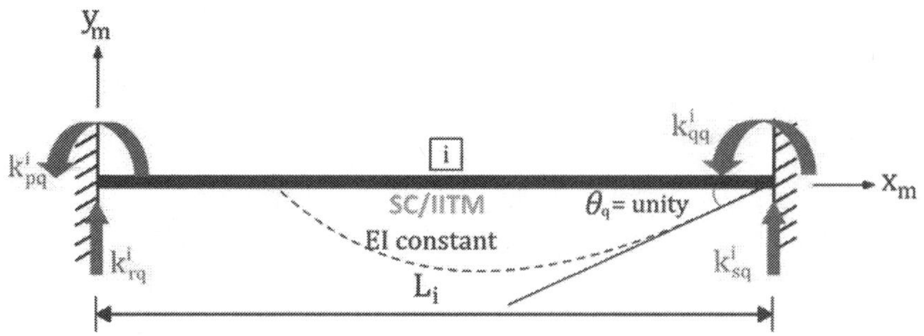

FIGURE 2.4 Fixed beam under unit rotation at the kth end.

Figures 2.5 to 2.6. The stiffness coefficients by applying unit displacement at the jth end are $k^i_{pr}, k^i_{qr}, k^i_{rr}, k^i_{sr}$. By applying unit displacement at the kth end, the following stiffness coefficients are obtained: $k^i_{ps}, k^i_{qs}, k^i_{rs}, k^i_{ss}$. A tangent can be drawn by connecting the deflected position of the beam at which the unit rotation is applied and the initial position of the beam at the other end. From this, one can say that the beam has undergone a rotation of $1/L_i$, where L_i is the length of the member. The rotation at the ends of the beam is equal to $1/L_i$.

For ith beam element experiencing arbitrary end displacements $(\theta_p, \delta_r, \theta_q, \delta_s)$, corresponding end reactions (moment, shear) are required to be estimated by maintaining the equilibrium of the restrained member. The governing equations are as follows:

$$m^i_p = k^i_{pp}\theta_p + k^i_{pq}\theta_q + k^i_{pr}\delta_r + k^i_{ps}\delta_s \tag{2.1a}$$

$$m^i_q = k^i_{qp}\theta_p + k^i_{qq}\theta_q + k^i_{qr}\delta_r + k^i_{qs}\delta_s \tag{2.1b}$$

$$p^i_r = k^i_{rp}\theta_p + k^i_{rq}\theta_q + k^i_{rr}\delta_r + k^i_{rs}\delta_s \tag{2.1c}$$

$$p^i_s = k^i_{sp}\theta_p + k^i_{sq}\theta_q + k^i_{sr}\delta_r + k^i_{ss}\delta_s \tag{2.1d}$$

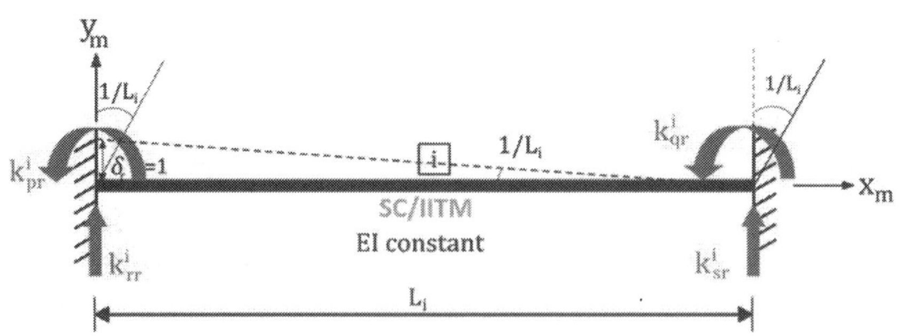

FIGURE 2.5 Fixed beam under unit displacement at jth end.

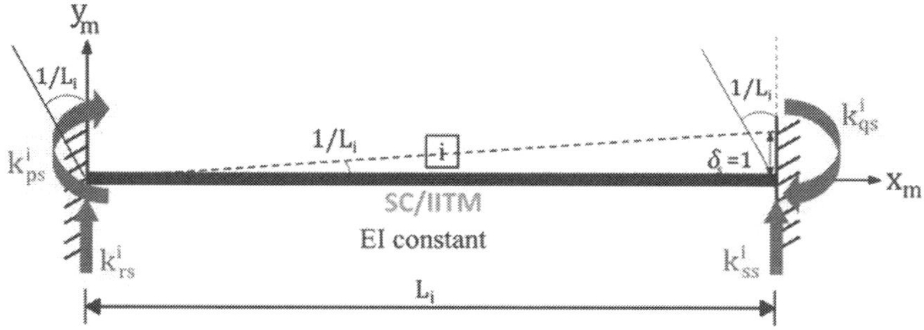

FIGURE 2.6 Fixed beam under unit displacement at the kth end.

In the above equations, the first subscript in the stiffness coefficients corresponds to the node at which the moment or force is measured, while the second subscript refers to the node at which the unit rotation or displacement is applied, respectively. Equations (2.1a, 2.1b) refer to the moments, while the rest refers to shear. The above equations give the end moments and shear for arbitrary unit displacements $(\theta_p, \theta_q, \delta_r, \delta_s)$ in the respective degrees of freedom. They can be generalized as follows:

$$\{m_i\} = [k]_i \{\delta_i\} \tag{2.2}$$

$$\{m_i\} = \begin{Bmatrix} m_p \\ m_q \\ p_r \\ p_s \end{Bmatrix} \tag{2.3a}$$

$$\{\delta_i\} = \begin{Bmatrix} \theta_p \\ \theta_q \\ \delta_r \\ \delta_s \end{Bmatrix} \tag{2.3b}$$

$$[k]_i = \begin{bmatrix} k_{pp} & k_{pq} & k_{pr} & k_{ps} \\ k_{qp} & k_{qq} & k_{qr} & k_{qs} \\ k_{rp} & k_{rq} & k_{rr} & k_{rs} \\ k_{sp} & k_{sq} & k_{sr} & k_{ss} \end{bmatrix} \tag{2.4}$$

Let us consider ith member of any structure as a modular element for analysis. The rotational coefficients for the ith member, namely $(k_{pp}, k_{qp}, k_{pq}, k_{qq})$ are to be evaluated. The end shear, arising from the rotations, can then be computed using the

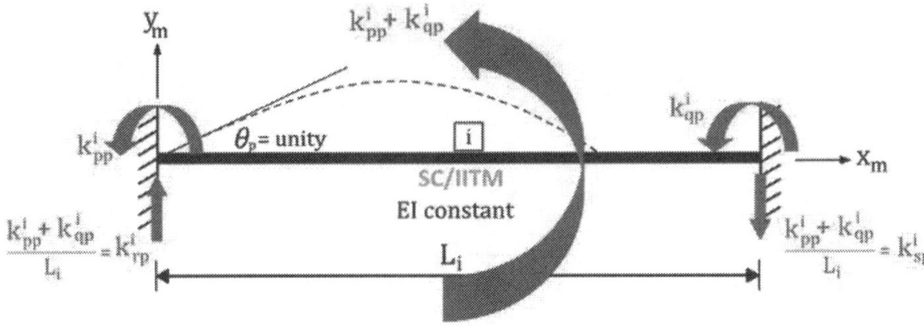

FIGURE 2.7 Stiffness coefficients for unit rotation.

following expressions. Figure 2.7 shows the stiffness coefficients under unit rotation at the jth end of the member.

$$k_{rp} = \frac{k_{pp} + k_{qp}}{L_i} \tag{2.5a}$$

$$k_{sp} = -\frac{k_{pp} + k_{qp}}{L_i} \tag{2.5b}$$

Similarly, for unit rotation applied at the kth end, the rotational coefficients are given below:

$$k_{rq} = \frac{k_{pq} + k_{qq}}{L_i} \tag{2.5c}$$

$$k_{sq} = -\frac{k_{pq} + k_{qq}}{L_i} \tag{2.5d}$$

For unit displacement applied at the jth end, the rotational coefficients are given by:

$$k_{pr} = \frac{k_{pp} + k_{pq}}{L_i} \tag{2.6a}$$

$$k_{qr} = \frac{k_{qp} + k_{qq}}{L_i} \tag{2.6b}$$

$$k_{rr} = \frac{k_{pr} + k_{qr}}{L_i} = \frac{1}{L_i}\left\{\frac{k_{pp} + k_{pq}}{L_i} + \frac{k_{qp} + k_{qq}}{L_i}\right\} = \frac{k_{pp} + k_{pq} + k_{qp} + k_{qq}}{(L_i)^2} \tag{2.6c}$$

$$k_{sr} = -\frac{k_{pr} + k_{qr}}{L_i} = -\frac{1}{L_i}\left\{\frac{k_{pp} + k_{pq}}{L_i} + \frac{k_{qp} + k_{qq}}{L_i}\right\} = -\frac{k_{pp} + k_{pq} + k_{qp} + k_{qq}}{(L_i)^2} \tag{2.6d}$$

For unit displacement applied at the kth end, the rotational coefficients are given as:

$$k_{ps} = -\frac{k_{pp} + k_{pq}}{L_i} \tag{2.7a}$$

$$k_{qs} = -\frac{k_{qp} + k_{qq}}{L_i} \tag{2.7b}$$

$$k_{rs} = -\left\{\frac{k_{ps} + k_{qs}}{L_i}\right\} = -\frac{1}{L_i}\left\{\frac{k_{pp} + k_{pq}}{L_i} + \frac{k_{qp} + k_{qq}}{L_i}\right\} = -\frac{k_{pp} + k_{pq} + k_{qp} + k_{qq}}{(L_i)^2} \tag{2.7c}$$

$$k_{ss} = \frac{k_{pr} + k_{qr}}{L_i} = \frac{1}{L_i}\left\{\frac{k_{pp} + k_{pq}}{L_i} + \frac{k_{qp} + k_{qq}}{L_i}\right\} = \frac{k_{pp} + k_{pq} + k_{qp} + k_{qq}}{(L_i)^2} \tag{2.7d}$$

From the above equations, it can be seen that the end shear coefficients are expressed in terms of rotational coefficients. Thus, by evaluating the four rotational coefficients, the stiffness matrix with sixteen stiffness coefficients can be framed and is given as follows:

$$[K] = \begin{bmatrix} k_{pp} & k_{pq} & \frac{k_{pp} + k_{pq}}{L} & -\left(\frac{k_{pp} + k_{pq}}{L}\right) \\ k_{qp} & k_{qq} & \frac{k_{qp} + k_{qq}}{L} & -\left(\frac{k_{qp} + k_{qq}}{L}\right) \\ \frac{k_{pp} + k_{pq}}{L} & \frac{k_{pq} + k_{qq}}{L} & \frac{k_{pp} + k_{pq} + k_{qp} + k_{qq}}{L^2} & -\frac{k_{pp} + k_{pq} + k_{qp} + k_{qq}}{L^2} \\ -\left(\frac{k_{pp} + k_{pq}}{L}\right) & -\left(\frac{k_{pq} + k_{qq}}{L}\right) & -\left(\frac{k_{pp} + k_{pq} + k_{qp} + k_{qq}}{L^2}\right) & \left(\frac{k_{pp} + k_{pq} + k_{qp} + k_{qq}}{L^2}\right) \end{bmatrix} \tag{2.8}$$

Flexibility and stiffness matrices are related as follows:

$$[f][K] = [I] \tag{2.9}$$

Let $\begin{bmatrix} \delta_{jj} & \delta_{jk} \\ \delta_{kj} & \delta_{kk} \end{bmatrix}$ be the flexibility coefficients of a member with nodes j and k and $\begin{bmatrix} k_{pp} & k_{pq} \\ k_{qp} & k_{qq} \end{bmatrix}$ be the stiffness coefficients, then the following relationship holds well:

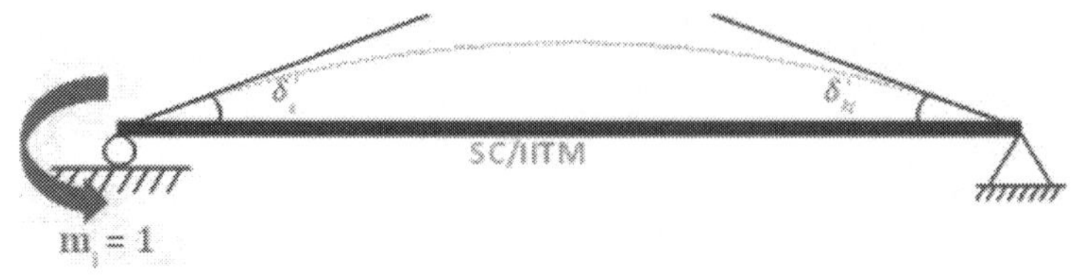

FIGURE 2.8 Beam under unit moment at jth end.

$$\begin{bmatrix} \delta_{jj} & \delta_{jk} \\ \delta_{kj} & \delta_{kk} \end{bmatrix} \begin{bmatrix} k_{pp} & k_{pq} \\ k_{qp} & k_{qq} \end{bmatrix} = \begin{bmatrix} 1 & 0 \\ 0 & 1 \end{bmatrix} \tag{2.10}$$

The stiffness matrix [k] can be evaluated using the following relationship:

$$[K] = [f]^{-1}[I] \tag{2.11}$$

Now, let us evaluate the flexibility coefficients to form the flexibility matrix. Then, the stiffness matrix is derived from the flexibility matrix using the above equation.

2.2 ESTIMATING ROTATION COEFFICIENTS

Consider a simply supported beam, as shown in Figure 2.8. The unit moment is applied at the jth end, and the flexibility coefficients are determined. As seen in Figure 2.9, flexibility coefficients are determined by applying unit displacement at the kth end. The flexibility coefficients $\left(\delta^i_{jj}, \delta^i_{kj}\right)$ define rotations at ends j and k, respectively, of the ith member, caused due to unit moment applied at the jth end. The first subscript refers to the degree of freedom where the displacement is measured, and the second subscript refers to the degree of freedom where the unit moment is applied; the superscript refers to the member identity, as it is the ith beam. Similarly, the flexibility coefficients $\left(\delta^i_{jk}, \delta^i_{kk}\right)$ define rotations at the jth and kth ends of the ith member due to unit moment applied at the kth end.

2.3 STIFFNESS MATRIX OF A BEAM ELEMENT

In this section, we shall derive the stiffness coefficients of the fixed beam. By neglecting axial deformation, the size of [K] shall be 4×4. As per the classical definition, the stiffness coefficient, k_{ij}, is the force in the ith degree of freedom for unit

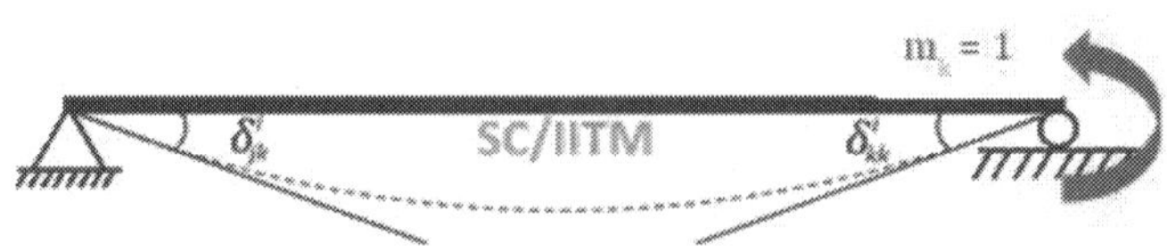

FIGURE 2.9 Beam under unit moment at the kth end.

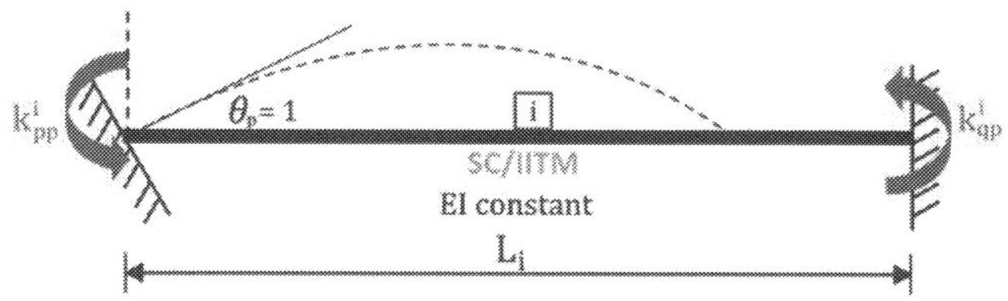

FIGURE 2.10 Fixed beam under unit rotation at pth degree of freedom.

displacement in the jth degree of freedom, keeping all other degrees of freedom as restrained. Let us apply unit rotation at pth degree of freedom ($\theta_p = 1$), and derive forces in all degrees of freedom. Please note that all other degrees of freedom are restrained from any displacement. By imposing this condition, we shall derive the first column of the [K] (Figure 2.10). Similarly, let us apply unit rotation at the qth degree of freedom ($\theta_q = 1$), and derive forces in all degrees of freedom. Please note that all other degrees of freedom are restrained from any displacement. By imposing this condition, we shall derive the second column of the [K] (Figure 2.11).

The stiffness coefficients $\left(k^i_{pp}, k^i_{qp}\right)$ define end moments required at jth and kth ends to maintain equilibrium when the jth end is subjected to unit rotation while the kth end is restrained. Similarly, the stiffness coefficients $\left(k^i_{pq}, k^i_{qq}\right)$ define end moments required at jth and kth ends to maintain equilibrium when the kth end is subjected to unit rotation while the jth end is restrained.

$$\begin{bmatrix} \delta_{jj} & \delta_{jk} \\ \delta_{kj} & \delta_{kk} \end{bmatrix} \begin{bmatrix} k_{pp} & k_{pq} \\ k_{qp} & k_{qq} \end{bmatrix} = \begin{bmatrix} 1 & 0 \\ 0 & 1 \end{bmatrix} \tag{2.12}$$

$$k^i_{pp}\,\delta^i_{jj} + k^i_{qp}\,\delta^i_{jk} = 1 \tag{2.13}$$

$$k^i_{pq}\,\delta^i_{kj} + k^i_{qq}\,\delta^i_{kk} = 1 \tag{2.14}$$

Let us denote the flexibility matrix as $[f_r]$ and the stiffness matrix as $[K_r]$. The subscript refers to the rotational degree of freedom. To estimate the flexibility matrix for the beam element, assume a simply supported beam with the unit moment at the jth end, as shown in Figure 2.12. The clockwise couple balances the anticlockwise moment. The bending moment diagram, as seen in Figure 2.13, with tension at the top and compression at the bottom, helps derive the couple.

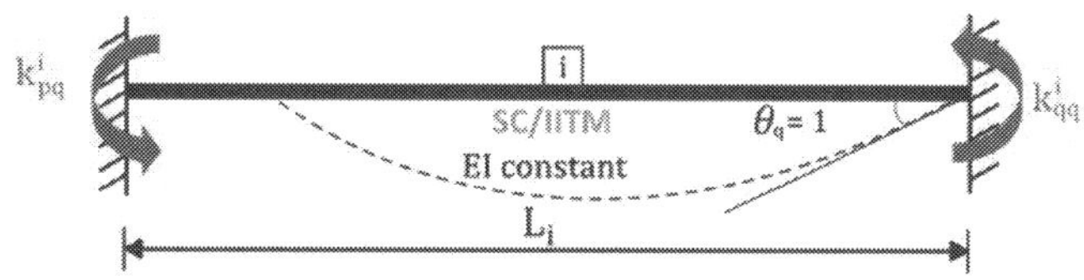

FIGURE 2.11 Fixed beam under unit rotation at the degree of freedom.

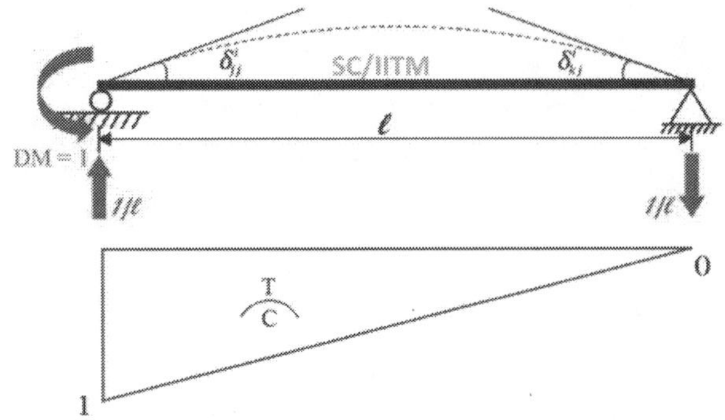

FIGURE 2.12 Beam under unit moment at jth end.

Let us replace the loading diagram with a conjugate beam, as shown in Figure 2.13. Taking a moment about A, we get

$$V_B = \left[\left\{\frac{1}{2}L_i\left(\frac{1}{EI}\right)\right\}\frac{1}{3}L_i\right]\frac{1}{L_i} = \frac{L_i}{6EI}(\text{downward}) \tag{2.15a}$$

$$V_A = \left\{\frac{1}{2}L_i\left(\frac{1}{EI}\right)\right\} - \frac{L_i}{6EI} = \frac{L_i}{3EI}(\text{upward}) \tag{2.15b}$$

The same procedure is followed for the other case to derive the following flexibility matrix.

$$f_r = \begin{bmatrix} \frac{L}{3EI} & -\frac{L}{6EI} \\ -\frac{L}{6EI} & \frac{L}{3EI} \end{bmatrix} \tag{2.16}$$

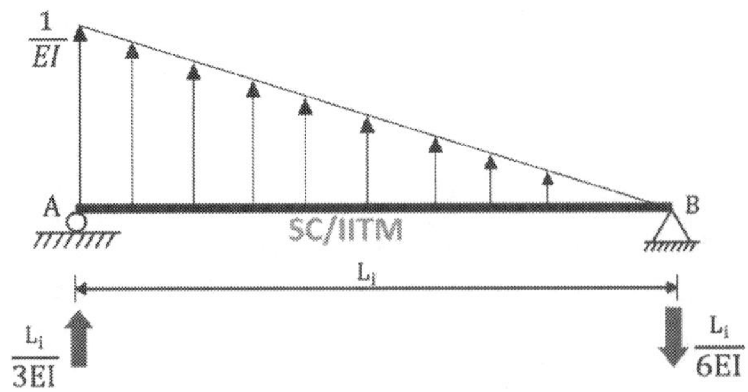

FIGURE 2.13 Conjugate beam.

$$k_r = \left[f_r\right]^{-1} = \frac{12(EI)^2}{L^2}\begin{bmatrix} \frac{L}{3EI} & \frac{L}{6EI} \\ \frac{L}{6EI} & \frac{L}{3EI} \end{bmatrix} = \begin{bmatrix} \frac{4EI}{L} & \frac{2EI}{L} \\ \frac{2EI}{L} & \frac{4EI}{L} \end{bmatrix} \tag{2.17}$$

Once the rotation coefficients are derived, using Eq. (2.8), [K] is derived as follows:

$$K_i = \begin{bmatrix} \frac{4EI}{L} & \frac{2EI}{L} & \frac{6EI}{L^2} & -\frac{6EI}{L^2} \\ \frac{2EI}{L} & \frac{4EI}{L} & \frac{6EI}{L^2} & -\frac{6EI}{L^2} \\ \frac{6EI}{L^2} & \frac{6EI}{L^2} & \frac{12EI}{L^3} & -\frac{12EI}{L^3} \\ -\frac{6EI}{L^2} & -\frac{6EI}{L^2} & -\frac{12EI}{L^3} & \frac{12EI}{L^3} \end{bmatrix} \tag{2.18}$$

2.4 BEAM ELEMENT WITH VARYING FLEXURAL RIGIDITY

Let us consider a beam of 300 mm wide and 5 m long with varying depth, as shown in Figure 2.14. As the depth varies along the span of the beam, there will also be a continuous variation of the flexural rigidity (EI) along the length. It is due to the continuous variation of the moment of inertia about the bending axis along the length.

The stiffness matrix is developed by neglecting the axial deformation. Let us divide the beam into ten parts of equal width, as shown in Figure 2.15. A different moment of inertia is assigned to each part. For the cross-section of the beam at the end with 400 mm depth, the moment of inertia about the bending axis is given as:

$$I_{zz} = \frac{300 \times 400^3}{12} = 1.6 \times 10^9\, mm^4 \tag{2.19}$$

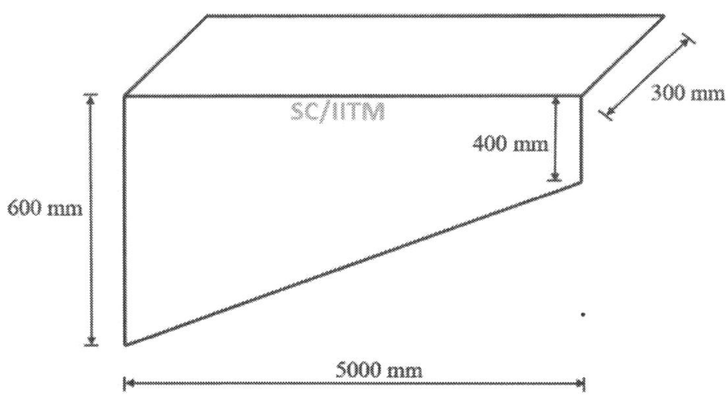

FIGURE 2.14 Beam with varying depth.

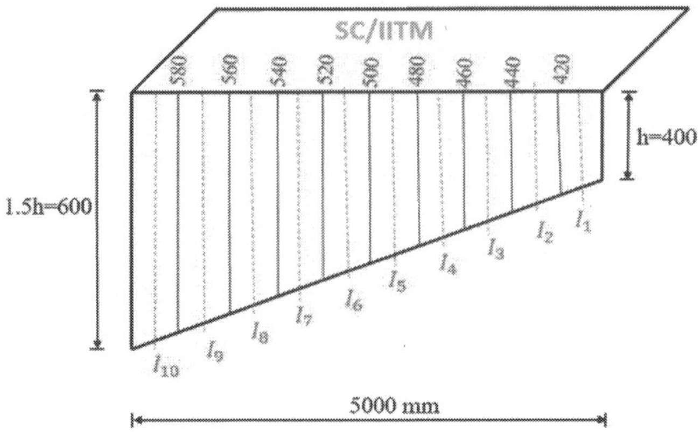

FIGURE 2.15 Beam with varying moments of inertia.

The moment of inertia of each strip can be computed for the known average depth of each strip. The depth of the beam at one end is 400 mm, and the other end is 600 mm. For h = 400, the depth at the other end can be assigned to 1.5 h. Hence, the average thickness of all the strips can be expressed in terms of 'h', as tabulated in Table 2.1. For strip-1, the average depth and moment of inertia are computed as follows:

Depth at one end of the strip-1 = 400 mm
Depth at the other end of the strip-1 = 420 mm
Average depth of strip-1 = 410 mm = 1.025 h
Moment of inertia of strip-1 = $I_1 = (1.025)^3 I = 1.077I$

TABLE 2.1
M/EI Ordinates of Strips under Unit Moment at jth End

Strip	Average Thickness	Moment of Inertia	Average Moment (M)	M/EI Ordinates
I_1	410 = 1.025 h	1.077I	0.05	0.046/EI
I_2	430 = 1.075 h	1.242I	0.15	0.121/EI
I_3	450 = 1.125 h	1.424I	0.25	0.176/EI
I_4	470 = 1.175 h	1.622I	0.35	0.216/EI
I_5	490 = 1.225 h	1.838I	0.45	0.245/EI
I_6	510 = 1.275 h	2.073I	0.55	0.265/EI
I_7	530 = 1.325 h	2.326I	0.65	0.279/EI
I_8	550 = 1.375 h	2.60I	0.75	0.288/EI
I_9	570 = 1.425 h	2.894I	0.85	0.294/EI
I_{10}	590 = 1.475 h	3.209I	0.95	0.296/EI

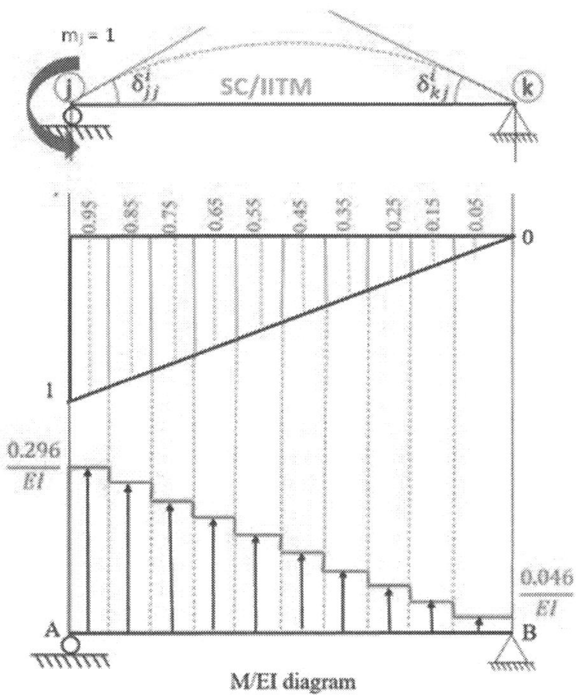

FIGURE 2.16 Beam under unit moment at jth end.

Similarly, the moment of inertia of all the strips can be calculated, as shown in Table 2.1. The corresponding values of (M/EI) ordinates for unit moment applied at jth end are shown in Figure 2.16.

By taking a moment about A and applying the equation of static equilibrium, we get:

$$v_B = \frac{1}{5}\frac{0.5}{EI}\{(0.296\times0.25)+(0.294\times0.75)+(0.288\times1.25)+(0.279\times1.75)$$
$$+(0.265\times2.25)+(0.245\times2.75)+(0.216\times3.25)+(0.176\times3.75)$$
$$+(0.121\times4.25)+(0.046\times4.75)\} = \frac{0.451}{EI}(\text{downward})$$

$$v_A = \frac{1}{EI}\{(0.296+0.294+0.288+0.279+0.265+0.245+0.216$$
$$+\ 0.176+0.121+0.0446)\times0.5-(0.451)\} = \frac{0.662}{EI}(\text{upward})$$

Thus, the first column of the flexibility matrix of the beam element is given by:

$$\left[f_r\right]_i = \frac{1}{EI}\begin{bmatrix} 0.662 & a_{12} \\ -0.451 & a_{22} \end{bmatrix} \tag{2.20}$$

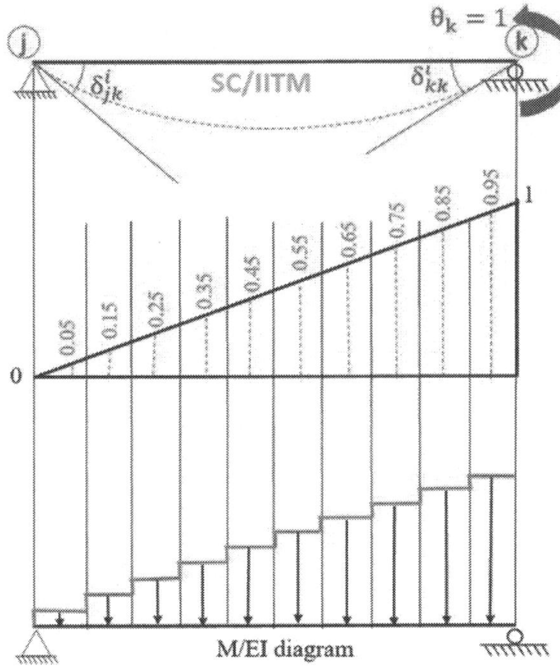

FIGURE 2.17 Beam under unit moment at kth end.

To derive the second column, a similar procedure is followed by applying a unit moment at kth end, as shown in Figure 2.17. The M/EI ordinates are listed in Table 2.2.

By taking a moment about A and applying the equation of static equilibrium, we get:

$$v_B = \frac{1}{5}\frac{0.5}{EI}\{(0.016\times 25)+(0.052\times 0.75)+(0.096\times 1.25)+(0.15\times 1.75)$$
$$+(0.217\times 2.25)+(0.293\times 2.75)+(0.401\times 3.25)+(0.527\times 3.75)$$
$$+(0.684\times 4.25)+(0.882\times 4.75)\} = \frac{1.210}{EI}(\text{upward})$$

$$v_A = \frac{1}{EI}\{(0.016+0.052+0.096+0.15+0.217+0.293+0.401+0.527+0.684+0.82)$$
$$\times 0.5 - 1.210\} = \frac{0.45}{EI}(\text{downward})$$

Therefore, the flexibility matrix is given by:

$$[f_r]_i = \frac{1}{EI}\begin{bmatrix} 0.662 & -0.450 \\ -0.451 & 1.210 \end{bmatrix} \tag{2.21}$$

TABLE 2.2
M/EI Ordinates of Strips under Unit Moment at kth End

Strip	Average Thickness of Strips	Moment of Inertia (I)	Average Moment (M)	M/EI Ordinates
I_1	410 = 1.025 h	1.077I	0.05	0.016/EI
I_2	430 = 1.075 h	1.242I	0.15	0.052/EI
I_3	450 = 1.125 h	1.424I	0.25	0.096/EI
I_4	470 = 1.175 h	1.622I	0.35	0.150/EI
I_5	490 = 1.225 h	1.838I	0.45	0.217/EI
I_6	510 = 1.275 h	2.073I	0.55	0.293/EI
I_7	530 = 1.325 h	2.326I	0.65	0.401/EI
I_8	550 = 1.375 h	2.60I	0.75	0.527/EI
I_9	570 = 1.425 h	2.894I	0.85	0.684/EI
I_{10}	590 = 1.475 h	3.209I	0.95	0.882/EI

Now, the stiffness matrix is obtained by inverting the flexibility matrix.

$$\left[K\right]_i = \left[f\right]_i^{-1} = \frac{EI}{0.599}\begin{bmatrix} 1.210 & 0.450 \\ 0.45 & 0.662 \end{bmatrix} \tag{2.22}$$

$$\left[K\right]_i = EI\begin{bmatrix} 2.202 & 0.751 \\ 0.750 & 1.105 \end{bmatrix} \tag{2.23}$$

Considering a fixed beam of 5 m long with varying cross-section, the stiffness matrix can be written with the known rotational coefficients as follows:

$$K_i = \begin{bmatrix} 2.02 & 0.751 & 0.554 & -0.554 \\ 0.751 & 1.105 & 0.371 & -0.371 \\ 0.554 & 0.371 & 0.185 & -0.185 \\ -0.554 & -0.371 & -0.185 & 0.185 \end{bmatrix} \tag{2.24}$$

2.5 PLANAR ORTHOGONAL STRUCTURES

A planar orthogonal structure is one where all members lie on the same plane and each member is orthogonal to its adjacent member. In the previous section, we derived the [K] for a fixed beam element. In this section, let us try to solve a planar orthogonal frame, with the basic element as a fixed beam module. Consider a single bay single-story frame as shown in Figure 2.18. Both ends of the frame are fixed, and the frame is subjected to some arbitrary loading. There are a few steps to be followed while formulating the stiffness method of analysis for solving such problems. There are three

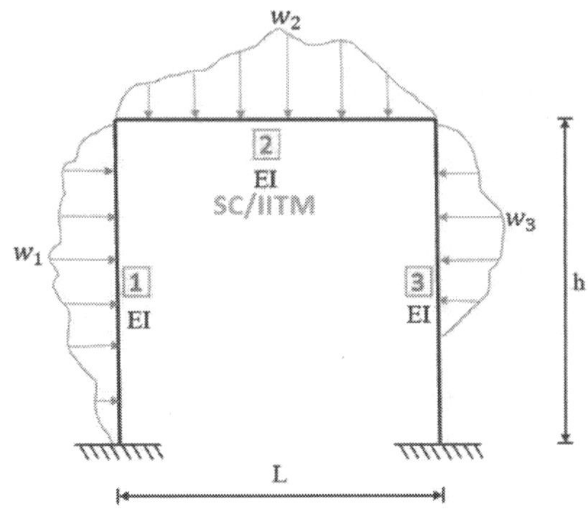

FIGURE 2.18 Planar orthogonal structure.

members and four joints in this frame. For every member, it is important to identify the joint numbering in accordance with the standard fixed beam. The standard module is a fixed beam whose degrees of freedom labels are shown in Figure 2.19.

Figure 2.20 shows the degrees of freedom of the frame. The degrees of freedom are grouped as unrestrained and restrained. The unrestrained degrees of freedom of the frame are marked as $(\theta_1, \theta_2, \delta_3)$ neglecting the axial deformation. The restrained degrees of freedom are $(\delta_4, \delta_5, \theta_6, \theta_7, \delta_8, \delta_9)$. Please note that the vertical displacement at nodes (1,2) is the same, confirming that the axial deformation in the column member is neglected; the same is true in the case of other members. Thus, the total number of degrees of freedom of the frame is 9, implying that the size of [K] is 9×9.

For a fixed beam used as a modular element, [K] of the modular element is given as:

$$K_i = \begin{bmatrix} \frac{4EI}{l} & \frac{2EI}{l} & \frac{6EI}{l^2} & -\frac{6EI}{l^2} \\ \frac{2EI}{l} & \frac{4EI}{l} & \frac{6EI}{l^2} & -\frac{6EI}{l^2} \\ \frac{6EI}{l^2} & \frac{6EI}{l^2} & \frac{12EI}{l^3} & -\frac{12EI}{l^3} \\ -\frac{6EI}{l^2} & -\frac{6EI}{l^2} & -\frac{12EI}{l^3} & \frac{12EI}{l^3} \end{bmatrix} \tag{2.25}$$

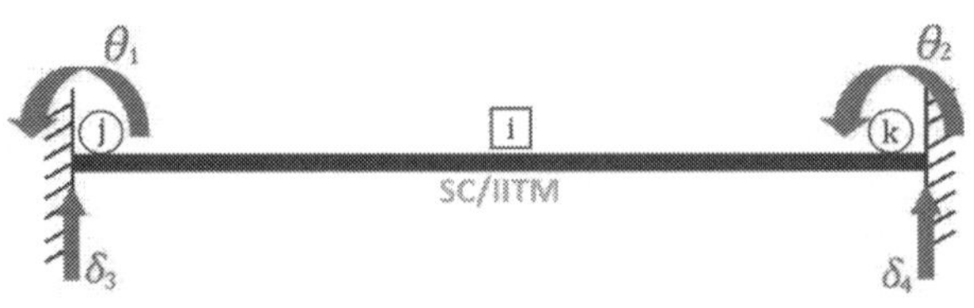

FIGURE 2.19 Modular fixed beam with degrees of freedom.

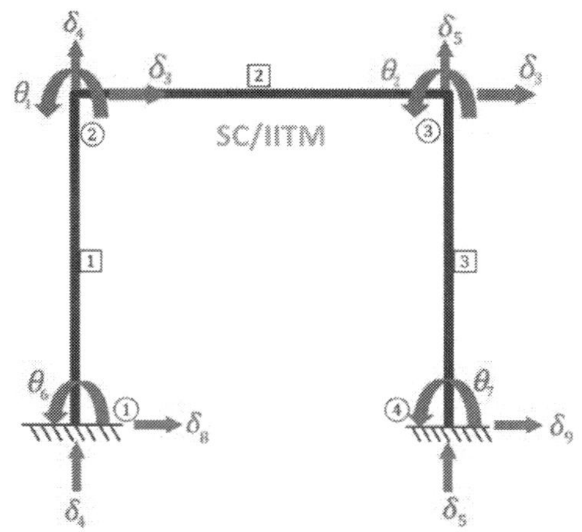

FIGURE 2.20 Degrees of freedom of the single bay single-story frame.

Using Eq. (2.25), the stiffness matrix of each member can be derived for the known values of (E, I, *l*). Once [k] of all three members is formulated, then these stiffness matrices can be assembled to obtain the [K] of the frame. This assembly is possible only if the labels of the degrees of freedom of each member are properly mapped to the frame. For example, labels of all members are identified in the same order as that of the modular element. Referring to Figure 2.19, labels of a modular element are in the following order:

(Rotation at jth end, rotation at kth end, displacement along +ve y-axis at jth end, displacement along +ve y-axis at kth end) amounting to $(\theta_1, \theta_2, \delta_3, \delta_4)$. The labels for the members are summarized in the following table.

Member	jth End	kth End	Labels
1	2	1	(1,6,3,8)
2	2	3	(1,2,4,5)
3	3	4	(2,7,3,9)

Once [K] of the frame is assembled, it can be partitioned as shown below:

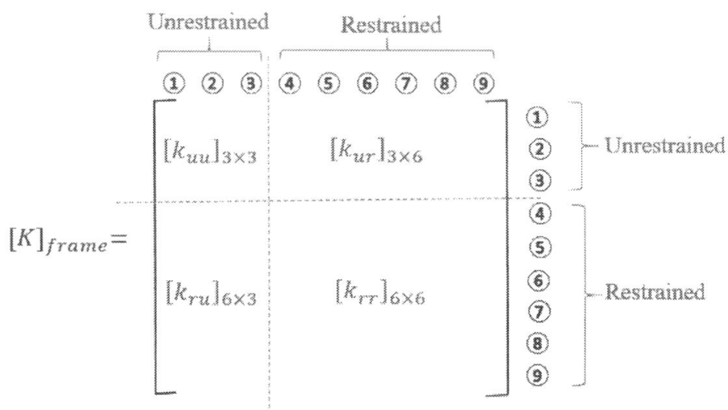

In the above matrix, partitioning is done at 3×3, as the frame has only (3) unrestrained degrees of freedom. Let us consider the stiffness matrix, formulated in terms of unrestrained and restrained sub-matrices, as shown above. Let the unrestrained and restrained displacements be denoted as (Δ_u, Δ_r), respectively. The compatibility equations can be written as follows:

$$\begin{bmatrix} [K_{uu}] & [K_{ur}] \\ [K_{ru}] & [K_{rr}] \end{bmatrix} \begin{Bmatrix} \Delta_u \\ \Delta_r \end{Bmatrix} = \begin{Bmatrix} J_{Lu} \\ J_{Lr} \end{Bmatrix} + \begin{Bmatrix} [0] \\ R_r \end{Bmatrix} \tag{2.26}$$

Where $\begin{Bmatrix} J_{Lu} \\ J_{Lr} \end{Bmatrix}$ is the load vector, partitioned as unrestrained and restrained, respectively. The subscripts (u, r) refer to unrestrained and restrained, respectively. By expanding Eq. (2.26) and substituting $(\Delta_r = 0)$, we get:

$$[k_{uu}]\{\Delta_u\} = \{J_L\}_u \tag{2.27}$$

$$\{\Delta_u\} = [k_{uu}]^{-1}\{J_L\}_u \tag{2.28}$$

Thus, the displacement vector $\{\Delta_u\}$ can be computed once we know the load vector, $\{J_L\}_u$. The joint loads are the consequences of the fixed end moments and shear at supports caused by the member loads. Table 2.3 summarizes the fixed end moments for different load cases with the sign convention as positive for the anticlockwise moment. Joint loads are the reversal of end moments and shear, as obtained from the fixed end moments. Table 2.4 summarizes the end moments caused by the joint displacements. The general equation to find the end moment of ith beam is given as:

$$[M]_i = k_i\delta_i + (FEM)_i \tag{2.29}$$

Where $(FEM)_i$ refers to the fixed end moment vector of the ith member.

2.6 SHEAR FORCE AND BENDING MOMENT

Shear force diagram (SFD) and bending moment diagram (BMD) are graphical representations of the variation of shear force and bending moment along the length of a beam. They are essential tools in structural analysis for the following reasons:

1. **Determining Critical Sections**
    - SFD and BMD help identify points of maximum shear force and bending moment, which are critical for the design and analysis of beams. These points typically experience the highest stress and are the most likely locations for failure.
2. **Assisting in Beam Design**
    - Engineers use the maximum bending moment to calculate the required section modulus, ensuring the beam can safely withstand applied loads.

TABLE 2.3
Fixed End Moments

Beam	M_F^{AB}	M_F^{BA}
Fixed beam with central concentrated load: 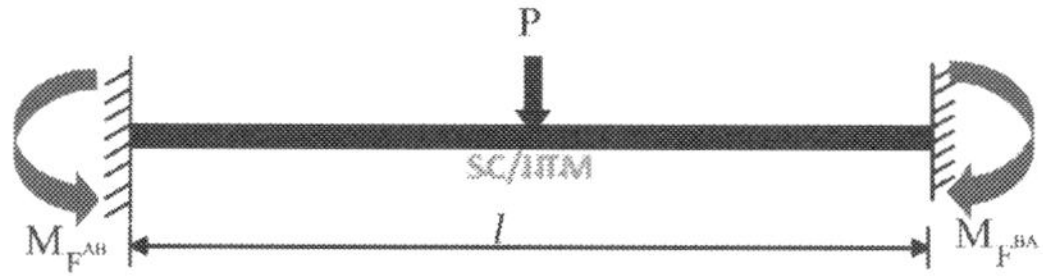	$\frac{Pl}{8}$	$\frac{Pl}{8}$
Fixed beam with uniformly distributed load: 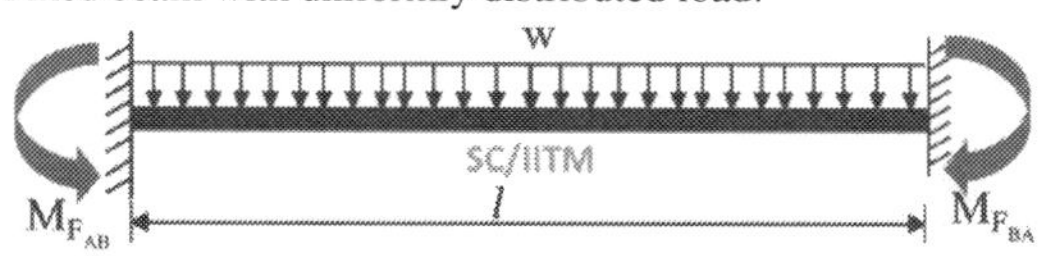	$\frac{wl^2}{12}$	$\frac{wl^2}{12}$
Fixed beam with eccentric concentrated load: 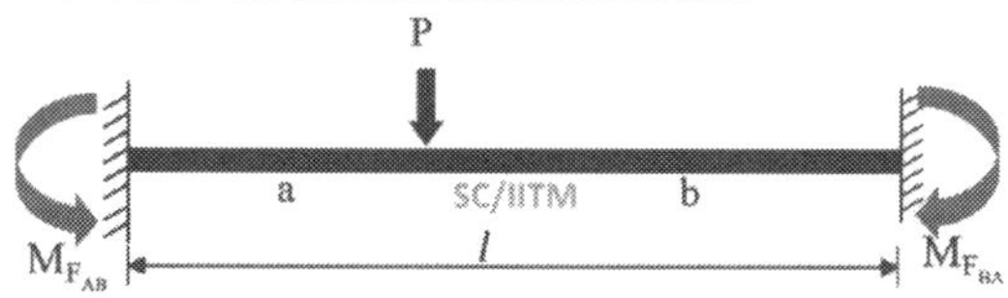	$\frac{pab^2}{l^2}$	$\frac{pba^2}{l^2}$
Fixed bean with uniformly varying load: 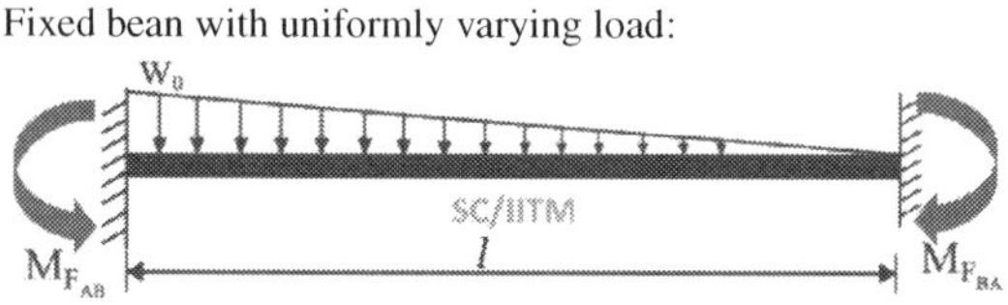	$\frac{W_o l^2}{20}$	$\frac{W_o l^2}{30}$
Fixed beam with triangular loading: 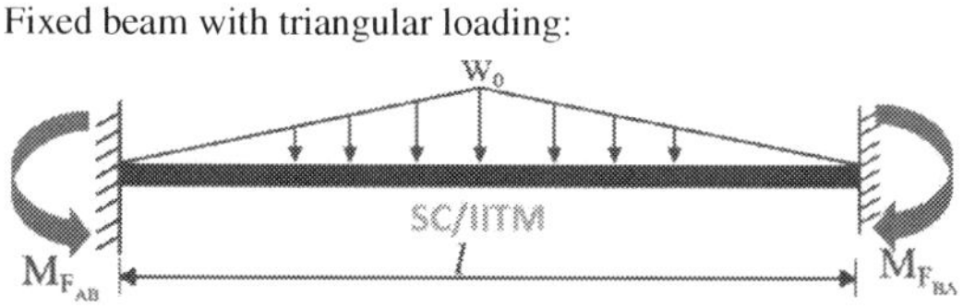	$\frac{5}{96}W_o l^2$	$\frac{5}{96}W_o l^2$
Fixed beam with eccentric anticlockwise moment: 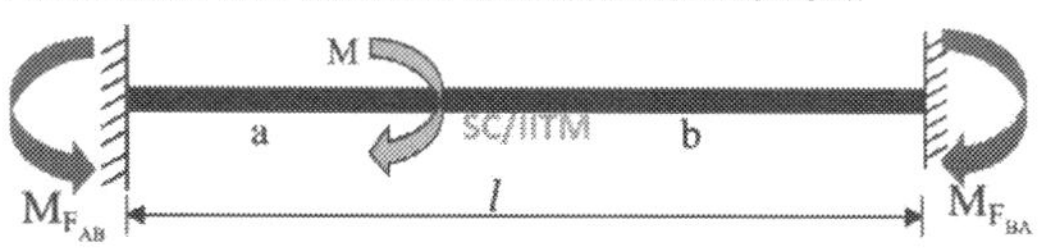	$\frac{M}{l^2}b(b-2a)$	$\frac{M}{l^2}a(2b-a)$
Fixed beam with central clockwise moment: 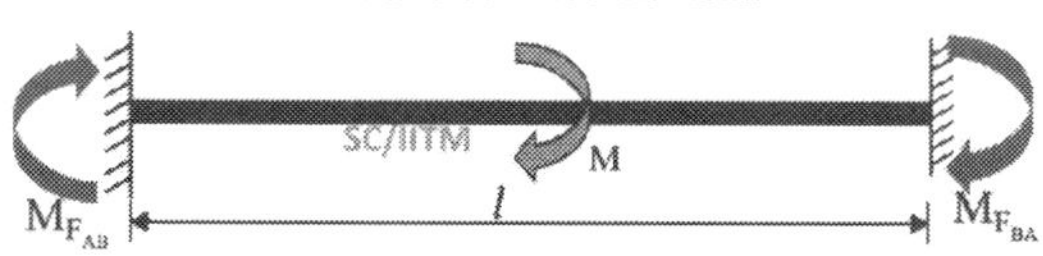	$\frac{M}{4}$	$\frac{M}{4}$

TABLE 2.4
End Moments Due to Joint Displacements

Beam	M_F^{AB}	M_F^{BA}
Beam with rotation at jth end: 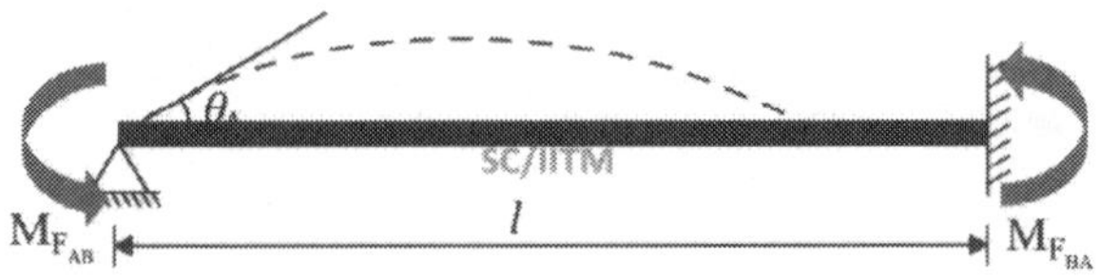	$\frac{4EI\theta_A}{l}$	$\frac{2EI\theta_A}{l}$
Beam with rotation at kth end: 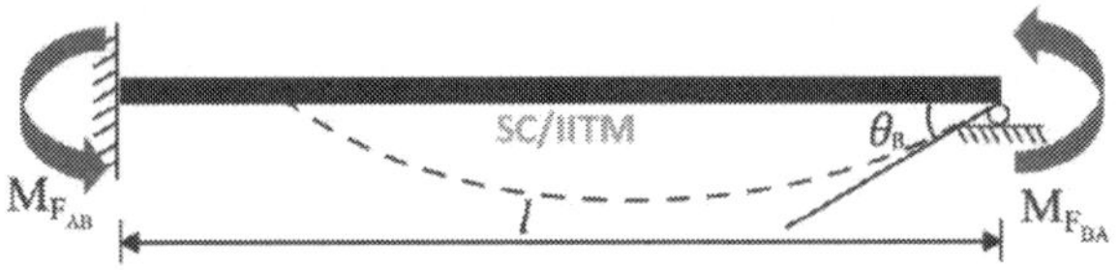	$\frac{2EI\theta_B}{l}$	$\frac{4EI\theta_B}{l}$
Simply supported beam with rotation: 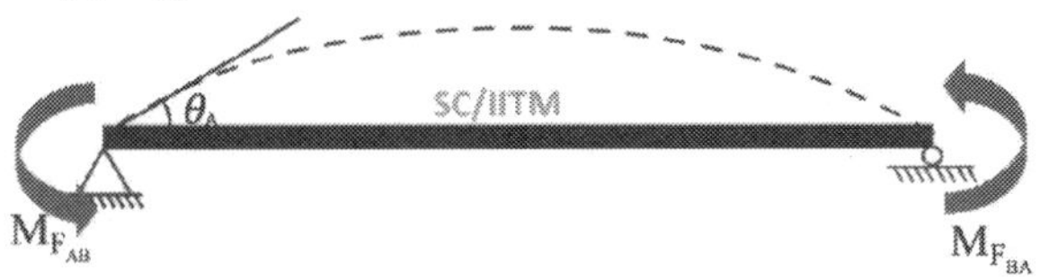	$\frac{3EI\theta_A}{l}$	NIL
Beam with settlement of support: 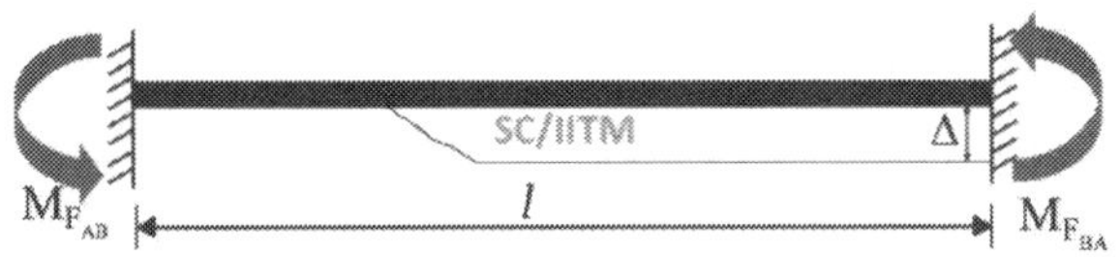	$\frac{6EI\Delta}{l^2}$	$\frac{6EI\Delta}{l^2}$
Beam with settlement of support: 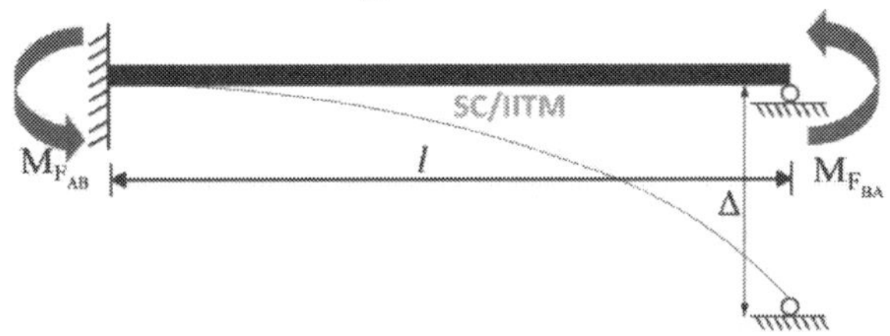	$\frac{3EI\Delta}{l^2}$	NIL

   - The shear force is used to determine the beam's shear strength, influencing web design in steel sections or shear reinforcement in concrete beams.
3. **Verifying Structural Safety**
   - SFD and BMD ensure that the selected material and cross-section can resist the calculated forces and moments without exceeding allowable stresses.

4. **Analyzing Structural Behavior**
    - These diagrams provide insights into how a beam behaves under loading. For instance:
        - Where the bending moment is zero, the beam changes curvature.
        - Sudden changes in SFD indicate concentrated forces, while linear changes represent uniformly distributed loads.
5. **Essential for Advanced Analysis**
    - SFD and BMD are often preliminary steps in more complex analyses like deflection calculations, stability checks, or dynamic response studies.

2.6.1 Steps to Draw Shear Force Diagram (SFD) and Bending Moment Diagram (BMD)

Step 1: Understand the Loading and Supports

- Identify the type of beam (simply supported, cantilever, fixed, etc.).
- Note all applied loads, including point loads, distributed loads, and moments.
- Determine the reactions at supports using equilibrium equations:

$$\boldsymbol{\sum F_y = 0, \sum M = 0}$$

Step 2: Compute Shear Force

- Divide the beam into segments based on the location of loads and supports.
- Start at one end of the beam (usually the left) and calculate the shear force at key points by summing vertical forces:

$$\mathrm{V} = \mathrm{V}_{\text{previous}} + (\text{Upward forces}) - (\text{Downward forces})$$

Step 3: Plot the Shear Force Diagram

- Plot the calculated shear force at each key point along the beam.
- **Key Characteristics:**
    - A concentrated load causes a sudden jump or drop in the shear force.
    - A uniformly distributed load (UDL) causes a linear variation in the shear force.

Step 4: Compute Bending Moment

- Use the relation between shear force and bending moment:

$$M_x = M_{x-1} + V.\Delta x$$

  Where M_x is the moment at a section, and Δx is the distance between sections.
- For point loads or moments:
    - Moment due to a point load: M = F·d (force times perpendicular distance).
    - Moment due to a UDL: Use integration to calculate.

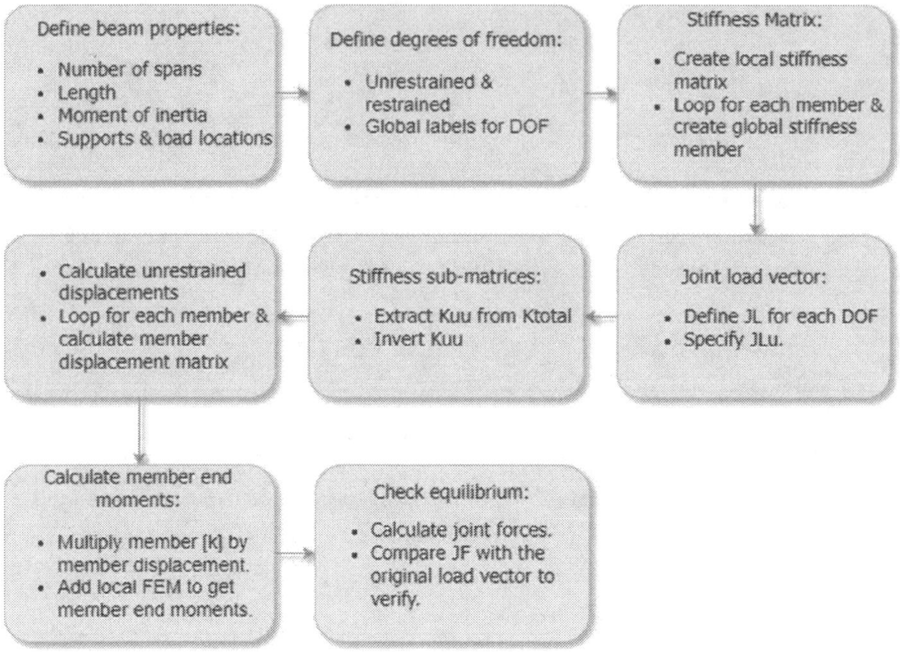

FIGURE 2.21 Steps to analyze a beam using MATLAB.

Step 5: Plot the Bending Moment Diagram

- Plot the bending moment values calculated at each segment.
- Key characteristics:
  - A point load causes a linear variation in the moment.
  - A UDL causes a parabolic variation.
  - Points of zero bending moment indicate locations of inflection (curvature change).

2.7 STEP-BY-STEP FRAMEWORK OF COMPUTER-AIDED ANALYSIS

The step-by-step framework to analyze a beam using MATLAB program is given in Figure 2.21.

2.8 CONTINUOUS BEAMS

2.8.1 Example Problem

Neglecting axial deformation, for a constant (EI) of the beam, determines the reactions and end moments of the beam shown:

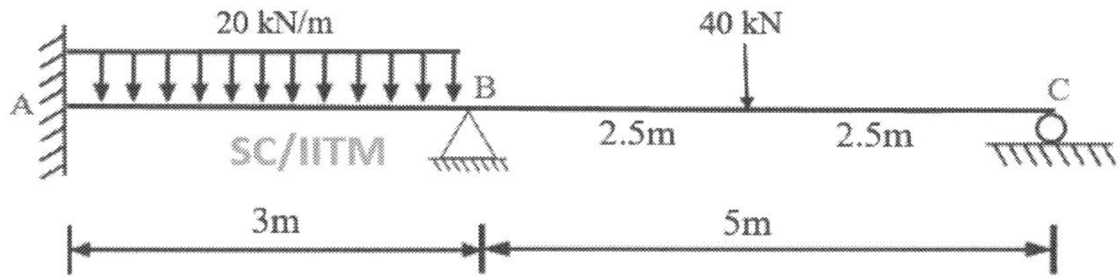

i. *Marking unrestrained and restrained degrees of freedom:*
The unrestrained and restrained degrees of freedom are marked as shown.

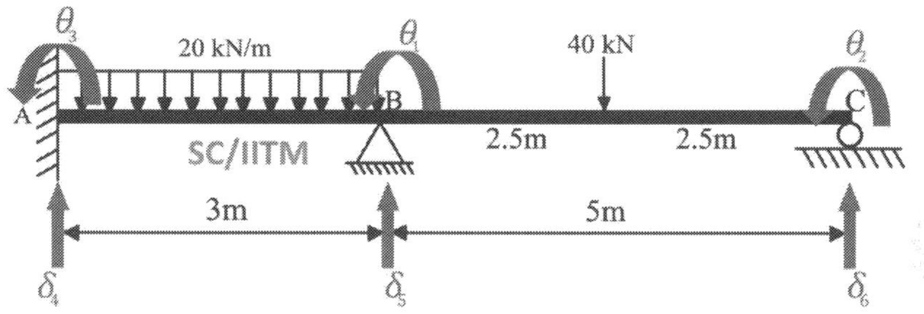

Unrestrained degrees-of-freedom = 2 [θ_1, θ_2]
Restrained degrees-of-freedom = 4 [θ_3, δ_4, δ_5, δ_6]

Thus, the unrestrained sub-matrix will be of size 2×2 while the size of [K] of the beam is 6×6. With the help of a modular member and a fixed beam, one can derive the [k] of each member and assemble them to derive [K] of the whole beam. The labels of each member are summarized below:

Member	jth End	kth End	Length (m)	Labels
AB	A	B	3	(3,1,4,5)
BC	B	C	4	(1,2,5,6)

ii. *Formulation of stiffness matrix:*
The stiffness matrix of a standard modular element, which is a fixed beam, is given as:

$$K_i = \begin{bmatrix} \frac{4EI}{l} & \frac{2EI}{l} & \frac{6EI}{l^2} & -\frac{6EI}{l^2} \\ \frac{2EI}{l} & \frac{4EI}{l} & \frac{6EI}{l^2} & -\frac{6EI}{l^2} \\ \frac{6EI}{l^2} & \frac{6EI}{l^2} & \frac{12EI}{l^3} & -\frac{12EI}{l^3} \\ -\frac{6EI}{l^2} & -\frac{6EI}{l^2} & -\frac{12EI}{l^3} & \frac{12EI}{l^3} \end{bmatrix}$$

The stiffness matrices of each member are given as follows (please pay attention to the labels of the rows and columns):

$$K_{AB} = EI\begin{bmatrix} 1.333 & 0.667 & 0.667 & -0.667 \\ 0.667 & 1.333 & 0.667 & -0.667 \\ 0.667 & 0.667 & 0.445 & -0.445 \\ -0.667 & -0.667 & -0.445 & 0.445 \end{bmatrix} \quad \text{(rows: 3, 1, 4, 5; columns: 3, 1, 4, 5)}$$

$$K_{BC} = EI\begin{bmatrix} 0.80 & 0.40 & 0.24 & -0.24 \\ 0.40 & 0.80 & 0.24 & -0.24 \\ 0.24 & 0.24 & 0.096 & -0.096 \\ -0.24 & -0.24 & -0.096 & 0.096 \end{bmatrix} \quad \text{(rows: 1, 2, 5, 6; columns: 1, 2, 5, 6)}$$

After assembling the respective rows and columns of the member stiffness matrices, the unrestrained stiffness matrix $[k_{uu}]$ of the continuous beam is given as follows (please note that the [K] of the continuous beam is partitioned at 2×2, as this problem has only two unrestrained degrees of freedom.

$$K_{UU} = EI\begin{bmatrix} 2.133 & 0.4 \\ 0.4 & 0.8 \end{bmatrix} \quad \text{(rows: 1, 2; columns: 1, 2)}$$

$$K_{UU}^{-1} = \frac{1}{1.546EI}\begin{bmatrix} 2.133 & 0.4 \\ 0.4 & 0.8 \end{bmatrix}$$

iii. *Calculation of fixed end moments:*

The fixed end moments and reactions of the members are shown below:

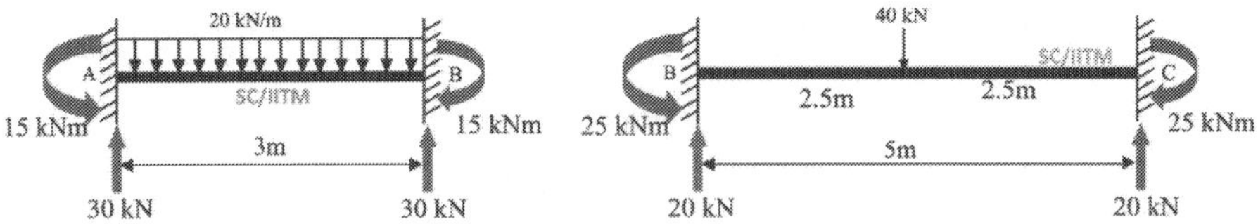

$$M^F_{AB} = \frac{wl^2}{12} = \frac{20\times 3^2}{12} = +15kNm$$

$$M^F_{BA} = -\frac{wl^2}{12} = -\frac{20\times 3^2}{12} = -15kNm$$

$$V_A = +30\,kN \text{ and } V_B = +30\,kN$$

$$M_{BC}^{F} = \frac{pl}{8} = \frac{40 \times 5}{8} = +25kNm$$

$$M_{CB}^{F} = -\frac{pl}{8} = -\frac{40 \times 5}{8} = -25kNm$$

$$V_B = +20\,kN \text{ and } V_C = +20\,kN$$

The fixed end moments for the complete beam are shown below:

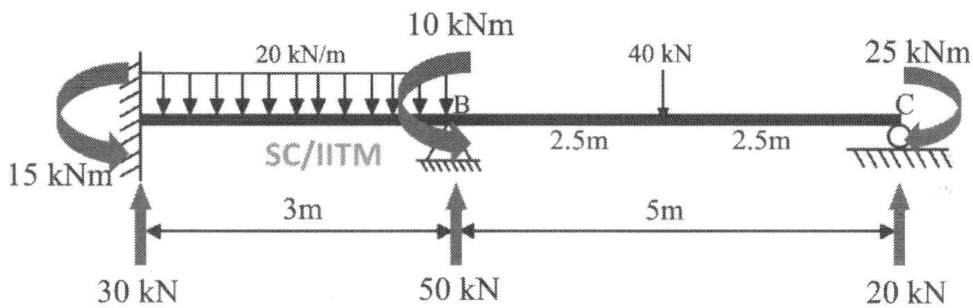

iv. *Calculation of joint load vectors:*
The joint loads will be the reversal of the fixed end moments and reactions. The joint load for the continuous beam is shown below:

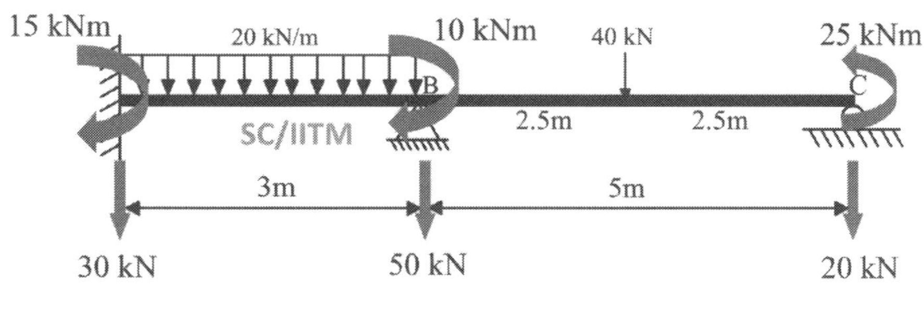

$$J_L = \begin{Bmatrix} -10 \\ 25 \\ -15 \\ -30 \\ -50 \\ -20 \end{Bmatrix} \begin{matrix} 1 \\ 2 \\ 3 \\ 4 \\ 5 \\ 6 \end{matrix}$$

The unrestrained joint load vector is given by,

$$J_{LU} = \begin{Bmatrix} \mathbf{-10} \\ \mathbf{25} \end{Bmatrix} \begin{matrix} 1 \\ 2 \end{matrix}$$

v. *Calculation of unrestrained displacements:*

$$\{\Delta_u\} = \begin{Bmatrix} \theta_1 \\ \theta_2 \end{Bmatrix} = \left[k_{uu}\right]^{-1} \{J_L\}_u = \frac{1}{1.546EI} \begin{Bmatrix} -18 \\ 57.325 \end{Bmatrix}$$

$$\{\Delta_r\} = \begin{Bmatrix} \theta_3 \\ \delta_4 \\ \delta_5 \\ \delta_6 \end{Bmatrix} = \{0\}$$

vi. *Calculation of end moment and shear:*

The general equation to find the end moment of ith member is as follows:

$$\left[M\right]_i = k_i \delta_i + \left(FEM\right)_i$$

$$M_{AB} = \left[k\right]_{AB} \{\delta\} + \{FEM\}$$

$$\begin{Bmatrix} M_3 \\ M_1 \\ V_4 \\ V_5 \end{Bmatrix} = EI \begin{bmatrix} 1.333 & 0.667 & 0.667 & -0.667 \\ 0.667 & 1.333 & 0.667 & -0.667 \\ 0.667 & 0.667 & 0.445 & -0.445 \\ -0.667 & -0.667 & -0.445 & 0.445 \end{bmatrix} \begin{Bmatrix} \theta_3 \\ \theta_1 \\ \delta_4 \\ \delta_5 \end{Bmatrix} + \begin{Bmatrix} -15 \\ -15 \\ +30 \\ +30 \end{Bmatrix} = \begin{Bmatrix} 7.234 \\ -30.52 \\ 22.234 \\ 37.766 \end{Bmatrix}$$

$$M_{BC} = \left[k\right]_{BC} \{\delta\} + \{FEM\}$$

$$\begin{Bmatrix} M_1 \\ M_2 \\ V_5 \\ V_6 \end{Bmatrix} = EI \begin{bmatrix} 0.8 & 0.4 & 0.24 & -0.24 \\ 0.4 & 0.8 & 0.24 & -0.24 \\ 0.24 & 0.24 & 0.096 & -0.096 \\ -0.24 & -0.24 & -0.096 & 0.096 \end{bmatrix} \begin{Bmatrix} \theta_1 \\ \theta_2 \\ \delta_5 \\ \delta_6 \end{Bmatrix} + \begin{Bmatrix} 25 \\ -25 \\ 20 \\ 20 \end{Bmatrix} = \begin{Bmatrix} 30.517 \\ 0 \\ 26.105 \\ 13.895 \end{Bmatrix}$$

The member end moments and shear are shown below:

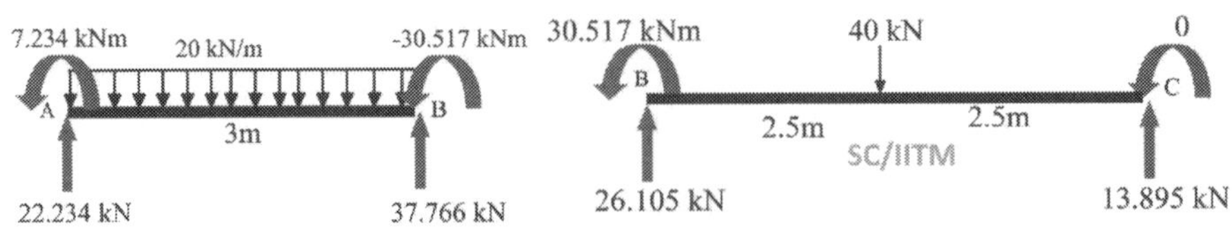

Check for span AB:

$\sum V = 0,$

$22.234 + 37.766 = 20 \times 3 = 60\,\text{kN}$

$M_A = (20 \times 3 \times 1.5) + 30.52 - 37.766(3) = 7.234\,\text{kNm}$

Hence, Ok.

Check for span BC:

$\sum V = 0,$

$(26.105 + 13.895) = 40\,\text{kN}$

$M_B = (40 \times 2.5) - (13.895 \times 5) = 30.52\,\text{kNm}$

Hence, Ok.

By superimposing the end moments of the members into the continuous beam, the final end moments and shear are obtained as shown below:

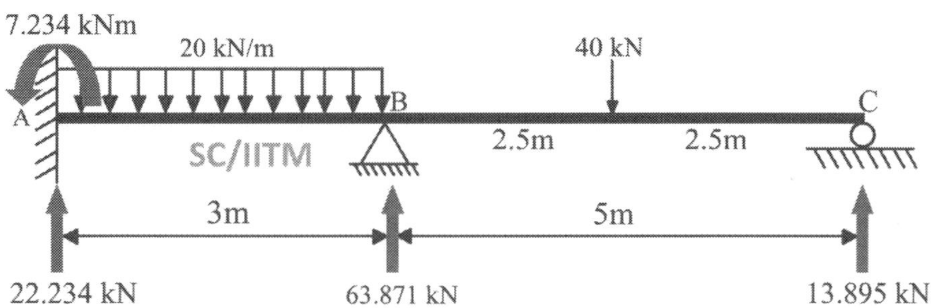

Check:

$$[k_{ru}]\{\Delta_u\} - \{J_l\}_r = \{R_r\}$$

$$[k_{ru}] = EI\begin{bmatrix} 0.667 & 0 \\ 0.667 & 0 \\ -0.427 & 0.24 \\ -0.24 & -0.24 \end{bmatrix}$$

$$\{\Delta_u\}\frac{1}{1.546EI}\begin{Bmatrix} -18 \\ 57.325 \end{Bmatrix}$$

$$\{J_L\}_r = \begin{Bmatrix} -15 \\ -30 \\ -50 \\ -20 \end{Bmatrix} \begin{matrix} 3 \\ 4 \\ 5 \\ 6 \end{matrix}$$

$$\{R_r\} = \begin{Bmatrix} M_3 \\ V_4 \\ V_5 \\ V_6 \end{Bmatrix} = \begin{Bmatrix} 7.234 \\ 22.234 \\ 63.81 \\ 13.895 \end{Bmatrix}. \text{Hence, Ok.}$$

2.8.2 Shear Force and Bending Moment Diagrams

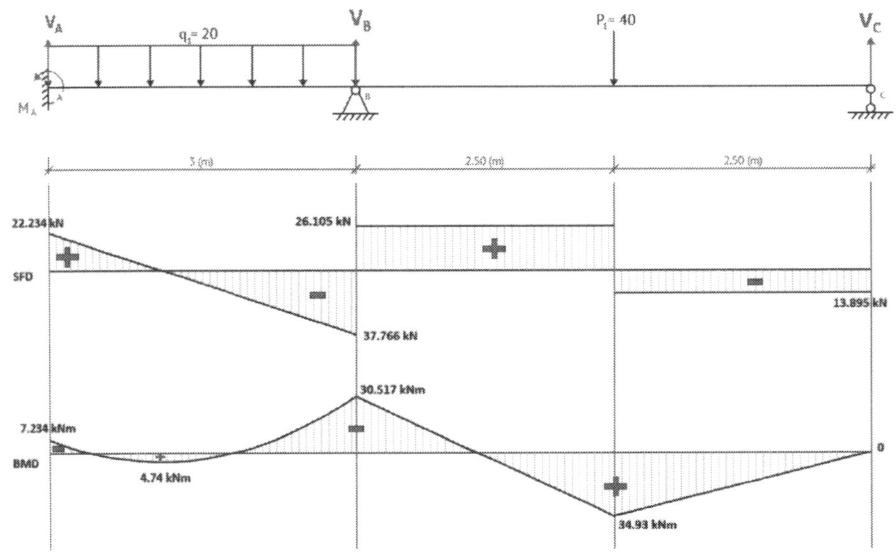

MATLAB program:

```
%% stiffness matrix method
% Input
clc;
clear;
n = 2; % number of members
I = [1 1]; %Moment of inertia in m4
L = [3 5]; % length in m
uu = 2; % Number of unrestrained degrees of freedom
ur = 4; % Number of restrained degrees of freedom
uul = [1 2]; % global labels of unrestrained dof
url = [3 4 5 6]; % global labels of restrained dof
l1 = [3 1 4 5]; % Global labels for member 1
```

```
l2 = [1 2 5 6]; % Global labels for member 2
l= [l1; l2];
dof = uu+ur;
Ktotal = zeros (dof);
fem1= [15 -15 30 30]; % Local Fixed end moments of member 1
fem2= [25 -25 20 20]; % Local Fixed end moments of member 2

%% Creation of joint load vector
jl= [-10; 25; -15; -30; -50; -20]; % values given in kN or kNm
jlu = [-10; 25]; % load vector in unrestrained dof

%% rotation coefficients for each member
rc1 = 4.*I./L;
rc2 = 2.*I./L;

%% stiffness matrix 4 by 4 (axial deformation neglected)
for i = 1:n
        Knew = zeros (dof);
        k1 = [rc1(i); rc2(i); (rc1(i)+rc2(i))/L(i); (-
(rc1(i)+rc2(i))/L(i))];
        k2 = [rc2(i); rc1(i); (rc1(i)+rc2(i))/L(i); (-
(rc1(i)+rc2(i))/L(i))];
        k3 = [(rc1(i)+rc2(i))/L(i); (rc1(i)+rc2(i))/L(i);
(2*(rc1(i)+rc2(i))/(L(i)^2)); (-2*(rc1(i)+rc2(i))/(L(i)^2))];
        k4 = -k3;
        K = [k1 k2 k3 k4];
        fprintf ('Member Number =');
        disp (i);
        fprintf ('Local Stiffness matrix of member, [K] = \n');
        disp (K);
        for p = 1:4
            for q = 1:4
                Knew((l(i,p)),(l(i,q))) =K(p,q);
            end
        end
        Ktotal = Ktotal + Knew;
        if i == 1
            Kg1=K;
        else
            Kg2 = K;
        end
end
fprintf ('Stiffness Matrix of complete structure, [Ktotal] = \n');
disp (Ktotal);
Kunr = zeros(uu);
for x=1:uu
    for y=1:uu
        Kunr(x,y)= Ktotal(x,y);
    end
end
fprintf ('Unrestrained Stiffness sub-matix, [Kuu] = \n');
disp (Kunr);
```

```
KuuInv= inv(Kunr);
fprintf ('Inverse of Unrestrained Stiffness sub-matix, [KuuInverse] = \n');
disp (KuuInv);
%% Calculation of displacements
delu = KuuInv*jlu;
fprintf ('Joint Load vector, [Jl] = \n');
disp (jl');
fprintf ('Unrestrained displacements, [DelU] = \n');
disp (delu');
delr = zeros (ur,1);
del = [delu; delr];
deli= zeros (4,1);
for i = 1:n
    for p = 1:4
        deli(p,1) = del((l(i,p)),1) ;
    end
    if i == 1
            delbar1 = deli;
            mbar1= (Kg1 * delbar1)+fem1';
            fprintf ('Member Number =');
            disp (i);
            fprintf ('Global displacement matrix [DeltaBar] = \n');
            disp (delbar1');
            fprintf ('Global End moment matrix [MBar] = \n');
            disp (mbar1');
    else
            delbar2 = deli;
            mbar2= (Kg2 * delbar2)+fem2';
            fprintf ('Member Number =');
            disp (i);
            fprintf ('Global displacement matrix [DeltaBar] = \n');
            disp (delbar2');
            fprintf ('Global End moment matrix [MBar] = \n');
            disp (mbar2');
     end
end
%% check
mbar = [mbar1'; mbar2'];
jf = zeros(dof,1);
for a=1:n
    for b=1:4 % size of k matrix
        d = l(a,b);
        jfnew = zeros(dof,1);
        jfnew(d,1)=mbar(a,b);
        jf=jf+jfnew;
    end
end
fprintf ('Joint forces = \n');
disp (jf');
```

MATLAB output:

```
Member Number =      1
Local Stiffness matrix of member, [K] =
    1.3333    0.6667    0.6667   -0.6667
    0.6667    1.3333    0.6667   -0.6667
    0.6667    0.6667    0.4444   -0.4444
   -0.6667   -0.6667   -0.4444    0.4444

Member Number =      2
Local Stiffness matrix of member, [K] =
    0.8000    0.4000    0.2400   -0.2400
    0.4000    0.8000    0.2400   -0.2400
    0.2400    0.2400    0.0960   -0.0960
   -0.2400   -0.2400   -0.0960    0.0960

Stiffness Matrix of complete structure, [Ktotal] =
    2.1333    0.4000    0.6667    0.6667   -0.4267   -0.2400
    0.4000    0.8000         0         0    0.2400   -0.2400
    0.6667         0    1.3333    0.6667   -0.6667         0
    0.6667         0    0.6667    0.4444   -0.4444         0
   -0.4267    0.2400   -0.6667   -0.4444    0.5404   -0.0960
   -0.2400   -0.2400         0         0   -0.0960    0.0960

Unrestrained Stiffness sub-matix, [Kuu] =
    2.1333    0.4000
    0.4000    0.8000
Inverse of Unrestrained Stiffness sub-matix, [KuuInverse] =
    0.5172   -0.2586
   -0.2586    1.3793

Joint Load vector, [Jl] =
   -10    25   -15   -30   -50   -20
Unrestrained displacements, [DelU] =
  -11.6379   37.0690

Member Number =      1
Global displacement matrix [DeltaBar] =
         0  -11.6379         0         0
Global End moment matrix [MBar] =
    7.2414  -30.5172   22.2414   37.7586

Member Number =      2
Global displacement matrix [DeltaBar] =
  -11.6379   37.0690         0         0
Global End moment matrix [MBar] =
   30.5172   -0.0000   26.1034   13.8966

Joint forces =
         0   -0.0000    7.2414   22.2414   63.8621   13.8966
```

2.9 PLANAR ORTHOGONAL FRAME

2.9.1 Example Problem

Neglecting the axial deformation, analyze the single bay single-storey frame shown below:

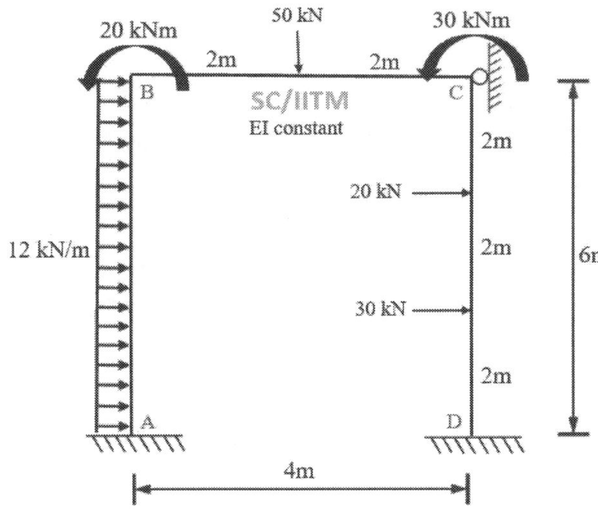

i. *Marking unrestrained and restrained degrees-of-freedom:*
   The unrestrained and restrained degrees of freedom are marked as shown in Figure 2.22:
   Thus, the size of the stiffness matrix will be 9 × 9.
   Unrestrained degrees-of-freedom = 2 [θ_1, θ_2]
   Restrained degrees-of-freedom = 7 [θ_3, θ_6, δ_4, δ_5, δ_7, δ_8, δ_9]

Thus, the unrestrained sub-matrix will be of size 2 × 2. The size of the total stiffness matrix will be 9 × 9. The element stiffness matrix will be 4 × 4, neglecting axial deformation.

$$[K]=\begin{bmatrix}[k_{uu}]_{2\times2} & [k_{ur}]_{2\times7}\\ [k_{ru}]_{7\times2} & [k_{rr}]_{7\times7}\end{bmatrix}$$

The orientation and the labels of the members are given in the following table. Please note that for column member AB, jth end of the member (origin of the member) is chosen at B. Hence, the x-axis will project toward A, and the y-axis will be marked anticlockwise 90° to the x-axis. It is important to note that only for this choice of origin, the degrees of freedom shall

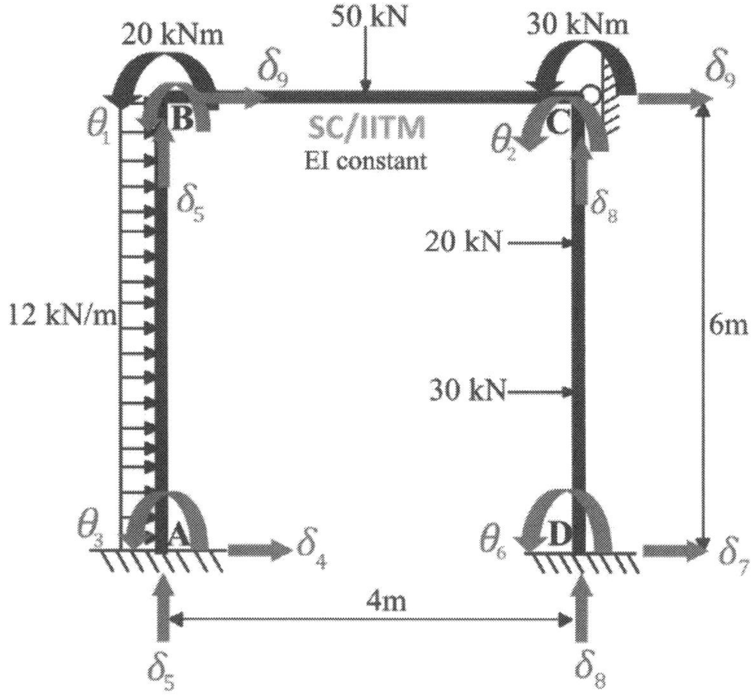

FIGURE 2.22 Unrestrained and restrained degrees of freedom.

remain positive along the y-axis of the member AB. This is also true for the column member CD. However, for the beam member, BC, the origin (jth end) is chosen at B, which is conventional.

Member	jth End	kth End	Labels
BA	B	A	(1,3,9,4)
BC	B	C	(1,2,5,8)
CD	C	D	(2,6,9,7)

$$\{\Delta\}_{9\times1} = \begin{Bmatrix} \theta_1 \\ \theta_2 \\ \theta_3 \\ \delta_4 \\ \delta_5 \\ \delta_6 \\ \delta_7 \\ \delta_8 \\ \delta_9 \end{Bmatrix} = \begin{Bmatrix} [\Delta_u]_{2\times1} \\ [\Delta_r]_{7\times1} \end{Bmatrix}$$

ii. *Formulation of stiffness matrix:*
The stiffness matrix is given by:

$$K_i = \begin{bmatrix} \frac{4EI}{l} & \frac{2EI}{l} & \frac{6EI}{l^2} & -\frac{6EI}{l^2} \\ \frac{2EI}{l} & \frac{4EI}{l} & \frac{6EI}{l^2} & -\frac{6EI}{l^2} \\ \frac{6EI}{l^2} & \frac{6EI}{l^2} & \frac{12EI}{l^3} & -\frac{12EI}{l^3} \\ -\frac{6EI}{l^2} & -\frac{6EI}{l^2} & -\frac{12EI}{l^3} & \frac{12EI}{l^3} \end{bmatrix}$$

By substituting the respective length and (EI) values of each member, the element stiffness matrix is given as follows:

$$\begin{matrix} & 1 & 3 & 9 & 4 & \\ K_{AB} = EI & \left[\begin{matrix} 1.333 \\ 0.667 \\ 0.333 \\ -0.333 \end{matrix}\right. & \begin{matrix} 0.667 \\ 1.333 \\ 0.333 \\ -0.333 \end{matrix} & \begin{matrix} 0.333 \\ 0.333 \\ 0.111 \\ -0.111 \end{matrix} & \left.\begin{matrix} -0.333 \\ -0.333 \\ -0.111 \\ 0.111 \end{matrix}\right] & \begin{matrix} 1 \\ 3 \\ 9 \\ 4 \end{matrix} \end{matrix}$$

$$\begin{matrix} & 1 & 2 & 5 & 8 & \\ K_{BC} = EI & \left[\begin{matrix} 1 \\ 0.5 \\ 0.375 \\ -0.375 \end{matrix}\right. & \begin{matrix} 0.5 \\ 1 \\ 0.375 \\ -0.375 \end{matrix} & \begin{matrix} 0.375 \\ 0.375 \\ 0.188 \\ -0.188 \end{matrix} & \left.\begin{matrix} -0.375 \\ -0.375 \\ -0.188 \\ 0.188 \end{matrix}\right] & \begin{matrix} 1 \\ 2 \\ 5 \\ 8 \end{matrix} \end{matrix}$$

$$\begin{matrix} & 2 & 6 & 9 & 7 & \\ K_{CD} = EI & \left[\begin{matrix} 1.333 \\ 0.667 \\ 0.333 \\ -0.333 \end{matrix}\right. & \begin{matrix} 0.667 \\ 1.333 \\ 0.333 \\ -0.333 \end{matrix} & \begin{matrix} 0.333 \\ 0.333 \\ 0.111 \\ -0.111 \end{matrix} & \left.\begin{matrix} -0.333 \\ -0.333 \\ -0.111 \\ 0.111 \end{matrix}\right] & \begin{matrix} 2 \\ 6 \\ 9 \\ 7 \end{matrix} \end{matrix}$$

It is important to note the labels, as marked in the rows and columns of each matrix. This is useful to assemble the [K] of the complete frame. However, to solve the problem, we need to estimate only the unrestrained displacements, as the restrained displacements are nil. Hence, from the assembly of [K] of the complete frame, we need to partition the matrix to obtain $[k_{uu}]$.

$$\begin{matrix} & 1 & 2 & \\ K_{UU} = EI & \left[\begin{matrix} 2.333 \\ 0.5 \end{matrix}\right. & \left.\begin{matrix} 0.5 \\ 2.333 \end{matrix}\right] & \begin{matrix} 1 \\ 2 \end{matrix} \end{matrix}$$

$$K_{UU}^{-1} = \frac{1}{5.193EI}\begin{bmatrix} 2.333 & -0.5 \\ -0.5 & 2.333 \end{bmatrix}$$

iii. *Calculation of fixed end moments:*
The fixed end moments and reactions of the members are given below:

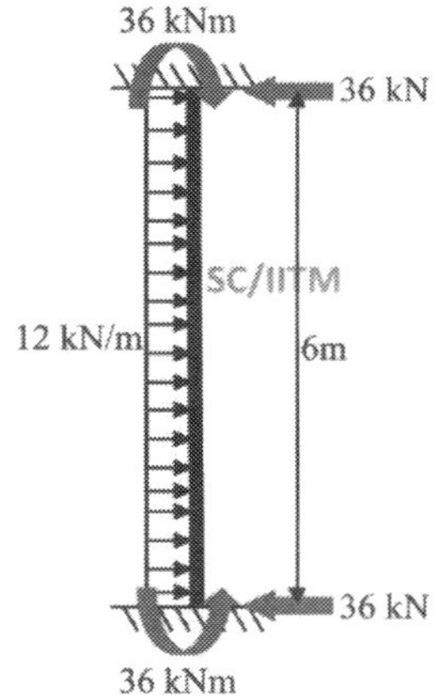

$$M_{AB}^{F} = -\frac{wl^2}{12} = -\frac{12\times6^2}{12} = -36\,\text{kNm}$$

$$M_{BA}^{F} = +36\,\text{kNm}$$

$$V_A = -36\,\text{kN};\ V_B = -36\,\text{kN}$$

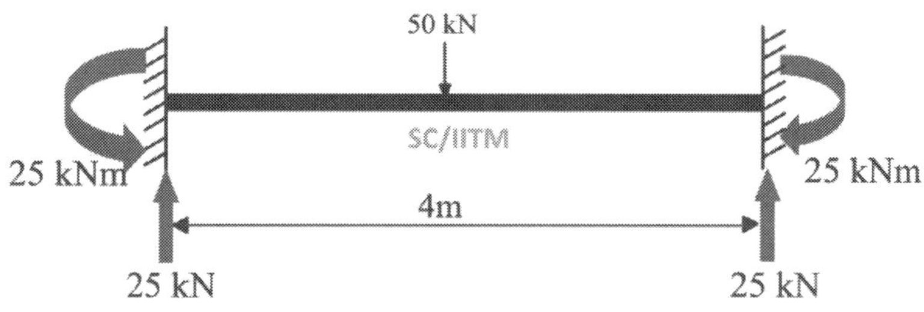

$$M_{BC}^{F} = \frac{pl}{8} = \frac{50\times4}{8} = +25\,\text{kNm}$$

$$M_{CB}^{F} = -25\,\text{kNm}$$

$$V_B = +25\,\text{kN};\ V_C = +25\,\text{kN}$$

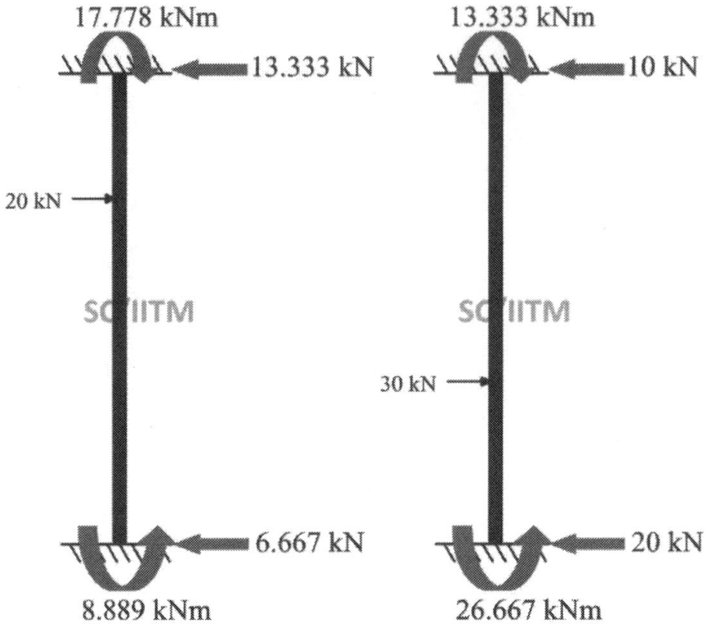

For 20 kN load,

$$M_{CD}^{F} = \frac{pab^2}{l^2} = \frac{20 \times 2 \times 4^2}{6^2} = +17.778\,\text{kNm}$$

$$M_{DC}^{F} = \frac{pa^2b}{l^2} = \frac{20 \times 2^2 \times 4}{6^2} = -8.889\,\text{kNm}$$

$$V_C = +13.333\,\text{kN}$$

$$V_D = +6.667\,\text{kN}$$

For 30 kN load,

$$M_{CD}^{F} = \frac{pab^2}{l^2} = \frac{30 \times 4 \times 2^2}{6^2} = +13.333\,\text{kNm}$$

$$M_{DC}^{F} = \frac{pa^2b}{l^2} = \frac{30 \times 2 \ \times 4^2}{6^2} = -26.667\,\text{kNm}$$

$$V_C = +10\,\text{kN}$$

$$V_D = +20\,\text{kN}$$

Thus, for the member CD, fixed end moments and reactions are given by,

$$M_{CD}^{F} = -17.778 - 13.333 = -31.111\,\text{kNm}$$

$$M_{DC}^{F} = +8.889 + 26.667 = +35.556\,\text{kNm}$$

$$V_C = -13.333 - 10 = -23.333\,\text{kN}$$

$$V_D = -6.667 - 20 = -26.667\,\text{kN}$$

Thus, for the whole beam, the fixed end moments are written as follows:

$$AB = \begin{Bmatrix} -36 \\ +36 \\ -36 \\ -36 \end{Bmatrix}, BC = \begin{Bmatrix} +25 \\ -25 \\ +25 \\ +25 \end{Bmatrix}, CD = \begin{Bmatrix} -31.111 \\ +35.556 \\ -23.333 \\ -26.667 \end{Bmatrix}$$

iv. *Calculation of joint load vectors:*

$$\{J_L\}_{9x1} = \begin{Bmatrix} \mathbf{31} \\ \mathbf{86.111} \\ \mathbf{-36} \\ \mathbf{36} \\ \mathbf{-25} \\ \mathbf{-35.556} \\ \mathbf{26.667} \\ \mathbf{-25} \\ \mathbf{59.333} \end{Bmatrix} \begin{matrix} 1 \\ 2 \\ 3 \\ 4 \\ 5 \\ 6 \\ 7 \\ 8 \\ 9 \end{matrix}$$

The joint load vector can be split into unrestrained and restrained degrees of freedom.

$$\{J_L\} = \begin{Bmatrix} (J_L)_{u} \\ {}_{2\times1} \\ --- \\ (J_L)_{r} \\ {}_{7\times1} \end{Bmatrix}_{9\times1}$$

Thus,

$$J_{LU} = \begin{Bmatrix} \mathbf{31} \\ \mathbf{86.111} \end{Bmatrix} \begin{matrix} 1 \\ 2 \end{matrix}$$

v. *Calculation of displacements:*

$$\{\Delta_u\} = [k_{uu}]^{-1}\{J_L\}_u$$

$$\begin{Bmatrix} \theta_1 \\ \theta_2 \end{Bmatrix} = \frac{1}{5.193EI}\begin{bmatrix} 2.333 & -0.5 \\ -0.5 & 2.333 \end{bmatrix}\begin{Bmatrix} 31 \\ 86.111 \end{Bmatrix} = \frac{1}{5.193EI}\begin{Bmatrix} 29.268 \\ 185.397 \end{Bmatrix} radians$$

vi. *Calculation of end moment and shear:*
The general equation to find the end moment of ith beam is,

$$[M]_i = k_i\delta_i + (FEM)_i$$

The equation can be rewritten element-wise as follows:

$$M_{AB} = [k]_{AB}\{\delta\} + \{FEM\}$$

$$\begin{Bmatrix} M_1 \\ M_3 \\ V_9 \\ V_4 \end{Bmatrix} = EI\begin{bmatrix} 1.333 & 0.667 & 0.333 & -0.333 \\ 0.667 & 1.333 & 0.333 & -0.333 \\ 0.333 & 0.333 & 0.111 & -0.111 \\ -0.333 & -0.333 & -0.111 & 0.111 \end{bmatrix}\begin{Bmatrix} \theta_1 \\ \theta_3 \\ \delta_9 \\ \delta_4 \end{Bmatrix} + \begin{Bmatrix} -36 \\ +36 \\ -36 \\ +36 \end{Bmatrix}$$

We know that, $\theta_1 = \frac{29.268}{5.193\text{EI}}$, $\theta_3 = \delta_9 = \delta_4 = 0$.

$$\text{Thus,}\begin{Bmatrix} M_1 \\ M_3 \\ V_9 \\ V_4 \end{Bmatrix} = \begin{Bmatrix} -28.487 \\ 39.759 \\ 34.123 \\ -34.123 \end{Bmatrix}$$

$$M_{BC} = [k]_{BC}\{\delta\} + \{FEM\}$$

$$\begin{Bmatrix} M_1 \\ M_2 \\ V_5 \\ V_8 \end{Bmatrix} = EI\begin{bmatrix} 1 & 0.5 & 0.375 & -0.375 \\ 0.5 & 1 & 0.375 & -0.375 \\ 0.375 & 0.375 & 0.188 & -0.188 \\ -0.375 & -0.375 & -0.188 & 0.188 \end{bmatrix}\begin{Bmatrix} \theta_1 \\ \theta_2 \\ \delta_5 \\ \delta_8 \end{Bmatrix} + \begin{Bmatrix} 25 \\ -25 \\ 25 \\ 25 \end{Bmatrix} = \begin{Bmatrix} 48.487 \\ 13.519 \\ 40.502 \\ 9.498 \end{Bmatrix}$$

Similarly for the member CD,

$$M_{CD} = [k]_{CD}\{\delta\} + \{FEM\}$$

$$\begin{Bmatrix} M_2 \\ M_6 \\ V_9 \\ V_7 \end{Bmatrix} = EI \begin{bmatrix} 1.333 & 0.667 & 0.333 & -0.333 \\ 0.667 & 1.333 & 0.333 & -0.333 \\ 0.333 & 0.333 & 0.111 & -0.111 \\ -0.333 & -0.333 & -0.111 & 0.111 \end{bmatrix} \begin{Bmatrix} \theta_2 \\ \theta_6 \\ \delta_9 \\ \delta_7 \end{Bmatrix} + \begin{Bmatrix} -31.111 \\ 35.556 \\ -23.333 \\ -26.667 \end{Bmatrix} = \begin{Bmatrix} 16.479 \\ 59.369 \\ -11.444 \\ -38.556 \end{Bmatrix}$$

The member end moments and shear are shown in the figure below:

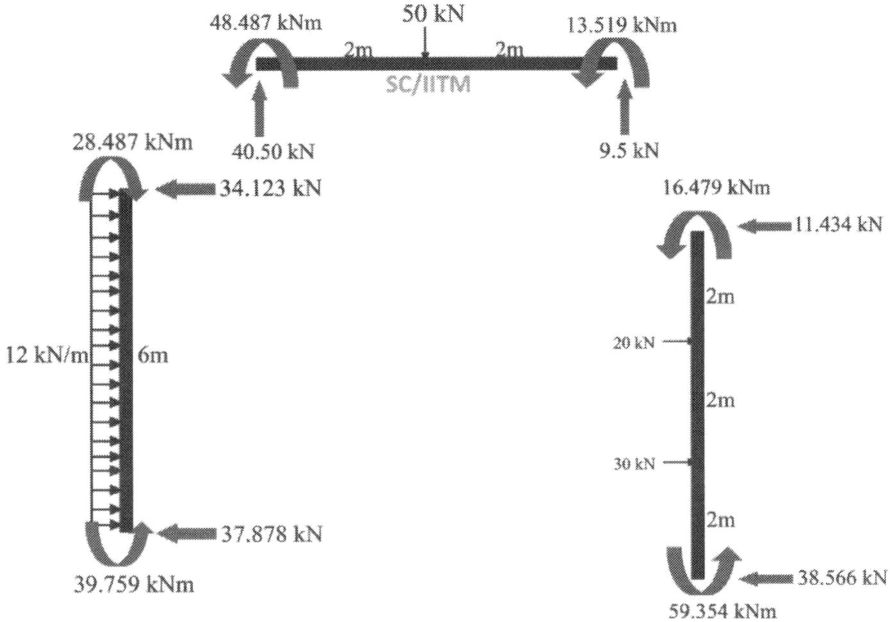

Check:
In member AB,

$\sum F_H = 0$

$(12\times6)-34.123-37.877=0$

$\sum M_A = 0$

$+28.487+(12\times6\times3)-(34.123\times6)+39.75=0$

In member BC,

$\sum F_V = 0$

$40.502+9.798-50=0$

$$\sum M_B = 0$$

$$-13.159 - (9.498 \times 4) + (50 \times 2) + 98.489 = 0$$

In member CD,

$$\sum F_H = 0$$

$$20 + 30 - 11.444 - 38.556 = 0$$

$$\sum M_D = 0$$

$$+16.479 + (11.444 \times 6) - (20 \times 4) - (30 \times 2) + 59.369 = 0$$

Final end moments and shear are shown below:

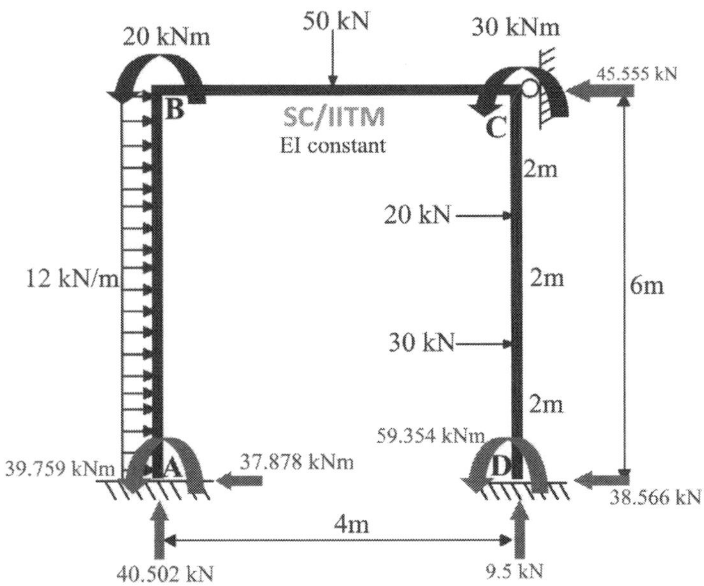

MATLAB program:

```
%% stiffness matrix method
% Input
clc;
clear;
n = 3; % number of members
I = [2 1 2]; %Moment of inertia in m4
L = [6 4 6]; % length in m
```

```
uu = 2; % Number of unrestrained degrees of freedom
ur = 7; % Number of restrained degrees of freedom
uul = [1 2]; % global labels of unrestrained dof
url = [3 4 5 6 7 8 9]; % global labels of restrained dof
l1 = [1 3 9 4]; % Global labels for member 1
l2 = [1 2 5 8]; % Global labels for member 2
l3 = [2 6 9 7]; % Global labels for member 3
l= [l1; l2; l3];
dof = uu+ur;
Ktotal = zeros (dof);
fem1= [-36 36 -36 -36]; % Local Fixed end moments of member 1
fem2= [25 -25 25 25]; % Local Fixed end moments of member 2
fem3= [-31.111 35.556 -23.333 -26.667]; % Local Fixed end moments
of member 3
%% rotation coefficients for each member
rc1 = 4.*I./L;
rc2 = 2.*I./L;
%% stiffness matrix 4 by 4 (axial deformation neglected)
for i = 1:n
        Knew = zeros (dof);
        k1 = [rc1(i); rc2(i); (rc1(i)+rc2(i))/L(i); (-
(rc1(i)+rc2(i))/L(i))];
        k2 = [rc2(i); rc1(i); (rc1(i)+rc2(i))/L(i); (-
(rc1(i)+rc2(i))/L(i))];
        k3 = [(rc1(i)+rc2(i))/L(i); (rc1(i)+rc2(i))/L(i);
(2*(rc1(i)+rc2(i))/(L(i)^2)); (-2*(rc1(i)+rc2(i))/(L(i)^2))];
        k4 = -k3;
        K = [k1 k2 k3 k4];
        fprintf ('Member Number =');
        disp (i);
        fprintf ('Local Stiffness matrix of member, [K] = \n');
        disp (K);
        for p = 1:4
            for q = 1:4
                Knew((l(i,p)),(l(i,q))) =K(p,q);
            end
        end
        Ktotal = Ktotal + Knew;
        if i == 1
            Kg1=K;
        elseif i == 2
            Kg2 = K;
        else
            Kg3=K;
        end
end
fprintf ('Stiffness Matrix of complete structure, [Ktotal] = \n');
disp (Ktotal);
Kunr = zeros(uu);
for x=1:uu
```

```
    for y=1:uu
        Kunr(x,y)= Ktotal(x,y);
    end
end
fprintf ('Unrestrained Stiffness sub-matrix, [Kuu] = \n');
disp (Kunr);
KuuInv= inv(Kunr);
fprintf ('Inverse of Unrestrained Stiffness sub-matrix,[KuuInverse]=
\n');
disp (KuuInv);
%% Creation of joint load vector
jl= [31; 86.111; -36; 36; -25; -35.556; 26.665; -25; 59.333]; % values
given in kN or kNm
jlu = [31; 86.111]; % load vector in unrestrained dof
delu = KuuInv*jlu;
fprintf ('Joint Load vector, [Jl] = \n');
disp (jl');
fprintf ('Unrestrained displacements, [DelU] = \n');
disp (delu');
delr = zeros (ur,1);
del = [delu; delr];
deli= zeros (4,1);
for i = 1:n
    for p = 1:4
        deli(p,1) = del((l(i,p)),1) ;
    end
    if i == 1
            delbar1 = deli;
            mbar1= (Kg1 * delbar1)+fem1';
            fprintf ('Member Number =');
            disp (i);
            fprintf ('Global displacement matrix [DeltaBar] = \n');
            disp (delbar1');
            fprintf ('Global End moment matrix [MBar] = \n');
            disp (mbar1');
        elseif i == 2
            delbar2 = deli;
            mbar2= (Kg2 * delbar2)+fem2';
            fprintf ('Member Number =');
            disp (i);
            fprintf ('Global displacement matrix [DeltaBar] = \n');
            disp (delbar2');
            fprintf ('Global End moment matrix [MBar] = \n');
            disp (mbar2');
        else
            delbar3 = deli;
            mbar3= (Kg3 * delbar3)+fem3';
            fprintf ('Member Number =');
            disp (i);
            fprintf ('Global displacement matrix [DeltaBar] = \n');
```

```
            disp (delbar3');
            fprintf ('Global End moment matrix [MBar] = \n');
            disp (mbar3');
       end
end
%% check
mbar = [mbar1'; mbar2'; mbar3'];
jf = zeros(dof,1);
for a=1:n
    for b=1:4 % size of k matrix
        d = l(a,b);
        jfnew = zeros(dof,1);
        jfnew(d,1)=mbar(a,b);
        jf=jf+jfnew;
    end
end
fprintf ('Joint forces = \n');
disp (jf');
```

MATLAB output:

```
Member Number =      1
Local Stiffness matrix of member, [K] =
    1.3333    0.6667    0.3333   -0.3333
    0.6667    1.3333    0.3333   -0.3333
    0.3333    0.3333    0.1111   -0.1111
   -0.3333   -0.3333   -0.1111    0.1111


Member Number =      2
Local Stiffness matrix of member, [K] =
    1.0000    0.5000    0.3750   -0.3750
    0.5000    1.0000    0.3750   -0.3750
    0.3750    0.3750    0.1875   -0.1875
   -0.3750   -0.3750   -0.1875    0.1875


Member Number =      3
Local Stiffness matrix of member, [K] =
    1.3333    0.6667    0.3333   -0.3333
    0.6667    1.3333    0.3333   -0.3333
    0.3333    0.3333    0.1111   -0.1111
   -0.3333   -0.3333   -0.1111    0.1111

Stiffness Matrix of complete structure, [Ktotal] =
 2.3333  0.5000  0.6667 -0.3333 0.3750       0       0 -0.3750  0.3333
 0.5000  2.3333       0       0 0.3750 0.6667 -0.3333 -0.3750  0.3333
 0.6667       0  1.3333 -0.3333      0       0       0       0  0.3333
 0.3333       0 -0.3333  0.1111      0       0       0       0 -0.1111
 0.3750  0.3750       0       0  0.1875       0       0 -0.1875       0
```

```
      0  0.6667       0       0       0  1.3333 -0.3333       0  0.3333
      0 -0.3333       0       0       0 -0.3333  0.1111       0 -0.1111
-0.3750 -0.3750       0       0 -0.1875       0       0  0.1875       0
 0.3333  0.3333  0.3333 -0.1111       0  0.3333 -0.1111       0  0.2222


Unrestrained Stiffness sub-matix, [Kuu] =
    2.3333    0.5000
    0.5000    2.3333
Inverse of Unrestrained Stiffness sub-matix, [KuuInverse] =
    0.4492   -0.0963
   -0.0963    0.4492


Joint Load vector, [Jl] =
  31.0000 86.1110 -36.0000 36.0000 -25.0000 -35.5560 26.6650 -25.0000
59.3330
Unrestrained displacements, [DelU] =
    5.6364   35.6969


Member Number =      1
Global displacement matrix [DeltaBar] =
    5.6364         0         0         0
Global End moment matrix [MBar] =
  -28.4848   39.7576  -34.1212  -37.8788


Member Number =      2
Global displacement matrix [DeltaBar] =
    5.6364   35.6969         0         0
Global End moment matrix [MBar] =
   48.4848   13.5151   40.5000    9.5000



Member Number =      3
Global displacement matrix [DeltaBar] =
   35.6969         0         0         0
Global End moment matrix [MBar] =
   16.4849   59.3539  -11.4340  -38.5660

Joint forces =

  20.0000   30.0000   39.7576   -37.8788   40.5000   59.3539   -38.5660   9.5000
-45.5552
```

2.10 STEP FRAME EXAMPLE PROBLEM

Analyze the frame shown below, neglecting the axial deformation. Note that column and beam members have different moments of inertia.

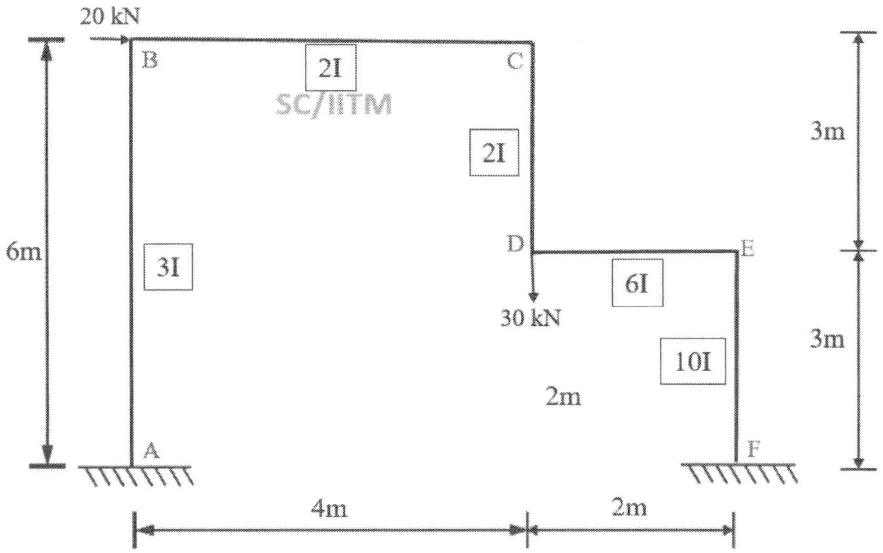

i. *Marking unrestrained and restrained degrees of freedom:*

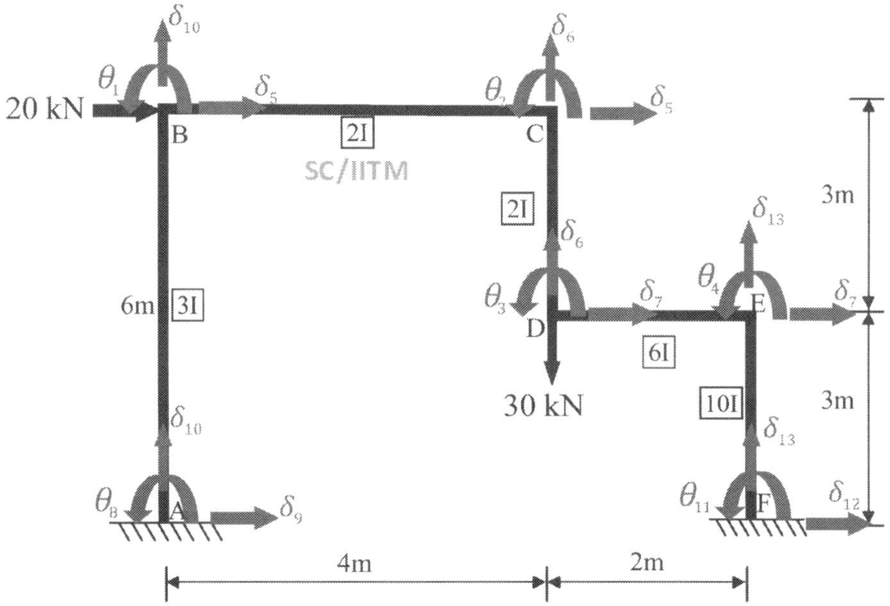

Total number of degrees-of-freedom is 13. Thus, the size of the stiffness matrix of the complete frame will be 13 × 13, while the member stiffness matrix shall remain as 4 × 4, as the axial deformation is neglected.

Unrestrained degrees-of-freedom = 7 $[\theta_1, \theta_2, \theta_3, \theta_4, \delta_5, \delta_6, \delta_7]$
Restrained degrees-of-freedom = 6 $[\theta_8, \theta_{11}, \delta_9, \delta_{10}, \delta_{12}, \delta_{13}]$

The orientation and the labels of the members are given in the following table. Please note that for column member AB, jth end of the member (origin of the member) is chosen at B. Hence, the x-axis will project toward A, and the y-axis will be marked anticlockwise 90° to the x-axis. It is important to note that only for this choice of origin, the degrees of freedom shall remain positive along the y-axis of the member AB. This is also true for all column members. However, for the beam members, say, for example, BC, origin (jth end) is chosen at B, which is conventional.

Member	jth End	kth End	Labels
BA	B	A	(1,8,5,9)
BC	B	C	(1,2,10,6)
CD	C	D	(2,3,5,7)
DE	D	E	(3,4,6,13)
EF	E	F	(4,11,7,12)

ii. *Formulation of stiffness matrix:*
The element stiffness matrix is given by:

$$K_i = \begin{bmatrix} \frac{4EI}{l} & \frac{2EI}{l} & \frac{6EI}{l^2} & -\frac{6EI}{l^2} \\ \frac{2EI}{l} & \frac{4EI}{l} & \frac{6EI}{l^2} & -\frac{6EI}{l^2} \\ \frac{6EI}{l^2} & \frac{6EI}{l^2} & \frac{12EI}{l^3} & -\frac{12EI}{l^3} \\ -\frac{6EI}{l^2} & -\frac{6EI}{l^2} & -\frac{12EI}{l^3} & \frac{12EI}{l^3} \end{bmatrix}$$

$$\begin{matrix} & 1 & 8 & 5 & 9 & \end{matrix}$$

$$K_{AB} = EI\begin{bmatrix} 2 & 1 & 0.5 & -0.5 \\ 1 & 2 & 0.5 & -0.5 \\ 0.5 & 0.5 & 0.167 & -0.167 \\ -0.5 & -0.5 & -0.167 & 0.167 \end{bmatrix} \begin{matrix} 1 \\ 8 \\ 5 \\ 9 \end{matrix}$$

$$\begin{matrix} & 1 & 2 & 10 & 6 & \end{matrix}$$

$$K_{BC} = EI\begin{bmatrix} 2 & 1 & 0.75 & -0.75 \\ 1 & 2 & 0.75 & -0.75 \\ 0.75 & 0.75 & 0.375 & -0.375 \\ -0.75 & -0.75 & -0.375 & 0.375 \end{bmatrix} \begin{matrix} 1 \\ 2 \\ 10 \\ 6 \end{matrix}$$

$$K_{CD} = EI \begin{bmatrix} 2.667 & 1.333 & 1.333 & -1.333 \\ 1.333 & 2.667 & 1.333 & -1.333 \\ 1.333 & 1.333 & 0.889 & -0.889 \\ -1.333 & -1.333 & -0.889 & 0.889 \end{bmatrix} \begin{matrix} 2 \\ 3 \\ 5 \\ 7 \end{matrix} \quad \text{(columns: 2, 3, 5, 7)}$$

$$K_{DE} = EI \begin{bmatrix} 12 & 6 & 9 & -9 \\ 6 & 12 & 9 & -9 \\ 9 & 9 & 9 & -9 \\ -9 & -9 & -9 & 9 \end{bmatrix} \begin{matrix} 3 \\ 4 \\ 6 \\ 13 \end{matrix} \quad \text{(columns: 3, 4, 6, 13)}$$

$$K_{EF} = EI \begin{bmatrix} 13.333 & 6.667 & 6.667 & -6.667 \\ 6.667 & 13.333 & 6.667 & -6.667 \\ 6.667 & 6.667 & 4.445 & -4.445 \\ -6.667 & -6.667 & -4.445 & 4.445 \end{bmatrix} \begin{matrix} 4 \\ 11 \\ 7 \\ 12 \end{matrix} \quad \text{(columns: 4, 11, 7, 12)}$$

It is important to note the labels, as marked in the rows and columns of each member matrix. This is useful to assemble the [K] of the complete frame. However, for solving the problem, we need to estimate only the unrestrained displacements, as the restrained displacements are nil. Hence, from the assembled [K] of the complete frame, we need to partition $[k_{uu}]$.

$$K_{UU} = EI \begin{bmatrix} 4 & 1 & 0 & 0 & 0.5 & -0.75 & 0 \\ 1 & 4.667 & 1.333 & 0 & 1.333 & -0.75 & -1.333 \\ 0 & 1.333 & 14.667 & 6 & 1.333 & 9 & -1.333 \\ 0 & 0 & 6 & 25.333 & 0 & 9 & 6.667 \\ 0.5 & 1.333 & 1.333 & 0 & 1.056 & 0 & -0.889 \\ -0.75 & -0.75 & 9 & 9 & 0 & 9.375 & 0 \\ 0 & -1.333 & -1.333 & 6.667 & -0.889 & 0 & 5.334 \end{bmatrix} \begin{matrix} 1 \\ 2 \\ 3 \\ 4 \\ 5 \\ 6 \\ 7 \end{matrix} \quad \text{(columns: 1, 2, 3, 4, 5, 6, 7)}$$

Since the size of the unrestrained stiffness matrix is 7 × 7. The inverse of the matrix is calculated using MATLAB programming, and the solution is obtained by MATLAB code written for solving the planar orthogonal structures.

$$K_{UU}^{-1} = \frac{1}{EI} \begin{bmatrix} 0.2758 & -0.0298 & -0.0053 & -0.0016 & -0.1070 & 0.0263 & -0.0246 \\ -0.0298 & 0.3843 & -0.0747 & -0.0808 & -0.2639 & 0.1776 & 0.1344 \\ -0.0053 & -0.0747 & 0.2963 & 0.1098 & -0.4029 & -0.3962 & -0.1490 \\ -0.0016 & -0.0808 & 0.1098 & 0.1914 & -0.269 & -0.2957 & -0.2768 \\ -0.107 & -0.2639 & -0.4029 & -0.2690 & 2.3068 & 0.6154 & 0.5540 \\ 0.0263 & 0.1776 & -0.3962 & -0.2957 & 0.6154 & 0.7873 & 0.4175 \\ -0.0246 & 0.1344 & -0.1490 & -0.2768 & 0.5540 & 0.4175 & 0.6221 \end{bmatrix}$$

iii. *Calculation of fixed end moments:*
As there are no member forces, the FEM vector of all members shall be a nil vector. Hence, the joint load vector can be written directly.

iv. *Calculation of joint load vectors:*

$$\{J_L\}_{13\,x\,1} = \begin{Bmatrix} \mathbf{0} \\ \mathbf{0} \\ \mathbf{0} \\ \mathbf{0} \\ \mathbf{20} \\ \mathbf{-30} \\ \mathbf{0} \\ \mathbf{0} \\ \mathbf{0} \\ \mathbf{0} \\ \mathbf{0} \\ \mathbf{0} \\ \mathbf{0} \end{Bmatrix} \begin{matrix} 1 \\ 2 \\ 3 \\ 4 \\ 5 \\ 6 \\ 7 \\ 8 \\ 9 \\ 10 \\ 11 \\ 12 \\ 13 \end{matrix}$$

The joint load vector can be split into unrestrained and restrained degrees of freedom.

$$\{J_L\} = \begin{Bmatrix} (J_L)_{u\atop 7\times1} \\ --- \\ (J_L)_{r\atop 6\times1} \end{Bmatrix}_{13\times1}$$

Then,

$$J_{LU} = \begin{Bmatrix} \mathbf{0} \\ \mathbf{0} \\ \mathbf{0} \\ \mathbf{0} \\ \mathbf{20} \\ \mathbf{-30} \\ \mathbf{0} \end{Bmatrix} \begin{matrix} 1 \\ 2 \\ 3 \\ 4 \\ 5 \\ 6 \\ 7 \end{matrix}$$

v. *Calculation of displacements:*

Now, $\{\Delta_u\} = [k_{uu}]^{-1}\{J_L\}_u$

$$\begin{Bmatrix} \theta_1 \\ \theta_2 \\ \theta_3 \\ \theta_4 \\ \delta_5 \\ \delta_6 \\ \delta_7 \end{Bmatrix} == \frac{1}{EI} \begin{Bmatrix} -2.9291 \\ -10.6047 \\ 3.8284 \\ 3.4921 \\ 27.6762 \\ -11.3104 \\ -1.4465 \end{Bmatrix}$$

vi. *Calculation of end moment and shear:*
The general equation to find the end moment of ith beam is given as:

$$[M]_i = k_i \delta_i + (FEM)_i$$

The equation can be rewritten element-wise as follows:

$$M_{AB} = [k]_{AB} \{\delta\} + \{FEM\}$$

$$\begin{Bmatrix} M_1 \\ M_8 \\ V_5 \\ V_9 \end{Bmatrix} = EI \begin{bmatrix} 2 & 1 & 0.5 & -0.5 \\ 1 & 2 & 0.5 & -0.5 \\ 0.5 & 0.5 & 0.167 & -0.167 \\ -0.5 & -0.5 & -0.167 & 0.167 \end{bmatrix} \begin{Bmatrix} \theta_1 \\ \theta_8 \\ \delta_5 \\ \delta_9 \end{Bmatrix} + 0$$

Thus, $\begin{Bmatrix} M_1 \\ M_8 \\ V_5 \\ V_9 \end{Bmatrix} = \begin{Bmatrix} 7.98 \\ 10.9091 \\ 3.1482 \\ -3.1482 \end{Bmatrix}$

$$M_{BC} = [k]_{BC} \{\delta\} + \{FEM\}$$

$$\begin{Bmatrix} M_1 \\ M_2 \\ V_{10} \\ V_6 \end{Bmatrix} = EI \begin{bmatrix} 2 & 1 & 0.75 & -0.75 \\ 1 & 2 & 0.75 & -0.75 \\ 0.75 & 0.75 & 0.375 & -0.375 \\ -0.75 & -0.75 & -0.375 & 0.375 \end{bmatrix} \begin{Bmatrix} \theta_1 \\ \theta_2 \\ \delta_{10} \\ \delta_6 \end{Bmatrix} + 0 = \begin{Bmatrix} -7.98 \\ -15.6556 \\ -5.9089 \\ 5.9089 \end{Bmatrix}$$

Similarly, for the member CD, $M_{CD} = [k]_{CD} \{\delta\} + \{FEM\}$

$$\begin{Bmatrix} M_2 \\ M_3 \\ V_5 \\ V_7 \end{Bmatrix} = EI \begin{bmatrix} 2.667 & 1.333 & 1.333 & -1.333 \\ 1.333 & 2.667 & 1.333 & -1.333 \\ 1.333 & 1.333 & 0.889 & -0.889 \\ -1.333 & -1.333 & -0.889 & 0.889 \end{bmatrix} \begin{Bmatrix} \theta_2 \\ \theta_3 \\ \delta_5 \\ \delta_7 \end{Bmatrix} + 0 = \begin{Bmatrix} 15.6556 \\ 34.8998 \\ 16.8518 \\ -16.8518 \end{Bmatrix}$$

$$M_{DE} = \left[k\right]_{DE} \left\{\delta\right\} + \left\{FEM\right\}$$

$$\begin{Bmatrix} M_3 \\ M_4 \\ V_6 \\ V_{13} \end{Bmatrix} = EI \begin{bmatrix} 12 & 6 & 9 & -9 \\ 6 & 12 & 9 & -9 \\ 9 & 9 & 9 & -9 \\ -9 & -9 & -9 & 9 \end{bmatrix} \begin{Bmatrix} \theta_3 \\ \theta_4 \\ \delta_6 \\ \delta_{13} \end{Bmatrix} + 0 = \begin{Bmatrix} -34.8998 \\ -36.9180 \\ -35.9089 \\ 35.9089 \end{Bmatrix}$$

$$M_{EF} = \left[k\right]_{EF} \left\{\delta\right\} + \left\{FEM\right\}$$

$$\begin{Bmatrix} M_4 \\ M_{11} \\ V_7 \\ V_{12} \end{Bmatrix} = EI \begin{bmatrix} 13.333 & 6.667 & 6.667 & -6.667 \\ 6.667 & 13.333 & 6.667 & -6.667 \\ 6.667 & 6.667 & 4.445 & -4.445 \\ -6.667 & -6.667 & -4.445 & 4.445 \end{bmatrix} \begin{Bmatrix} \theta_4 \\ \theta_{11} \\ \delta_7 \\ \delta_{12} \end{Bmatrix} + 0 = \begin{Bmatrix} 36.9180 \\ 13.6375 \\ 16.8518 \\ -16.8518 \end{Bmatrix}$$

The member and final end moments and shear are shown below:

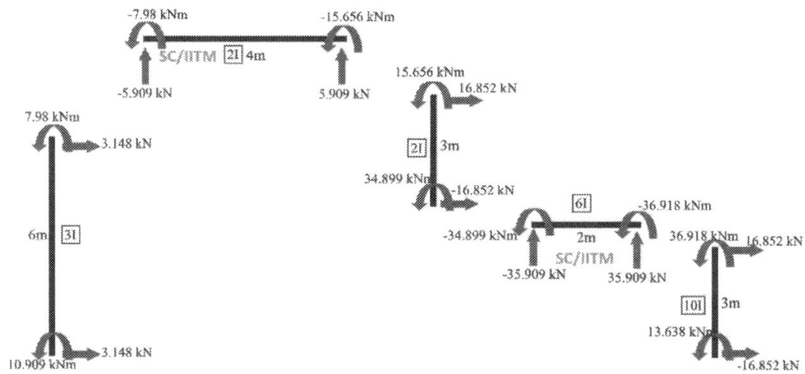

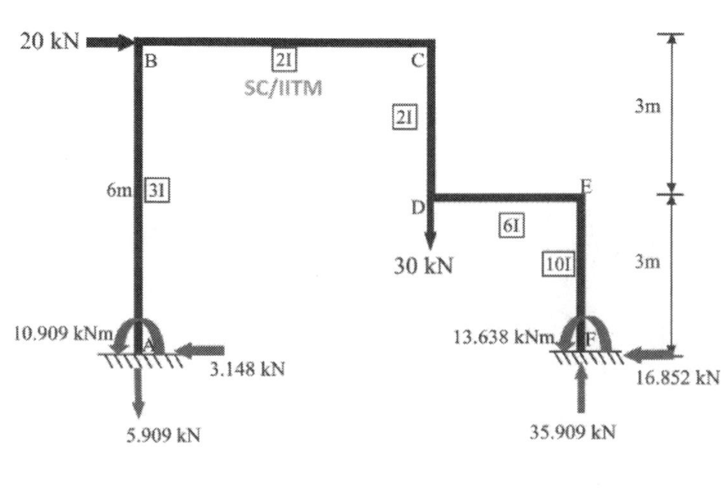

MATLAB program:

```
%% stiffness matrix method
% Input
clc;
clear;
n = 5; % number of members
I = [3 2 2 6 10]; %Moment of inertia in m4
L = [6 4 3 2 3]; % length in m
uu = 7; % Number of unrestrained degrees of freedom
ur = 6; % Number of restrained degrees of freedom
uul = [1 2 3 4 5 6 7]; % global labels of unrestrained dof
url = [8 9 10 11 12 13]; % global labels of restrained dof
l1 = [1 8 5 9]; % Global labels for member 1
l2 = [1 2 10 6]; % Global labels for member 2
l3 = [2 3 5 7]; % Global labels for member 3
l4 = [3 4 6 13]; % Global labels for member 4
l5 = [4 11 7 12]; % Global labels for member 5
l= [l1; l2; l3; l4; l5];
dof = uu+ur;
Ktotal = zeros (dof);
fem1= zeros (1,4); % Local Fixed end moments of member 1
fem2= zeros (1,4); % Local Fixed end moments of member 2
fem3= zeros (1,4); % Local Fixed end moments of member 3
fem4= zeros (1,4); % Local Fixed end moments of member 4
fem5= zeros (1,4); % Local Fixed end moments of member 5
%% rotation coefficients for each member
rc1 = 4.*I./L;
rc2 = 2.*I./L;
%% stiffness matrix 4 by 4 (axial deformation neglected)
for i = 1:n
        Knew = zeros (dof);
        k1 = [rc1(i); rc2(i); (rc1(i)+rc2(i))/L(i); (-
(rc1(i)+rc2(i))/L(i))];
        k2 = [rc2(i); rc1(i); (rc1(i)+rc2(i))/L(i); (-
(rc1(i)+rc2(i))/L(i))];
        k3 = [(rc1(i)+rc2(i))/L(i); (rc1(i)+rc2(i))/L(i);
(2*(rc1(i)+rc2(i))/(L(i)^2)); (-2*(rc1(i)+rc2(i))/(L(i)^2))];
        k4 = -k3;
        K = [k1 k2 k3 k4];
        fprintf ('Member Number =');
        disp (i);
        fprintf ('Local Stiffness matrix of member, [K] = \n');
        disp (K);
        for p = 1:4
            for q = 1:4
                Knew((l(i,p)),(l(i,q))) =K(p,q);
            end
        end
        Ktotal = Ktotal + Knew;
        if i == 1
```

```
            Kg1=K;
        elseif i == 2
            Kg2 = K;
        elseif i ==3
            Kg3=K;
        elseif i== 4
            Kg4 = K;
        else
            Kg5 = K;
        end
end
fprintf ('Stiffness Matrix of complete structure, [Ktotal] = \n');
disp (Ktotal);
Kunr = zeros(uu);
for x=1:uu
    for y=1:uu
        Kunr(x,y)= Ktotal(x,y);
    end
end
fprintf ('Unrestrained Stiffness sub-matix, [Kuu] = \n');
disp (Kunr);
KuuInv= inv(Kunr);
fprintf ('Inverse of Unrestrained Stiffness sub-matix,
[KuuInverse] = \n');
disp (KuuInv);
%% Creation of joint load vector
jl= [0; 0; 0; 0; 20; -30; 0; 0; 0; 0; 0; 0; 0]; % values given in kN or kNm
jlu = [0; 0; 0; 0; 20; -30; 0]; % load vector in unrestrained dof
delu = KuuInv*jlu;
fprintf ('Joint Load vector, [Jl] = \n');
disp (jl');
fprintf ('Unrestrained displacements, [DelU] = \n');
disp (delu');
delr = zeros (ur,1);
del = [delu; delr];
deli= zeros (4,1);
for i = 1:n
    for p = 1:4
        deli(p,1) = del((l(i,p)),1) ;
    end
    if i == 1
            delbar1 = deli;
            mbar1= (Kg1 * delbar1)+fem1';
            fprintf ('Member Number =');
            disp (i);
            fprintf ('Global displacement matrix [DeltaBar] = \n');
            disp (delbar1');
            fprintf ('Global End moment matrix [MBar] = \n');
            disp (mbar1');
        elseif i == 2
            delbar2 = deli;
```

```
            mbar2= (Kg2 * delbar2)+fem2';
            fprintf ('Member Number =');
            disp (i);
            fprintf ('Global displacement matrix [DeltaBar] = \n');
            disp (delbar2');
            fprintf ('Global End moment matrix [MBar] = \n');
            disp (mbar2');
        elseif i == 3
            delbar3 = deli;
            mbar3= (Kg3 * delbar3)+fem3';
            fprintf ('Member Number =');
            disp (i);
            fprintf ('Global displacement matrix [DeltaBar] = \n');
            disp (delbar3');
            fprintf ('Global End moment matrix [MBar] = \n');
            disp (mbar3');
        elseif i == 4
            delbar4 = deli;
            mbar4= (Kg4 * delbar4)+fem4';
            fprintf ('Member Number =');
            disp (i);
            fprintf ('Global displacement matrix [DeltaBar] = \n');
            disp (delbar4');
            fprintf ('Global End moment matrix [MBar] = \n');
            disp (mbar4');
        else
            delbar5 = deli;
            mbar5= (Kg5 * delbar5)+fem5';
            fprintf ('Member Number =');
            disp (i);
            fprintf ('Global displacement matrix [DeltaBar] = \n');
            disp (delbar5');
            fprintf ('Global End moment matrix [MBar] = \n');
            disp (mbar5');
    end
end
%% check
mbar = [mbar1'; mbar2'; mbar3'; mbar4'; mbar5'];
jf = zeros(dof,1);
for a=1:n
    for b=1:4 % size of k matrix
        d = l(a,b);
        jfnew = zeros(dof,1);
        jfnew(d,1)=mbar(a,b);
        jf=jf+jfnew;
    end
end
fprintf ('Joint forces = \n');
disp (jf');
```

MATLAB output:

```
Member Number =      1
Local Stiffness matrix of member, [K] =
    2.0000    1.0000    0.5000   -0.5000
    1.0000    2.0000    0.5000   -0.5000
    0.5000    0.5000    0.1667   -0.1667
   -0.5000   -0.5000   -0.1667    0.1667

Member Number =      2
Local Stiffness matrix of member, [K] =
    2.0000    1.0000    0.7500   -0.7500
    1.0000    2.0000    0.7500   -0.7500
    0.7500    0.7500    0.3750   -0.3750
   -0.7500   -0.7500   -0.3750    0.3750

Member Number =      3
Local Stiffness matrix of member, [K] =
    2.6667    1.3333    1.3333   -1.3333
    1.3333    2.6667    1.3333   -1.3333
    1.3333    1.3333    0.8889   -0.8889
   -1.3333   -1.3333   -0.8889    0.8889

Member Number =      4
Local Stiffness matrix of member, [K] =
    12      6      9     -9
     6     12      9     -9
     9      9      9     -9
    -9     -9     -9      9

Member Number =      5
Local Stiffness matrix of member, [K] =
   13.3333    6.6667    6.6667   -6.6667
    6.6667   13.3333    6.6667   -6.6667
    6.6667    6.6667    4.4444   -4.4444
   -6.6667   -6.6667   -4.4444    4.4444

Stiffness Matrix of complete structure, [Ktotal] =
 4.0000 1.0000       0       0 0.5000 -0.7500       0 1.0000 -0.5000 0.7500
0
 1.0000 4.6667  1.3333       0 1.3333 -0.7500 -1.3333       0       0 0.7500
0
      0 1.3333 14.6667 6.0000 1.3333  9.0000 -1.3333       0       0      0
0
      0       0 6.0000 25.3333      0  9.0000  6.6667       0       0      0
6.6667
 0.5000 1.3333 1.3333       0 1.0556       0 -0.8889  0.5000 -0.1667      0
0
-0.7500 -0.7500 9.0000 9.0000       0  9.3750       0       0       0 0.3750
0
      0 -1.3333 -1.3333 6.6667 -0.8889       0 5.3333       0       0      0
6.6667
 1.0000       0       0       0  0.5000       0       0  2.0000 -0.5000      0
0
```

```
-0.5000       0        0       0  -0.1667       0        0 -0.5000  0.1667       0
0
 0.7500 0.7500       0        0         0 -0.3750       0        0        0 0.3750
0
      0       0        0 6.6667       0         0 6.6667        0        0       0
13.3333
      0       0        0 -6.6667      0         0 -4.4444       0        0       0
-6.6667
      0       0 -9.0000 -9.0000       0 -9.0000        0        0        0       0
0
  Columns 12 through 13
        0         0
        0         0
        0   -9.0000
  -6.6667   -9.0000
        0         0
        0   -9.0000
  -4.4444         0
        0         0
        0         0
        0         0
  -6.6667         0
   4.4444         0
        0    9.0000

Unrestrained Stiffness sub-matix, [Kuu] =
   4.0000   1.0000        0        0   0.5000  -0.7500        0
   1.0000   4.6667   1.3333        0   1.3333  -0.7500  -1.3333
        0   1.3333  14.6667   6.0000   1.3333   9.0000  -1.3333
        0        0   6.0000  25.3333        0   9.0000   6.6667
   0.5000   1.3333   1.3333        0   1.0556        0  -0.8889
  -0.7500  -0.7500   9.0000   9.0000        0   9.3750        0
        0  -1.3333  -1.3333   6.6667  -0.8889        0   5.3333
Inverse of Unrestrained Stiffness sub-matix, [KuuInverse] =
   0.2758  -0.0298  -0.0053  -0.0016  -0.1070   0.0263  -0.0246
  -0.0298   0.3843  -0.0747  -0.0808  -0.2639   0.1776   0.1344
  -0.0053  -0.0747   0.2963   0.1098  -0.4029  -0.3962  -0.1490
  -0.0016  -0.0808   0.1098   0.1914  -0.2690  -0.2957  -0.2768
  -0.1070  -0.2639  -0.4029  -0.2690   2.3068   0.6154   0.5540
   0.0263   0.1776  -0.3962  -0.2957   0.6154   0.7873   0.4175
  -0.0246   0.1344  -0.1490  -0.2768   0.5540   0.4175   0.6221

Joint Load vector, [Jl] =
    0    0    0    0   20   -30    0    0    0    0    0    0    0
Unrestrained displacements, [DelU] =
  -2.9291  -10.6047  3.8284  3.4921  27.6762 -11.3104   -1.4465

Member Number =      1
Global displacement matrix [DeltaBar] =
   -2.9291         0   27.6762         0
Global End moment matrix [MBar] =
    7.9800   10.9091    3.1482   -3.1482
```

```
Member Number =      2
Global displacement matrix [DeltaBar] =
   -2.9291  -10.6047         0  -11.3104
Global End moment matrix [MBar] =
   -7.9800  -15.6556   -5.9089    5.9089

Member Number =      3
Global displacement matrix [DeltaBar] =
  -10.6047    3.8284   27.6762   -1.4465
Global End moment matrix [MBar] =
   15.6556   34.8998   16.8518  -16.8518

Member Number =      4
Global displacement matrix [DeltaBar] =
    3.8284    3.4921  -11.3104         0
Global End moment matrix [MBar] =
  -34.8998  -36.9180  -35.9089   35.9089

Member Number =      5
Global displacement matrix [DeltaBar] =
    3.4921         0   -1.4465         0
Global End moment matrix [MBar] =
   36.9180   13.6375   16.8518  -16.8518

Joint forces =
0.0000  -0.0000  -0.0000  -0.0000  20.0000  -30.0000  -0.0000   10.9091
-3.1482   -5.9089   13.6375  -16.8518   35.9089
\
```

3 Planar Non-Orthogonal Structures

3.1 PLANAR NON-ORTHOGONAL STRUCTURE

As discussed in the previous chapter, the analysis of an orthogonal frame is easy, as no transformation is required between the local and global axis systems. But, even if one of the members is non-orthogonal to the adjacent, then the complexity increases due to two reasons: (i) axial displacements need to be accounted for, and (ii) the size of the problem increases as the [k] of the modular beam is 6×6. The method becomes more complicated when the members in the structure are not orthogonal to each other. It is a common geometric form in that the members are inclined to support the lateral loads; they also add to the architectural aesthetics, as seen in the earlier chapter on the structural form. Please note that the non-orthogonality does not arise from the initial imperfection of the column members; it is designed to remain non-orthogonal to counteract lateral loads primarily.

Let us consider the conventional equation for a complete structure:

$$[K]_c \{\Delta\}_c = \{J_L\}_c + [R]_c \tag{3.1}$$

The above equation is valid for the case where the local axis of every member is aligned with the global axis of reference chosen for the analysis. In such conditions, the stiffness matrix can be established in terms of sub-matrices by partitioning them and solving for displacements and reaction forces. But if the reference axes of the system do not align with the member axes, then the stiffness matrix of each member must be transformed to its equivalent value in the global frame of reference. It is one of the main complexities in solving a non-orthogonal frame problem with a stiffness matrix approach.

One of the advantages of the stiffness matrix approach is that the same analysis principles will be used with a minor modification/update to the development of the method. It is, therefore, expected that the MATLAB program used to solve the orthogonal frames and continuous beams may not be useful in solving the non-orthogonal frames. The member stiffness matrix shall include axial displacement degrees of freedom, thus making its size 6×6; please note that it is 4×4 for orthogonal frames and beams.

Let us now discuss the transformation process between two axes systems, as shown in Figure 3.1. It is important to establish a connectivity between these axes systems to solve the problem. One of good news is that the derivations of the member stiffness matrix, discussed before, are completely valid but need to be marginally modified to accommodate the axial displacement degrees of freedom.

DOI: 10.1201/9781003639152-3

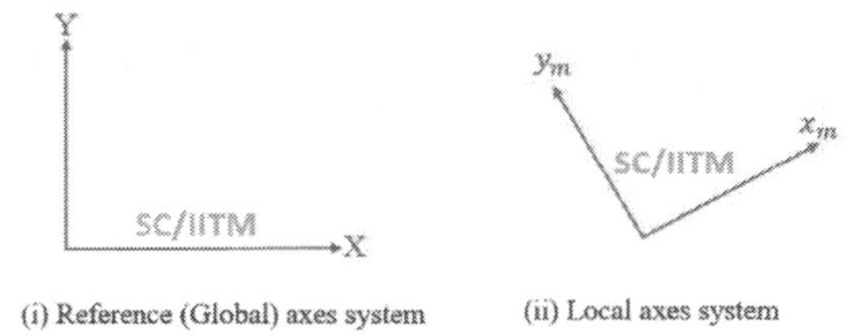

FIGURE 3.1 Global and local axis system.

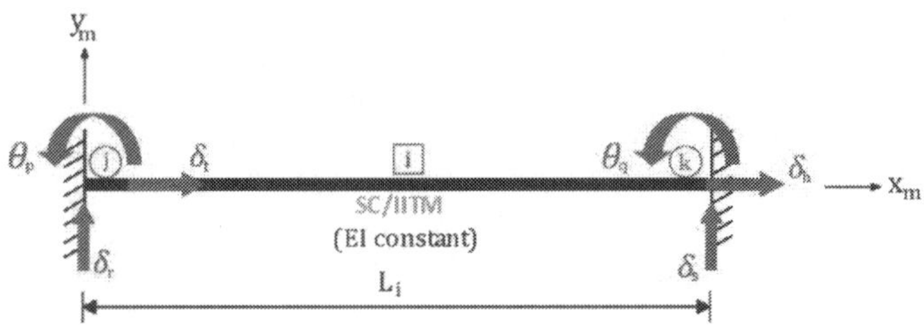

FIGURE 3.2 Modular beam with axial displacement.

Let us now consider a modular beam element with axial deformation, as shown in Figure 3.2. As there are six degrees of freedom, the member stiffness matrix will be size 6×6.

We already know the sub-matrix of [K] by neglecting axial deformation, and it is given as below:

$$[K]=\begin{bmatrix} k_{pp} & k_{pq} & \frac{k_{pp}+k_{pq}}{L} & -\left(\frac{k_{pp}+k_{pq}}{L}\right) \\ k_{qp} & k_{qq} & \frac{k_{qp}+k_{qq}}{L} & -\left(\frac{k_{qp}+k_{qq}}{L}\right) \\ \frac{k_{pp}+k_{pq}}{L} & \frac{k_{pq}+k_{qq}}{L} & \frac{k_{pp}+k_{pq}+k_{qp}+k_{qq}}{L^2} & -\frac{k_{pp}+k_{pq}+k_{qp}+k_{qq}}{L^2} \\ -\left(\frac{k_{pp}+k_{pq}}{L}\right) & -\left(\frac{k_{pq}+k_{qq}}{L}\right) & -\left(\frac{k_{pp}+k_{pq}+k_{qp}+k_{qq}}{L^2}\right) & \left(\frac{k_{pp}+k_{pq}+k_{qp}+k_{qq}}{L^2}\right) \end{bmatrix} \tag{3.2}$$

For developing the stiffness matrix for the non-orthogonal frame, consider the beam element with unit axial deformation at the jth end, as shown in Figure 3.3.

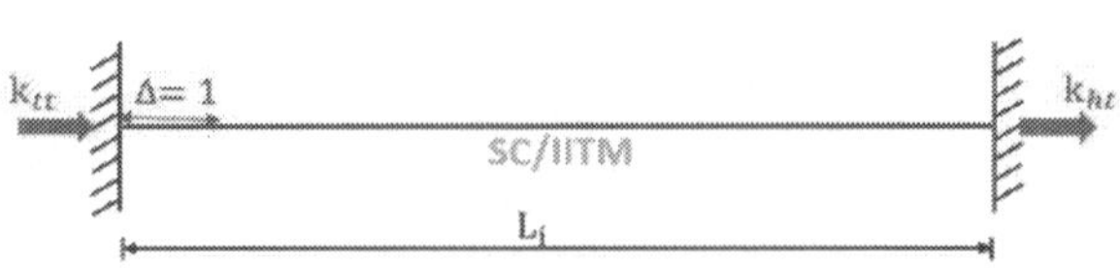

FIGURE 3.3 Beam with axial deformation at jth end.

k_{tt} is the force in the tth degree by giving unit displacement in tth degree. Whereas k_{ht} is the force in hth degree by giving unit displacement in tth degree.

$$k_{tt} = \frac{AE}{L}\Delta = \frac{AE}{L} \text{ for unit axial displacement} \tag{3.3a}$$

$$k_{ht} = -\frac{AE}{L} \tag{3.3b}$$

Similarly, assuming the beam element is undergoing axial deformation on the kth end, as shown in Figure 3.4.

$$k_{hh} = \frac{AE}{L}\Delta = \frac{AE}{L} \text{ for unit axial displacement} \tag{3.4a}$$

$$k_{th} = -\frac{AE}{L} \tag{3.4b}$$

Thus, the matrix stiffness matrix, including axial deformation, is given as:

$$[K]_i = \begin{bmatrix} \frac{4EI}{L} & \frac{2EI}{L} & \frac{6EI}{L^2} & -\frac{6EI}{L^2} & \begin{matrix}0\\0\end{matrix} & \begin{matrix}0\\0\end{matrix} \\ \frac{2EI}{L} & \frac{4EI}{L} & \frac{6EI}{L^2} & -\frac{6EI}{L^2} & \begin{matrix}0\\0\end{matrix} & \begin{matrix}0\\0\end{matrix} \\ \frac{6EI}{L^2} & \frac{6EI}{L^2} & \frac{12EI}{L^3} & -\frac{12EI}{L^3} & \frac{AE}{L} & -\frac{AE}{L} \\ -\frac{6EI}{L^2} & -\frac{6EI}{L^2} & -\frac{12EI}{L^3} & \frac{12EI}{L^3} & -\frac{AE}{L} & \frac{AE}{L} \end{bmatrix} \tag{3.5}$$

For example, consider an orthogonal and non-orthogonal frame, as shown in Figure 3.5. The local axes for all the members are marked, as shown. In the case of an orthogonal frame, one can observe that the member and the reference axes are mutually aligned. But in the case of a non-orthogonal frame, member axes are inclined (non-orthogonal) to the reference axes, demanding a transformation approach to solve the problem.

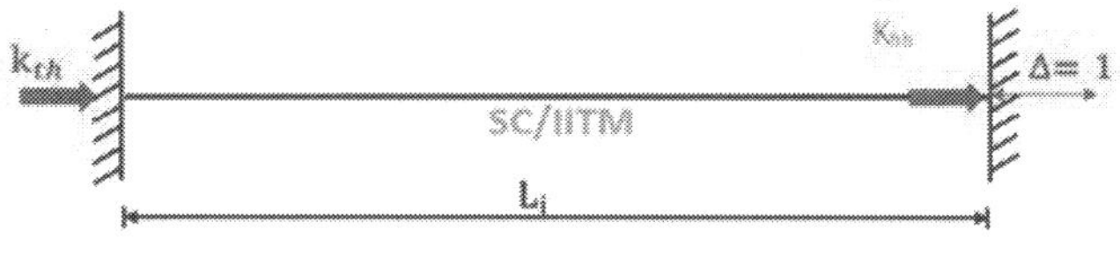

FIGURE 3.4 Beam with axial deformation at kth end.

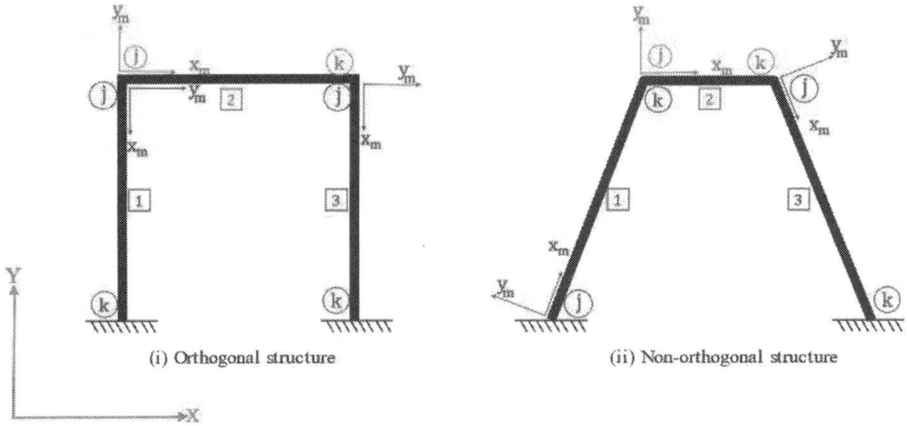

FIGURE 3.5 Global and local axes systems.

In the case of non-orthogonal frames, the orientation of the local axes (x_m-y_m) may be such that this cannot be aligned or mapped to the same orientation of the reference axes (X-Y). See, for example, members (1,3). Hence, the member stiffness matrix cannot be directly written concerning the reference axes system but is required to be transformed. Though the member stiffness matrix, developed for the local axes system, is still valid, the [K] will be transformed to become a global [K]. It is also interesting to note the style in which the degrees of freedom are marked for a non-orthogonal frame. See, for example, the gable frame shown in Figure 3.6. All

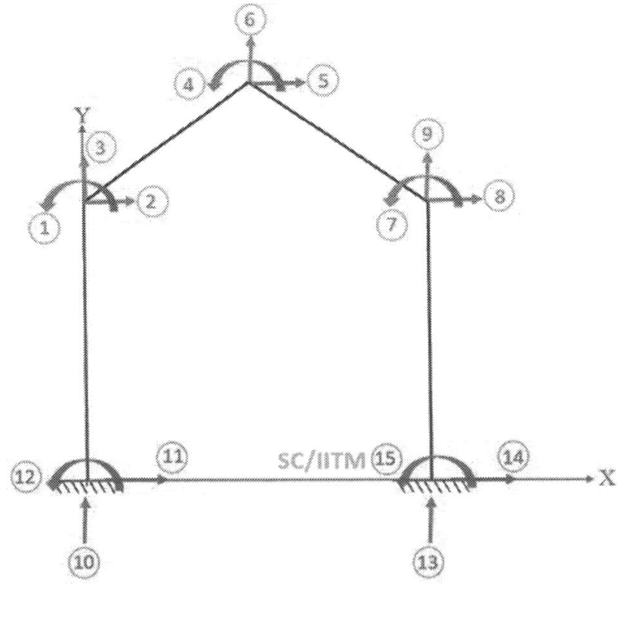

FIGURE 3.6 Degrees of freedom of non-orthogonal frame.

degrees of freedom, three at each node, are marked in alignment with the reference axes system, not about the local axes of the members.

3.2 TRANSFORMATION MATRIX

Let us consider two orthogonal sets of axes, such as (x_1-y_1) and (x_2-y_2), as shown in Figure 3.7.

$$V_1 = \overline{V_1}\cos\theta + \overline{V_2}\sin\theta \tag{3.6a}$$

$$V_2 = \overline{V_2}\cos\theta - \overline{V_1}\sin\theta \tag{3.6b}$$

The above equations can be written in matrix form as follows:

$$\begin{Bmatrix} V_1 \\ V_2 \end{Bmatrix} = \begin{bmatrix} \cos\theta & \sin\theta \\ -\sin\theta & \cos\theta \end{bmatrix} \begin{Bmatrix} \overline{V_1} \\ \overline{V_2} \end{Bmatrix} \tag{3.7}$$

It can also be expressed as follows:

$$\{V\} = [T]\{\overline{V}\} \tag{3.8}$$

where [T] is the transformation matrix.

Alternatively, one can also resolve (x_2-y_2) to (x_1-y_1) axes. In this case,

$$\overline{V_1} = V_1\cos\theta - V_2\sin\theta \tag{3.9a}$$

$$\overline{V_2} = V_2\cos\theta + V_1\sin\theta \tag{3.9b}$$

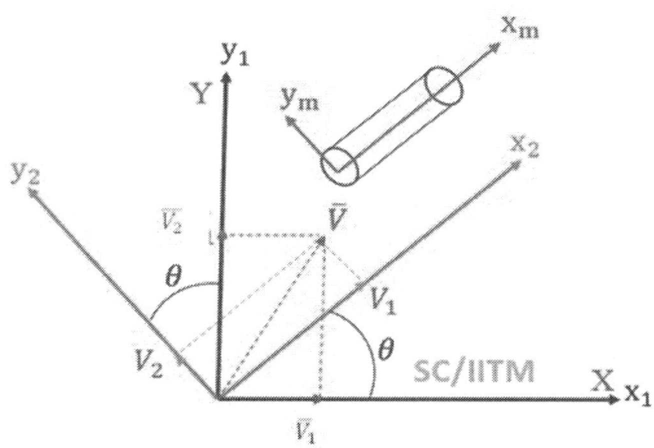

FIGURE 3.7 Two orthogonal sets of axes.

Expressing the above equations in matrix form,

$$\begin{Bmatrix} \bar{V}_1 \\ \bar{V}_2 \end{Bmatrix} = \begin{bmatrix} \cos\theta & -\sin\theta \\ \sin\theta & \cos\theta \end{bmatrix} \begin{Bmatrix} V_1 \\ V_2 \end{Bmatrix}$$

$$\{\bar{V}\} = [T]^T \{V\} \tag{3.10}$$

The transformation between the axes in both ways is valid. Thus, the transformation matrix is given by:

$$[T] = \begin{bmatrix} c & s \\ -s & c \end{bmatrix} \tag{3.11}$$

where, c represents cos(θ) and s represents sin(θ). The angle of rotation (θ) is the inclination or rotation of the x_m-axis of the member concerning the x-axis of the reference frame, measured in an anticlockwise manner. The inverse of the transformation matrix is given as follows:

$$[T]^{-1} = \frac{1}{\left(c^2 + s^2\right)} \begin{bmatrix} c & -s \\ s & c \end{bmatrix} = [T]^T \tag{3.12}$$

Thus, the inverse of the transformation matrix is equal to the transpose of the matrix, terming such matrix as an orthogonal matrix.

3.3 TRANSFORMATION OF END MOMENTS, REACTIONS, AND DISPLACEMENTS

Let us now consider a beam, arbitrarily oriented to the reference axes at an angle θ, as shown in Figure 3.8. Let us map all the degrees of freedom to the reference axes framework.

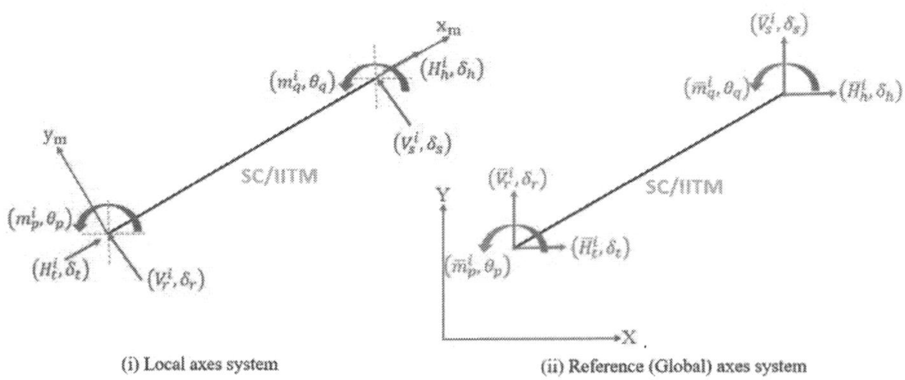

FIGURE 3.8 Transformation of end moments, reactions, and displacements.

The displacements along both local and reference (global) axes are given as follows:

Displacements along the reference axes

$$\{\bar{\Delta}\} = \begin{Bmatrix} \bar{m}_p \\ \bar{m}_q \\ \bar{V}_r \\ \bar{V}_s \\ \bar{H}_t \\ \bar{H}_h \end{Bmatrix} \tag{3.13a}$$

$$\{\Delta\} = \begin{Bmatrix} m_p \\ m_q \\ V_r \\ V_s \\ H_t \\ H_h \end{Bmatrix} \tag{3.13b}$$

Concerning Figure 3.8, the moments and forces on the reference axes can be connected to the local axis system using the following relationship:

$$\begin{Bmatrix} \bar{m}_p \\ \bar{m}_q \\ \bar{V}_r \\ \bar{V}_s \\ \bar{H}_t \\ \bar{H}_h \end{Bmatrix} = \begin{bmatrix} 1 & 0 & 0 & 0 & 0 & 0 \\ 0 & 1 & 0 & 0 & 0 & 0 \\ 0 & 0 & cos\theta & 0 & sin\theta & 0 \\ 0 & 0 & 0 & cos\theta & 0 & sin\theta \\ 0 & 0 & -sin\theta & 0 & cos\theta & 0 \\ 0 & 0 & 0 & -sin\theta & 0 & cos\theta \end{bmatrix} \begin{Bmatrix} m_p \\ m_q \\ V_r \\ V_s \\ H_t \\ H_h \end{Bmatrix} \tag{3.14a}$$

Equation (3.14a) can also be written as follows:

$$\{\bar{m}_i\} = [T]^T \{m_i\} \tag{3.14b}$$

As the transformation matrix is orthogonal (see Eq. (3.12) for more explanation), the following relationship also holds:

$$\begin{Bmatrix} m_p \\ m_q \\ V_r \\ V_s \\ H_t \\ H_h \end{Bmatrix} = \begin{bmatrix} 1 & 0 & 0 & 0 & 0 & 0 \\ 0 & 1 & 0 & 0 & 0 & 0 \\ 0 & 0 & cos\theta & 0 & -sin\theta & 0 \\ 0 & 0 & 0 & cos\theta & 0 & -sin\theta \\ 0 & 0 & sin\theta & 0 & cos\theta & 0 \\ 0 & 0 & 0 & sin\theta & 0 & cos\theta \end{bmatrix} \begin{Bmatrix} \bar{m}_p \\ \bar{m}_q \\ \bar{V}_r \\ \bar{V}_s \\ \bar{H}_t \\ \bar{H}_h \end{Bmatrix} \tag{3.15a}$$

$$\{m_i\} = [T]\{\bar{m}_i\} \tag{3.15b}$$

Thus, the following set of equations can be used to connect the displacements, which are similar to Eqs. (3.15b and 3.16b):

$$\{\bar{\delta}_i\} = [T]_i^T \{\delta\}_i \tag{3.16a}$$

$$\{\delta\}_i = [T]_i \{\bar{\delta}_i\} \tag{3.16b}$$

where,

$$\{\delta\}_i = \begin{Bmatrix} \theta_p \\ \theta_q \\ \delta_r \\ \delta_s \\ \delta_t \\ \delta_h \end{Bmatrix}, \{\bar{\delta}\}_i = \begin{Bmatrix} \bar{\theta}_p \\ \bar{\theta}_q \\ \bar{\delta}_r \\ \bar{\delta}_s \\ \bar{\delta}_t \\ \bar{\delta}_h \end{Bmatrix} \tag{3.17}$$

Further, to simplify the transformation process, it is necessary to get rid of the angle, as it involves a sign convention. Hence, the direction cosines can also be written in terms of the coordinates of the jth and kth ends of a member. Let us consider an arbitrarily oriented member concerning the reference axes as shown in Figure 3.9. For the length of the member (L_i), the direction cosines can be expressed as follows:

$$cos\theta = \frac{X_k - X_j}{L_i} \tag{3.18a}$$

$$sin\theta = \frac{Y_k - Y_j}{L_i} \tag{3.18b}$$

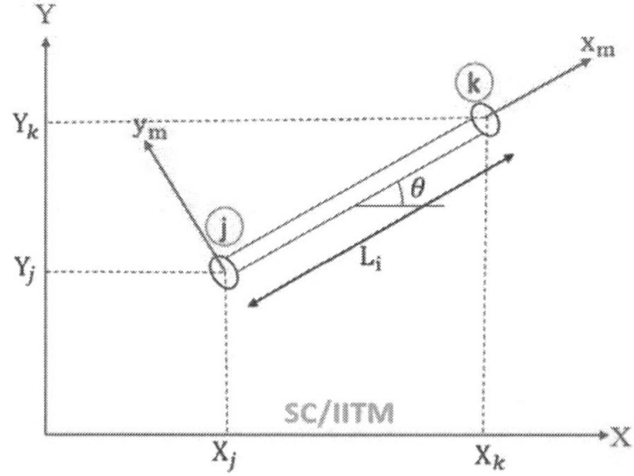

FIGURE 3.9 Transformation of member geometry.

$$L_i = \sqrt{\left(X_k - X_j\right)^2 + \left(Y_k - Y_j\right)^2} \tag{3.18c}$$

The following force-displacement relationships are also true:

$$\{m_i\} = [T]\{\overline{m_i}\} \tag{3.19a}$$

$$\{\delta_i\} = [T]\{\overline{\delta_i}\} \tag{3.19b}$$

$$\{m_i\} = [K]\{\delta_i\} \tag{3.19c}$$

Substituting Eqs. (3.19a and 3.19b) in Eq. (3.19c), we get:

$$[T]\{\overline{m_i}\} = [K][T]\{\overline{\delta_i}\} \tag{3.20a}$$

$$\{\overline{m_i}\} = [T]^T [K][T]\{\overline{\delta_i}\} \tag{3.20b}$$

The above equation is true as $[T]^T = [T]^{-1}$. Hence,

$$\{\overline{m_i}\} = [\overline{K_i}]\{\overline{\delta_i}\} \tag{3.21}$$

Where

$$[\overline{K_i}] = [T_i]^T [K_i][T_i] \tag{3.22}$$

3.4 LOAD VECTOR AND END MOMENTS

The joint loads should be computed for the reference axes only. As mentioned earlier, the joint loads are the reversal of the fixed end moments (FEM) caused by the member loads. Thus,

$$\{\overline{FEM_i}\} = [T_i]^T \{FEM_i\} \tag{3.23a}$$

$$\{\overline{Joint\,Load_i}\} = [T_i]^T \{Joint\,load_i\} \tag{3.23b}$$

$$[\overline{FEM}]_i = [T]_i^T [FEM]_i \tag{3.24}$$

Then, the unrestrained displacement is computed using the following relationship:

$$\{\overline{\delta_u}\} = [\overline{K_{uu}}]^{-1} \{\overline{J_L}\}_u \tag{3.25}$$

The end moments and shear of every member are obtained by the following equations:

$$\{M\}_i = [K]_i \{\delta_i\} + \{FEM\}_i$$
$$\{\bar{M}\}_i = [\bar{K}]_i \{\bar{\delta}_i\} + \{F\bar{E}M\}_i \quad (3.26)$$

3.5 EXAMPLE PROBLEMS USING MATLAB

Example 3.1:

Analyze the planar non-orthogonal structure shown below using the stiffness method.

Given data: Moment of inertia about bending axis, $I_{zz} = 0.0016\ m^4$, $A = 0.120\ m^2$, and E is constant.

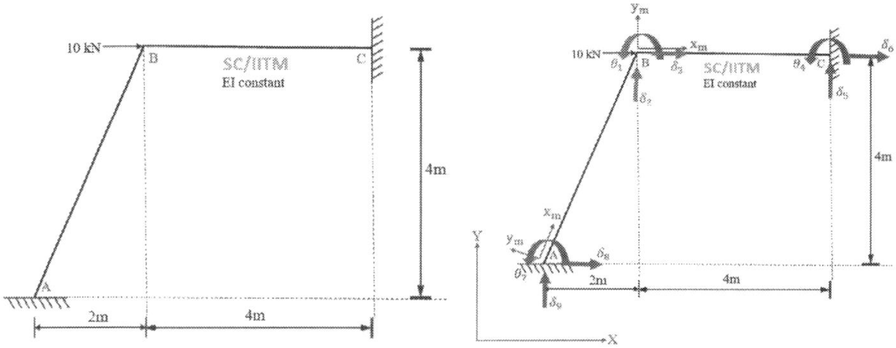

SOLUTION

Step 1: Marking global axes, local axes of each member, and degrees of freedom

Mark the local and global axes, and mark the unrestrained and restrained degrees of freedom, as shown above. Please check that all degrees of freedom are to be marked concerning the global axes and not the local axes of the member. Further, it also ensures that there are three degrees of freedom in each node. The inclined (non-orthogonal) member (AB) has an inclination of θ (= $\tan^{-1}(4/2)$ = 63.435°), which is positive. This is because the x_m-axis of the member (AB) is at counterclockwise 63.435° with the global X-axis. In the case of member (BC), this angle is zero (and not 180°) as the jth of the member (BC) is at B. Had the jth end of the member been assumed at C, then the x_m-axis of the (BC) would point toward B, making an angle of 180° with the X-axis.

Step 2: Calculation of transformation matrix coefficients and global labels

The unrestrained degrees of freedom are 3 (θ_1, δ_2, δ_3)

The restrained degrees of freedom: 6 (θ_4, δ_5, δ_6, θ_7, δ_8, δ_9)

Thus, the size of the total stiffness matrix will be 9×9, in which the sub-matrix for the unrestrained degrees-of-freedom will be of size 3×3.

Member Number	Ends j	k	Length (m)	θ (degrees)	C_x	C_y	Global Labels
1	A	B	4.472	+63.435	0.447	0.894	(7,1,9,2,8,3)
2	B	C	4	0	1	0	(1,4,2,5,3,6)

Step 3: Calculation of local stiffness matrix

The stiffness matrix for the standard beam element, including the axial deformation, is given by,

$$[K]_i = \begin{bmatrix} \frac{4EI}{l} & \frac{2EI}{l} & \frac{6EI}{l^2} & -\frac{6EI}{l^2} & 0 & 0 \\ \frac{2EI}{l} & \frac{4EI}{l} & \frac{6EI}{l^2} & -\frac{6EI}{l^2} & 0 & 0 \\ \frac{6EI}{l^2} & \frac{6EI}{l^2} & \frac{12EI}{l^3} & -\frac{12EI}{l^3} & 0 & 0 \\ -\frac{6EI}{l^2} & -\frac{6EI}{l^2} & -\frac{12EI}{l^3} & \frac{12EI}{l^3} & 0 & 0 \\ 0 & 0 & 0 & 0 & \frac{AE}{l} & -\frac{AE}{l} \\ 0 & 0 & 0 & 0 & -\frac{AE}{l} & \frac{AE}{l} \end{bmatrix}$$

$$\begin{matrix} & 7 & 1 & 9 & 2 & 8 & 3 & \\ K_{AB} = E & \left[\begin{matrix} 0.0014 & 0.0007 & 0.0005 & -0.0005 & 0 & 0 \\ 0.0007 & 0.0014 & 0.0005 & -0.0005 & 0 & 0 \\ 0.0005 & 0.0005 & 0.0002 & -0.0002 & 0 & 0 \\ -0.0005 & -0.0005 & -0.0002 & 0.0002 & 0 & 0 \\ 0 & 0 & 0 & 0 & 0.0268 & -0.0268 \\ 0 & 0 & 0 & 0 & -0.0268 & 0.0268 \end{matrix}\right. & \left.\begin{matrix} \\ \\ \\ \\ \\ \\ \end{matrix}\right] \begin{matrix} 7 \\ 1 \\ 9 \\ 2 \\ 8 \\ 3 \end{matrix} \end{matrix}$$

$$\begin{matrix} & 1 & 4 & 2 & 5 & 3 & 6 & \\ K_{BC} = E & \left[\begin{matrix} 0.0016 & 0.0008 & 0.0006 & -0.0006 & 0 & 0 \\ 0.0008 & 0.0016 & 0.0006 & -0.0006 & 0 & 0 \\ 0.0006 & 0.0006 & 0.0003 & -0.0003 & 0 & 0 \\ -0.0006 & -0.0006 & -0.0003 & 0.0003 & 0 & 0 \\ 0 & 0 & 0 & 0 & 0.0300 & -0.0300 \\ 0 & 0 & 0 & 0 & -0.0300 & 0.0300 \end{matrix}\right. & \left.\begin{matrix} \\ \\ \\ \\ \\ \\ \end{matrix}\right] \begin{matrix} 1 \\ 4 \\ 2 \\ 5 \\ 3 \\ 6 \end{matrix} \end{matrix}$$

Step 4: Calculation of transformation matrix

The transformation matrix for any member 'i' is given by,

$$[T]=\begin{bmatrix}1&0&0&0&0&0\\0&1&0&0&0&0\\0&0&C_x&0&-C_y&0\\0&0&0&C_x&0&-C_y\\0&0&C_y&0&C_x&0\\0&0&0&C_y&0&C_x\end{bmatrix}$$

$$[T]_{AB}=\begin{bmatrix}1&0&0&0&0&0\\0&1&0&0&0&0\\0&0&0.4472&0&-0.8944&0\\0&0&0&0.4472&0&-0.8944\\0&0&0.0944&0&0.4472&0\\0&0&0&0.8944&0&0.4472\end{bmatrix}$$

$$[T]_{BC}=\begin{bmatrix}1&0&0&0&0&0\\0&1&0&0&0&0\\0&0&1&0&0&0\\0&0&0&1&0&0\\0&0&0&0&1&0\\0&0&0&0&0&1\end{bmatrix}$$

Step 5: Estimation of joint load vector

The frame is subjected to only the joint load; no loads on any members. Hence, {FEM} will be a null vector. Thus, the joint load vector can be written directly as follows:

$$\{\boldsymbol{J_L}\}_{9x1}=\begin{Bmatrix}\mathbf{0}\\\mathbf{0}\\\mathbf{10}\\\mathbf{0}\\\mathbf{0}\\\mathbf{0}\\\mathbf{0}\\\mathbf{0}\\\mathbf{0}\end{Bmatrix}\begin{matrix}1\\2\\3\\4\\5\\6\\7\\8\\9\end{matrix}$$

Thus, the joint load vector in unrestrained degrees of freedom is given by:

$$\overline{\{\boldsymbol{J_{LU}}\}}_{3x1}=\begin{Bmatrix}\mathbf{0}\\\mathbf{0}\\\mathbf{10}\end{Bmatrix}\begin{matrix}1\\2\\3\end{matrix}$$

Step 6: Calculation of global stiffness matrix

The stiffness matrix of every member with respect to the global axes system is obtained by the following equation:

$$\{\overline{K}\}_i = [T]_i^T \, [K]_i \, [T]_i$$

$$\overline{K_{AB}} = E\times10^{-4}\begin{array}{c} \begin{array}{cccccc} 7 & 1 & 9 & 2 & 8 & 3 \end{array} \\ \begin{bmatrix} 14 & 7 & 2 & -2 & -4 & 4 \\ 7 & 14 & 2 & -2 & -4 & 4 \\ 2 & 2 & 215 & -215 & 106 & -106 \\ -2 & -2 & -215 & 215 & 106 & 106 \\ -4 & -4 & 106 & -106 & 55 & -55 \\ 4 & 4 & 106 & 106 & -55 & 55 \end{bmatrix} \end{array} \begin{array}{c} 7 \\ 1 \\ 9 \\ 2 \\ 8 \\ 3 \end{array}$$

$$\overline{K_{BC}} = E\times10^{-4}\begin{array}{c} \begin{array}{cccccc} 1 & 4 & 2 & 5 & 3 & 6 \end{array} \\ \begin{bmatrix} 16 & 8 & 6 & -6 & 0 & 0 \\ 8 & 16 & 6 & -6 & 0 & 0 \\ 6 & 6 & 3 & -3 & 0 & 0 \\ -6 & -6 & -3 & 3 & 0 & 0 \\ 0 & 0 & 0 & 0 & 300 & -300 \\ 0 & 0 & 0 & 0 & -300 & 300 \end{bmatrix} \end{array} \begin{array}{c} 1 \\ 4 \\ 2 \\ 5 \\ 3 \\ 6 \end{array}$$

These matrices are then assembled to get the total stiffness matrix, which will be a 9×9 matrix with sub-matrices of unrestrained and restrained degrees of freedom.

$$\overline{K_{TOTAL}} = E\times10^{-4}\begin{array}{c} \begin{array}{ccccccccc} 1 & 2 & 3 & 4 & 5 & 6 & 7 & 8 & 9 \end{array} \\ \begin{bmatrix} 38 & 4 & 4 & 8 & -6 & 0 & 7 & -4 & 2 \\ 4 & 218 & 106 & 6 & -3 & 0 & -2 & -106 & -215 \\ 4 & 106 & 355 & 0 & 0 & -300 & 4 & -55 & -106 \\ 8 & 6 & 0 & 16 & -6 & 0 & 0 & 0 & 0 \\ -6 & -3 & 0 & -6 & 3 & 0 & 0 & 0 & 0 \\ 0 & 0 & -300 & 0 & 0 & 300 & 0 & 0 & 0 \\ 7 & -2 & 4 & 0 & 0 & 0 & 14 & -4 & 2 \\ -4 & -106 & -55 & 0 & 0 & 0 & -4 & 55 & 106 \\ 2 & -215 & -106 & 0 & 0 & 0 & 2 & 106 & 215 \end{bmatrix} \end{array} \begin{array}{c} 1 \\ 2 \\ 3 \\ 4 \\ 5 \\ 6 \\ 7 \\ 8 \\ 9 \end{array}$$

$$\overline{\{K_{UU}\}} = E\times10^{-4}\begin{array}{c} \begin{array}{ccc} 1 & 2 & 3 \end{array} \\ \begin{bmatrix} 30 & 4 & 4 \\ 4 & 218 & 106 \\ 4 & 106 & 355 \end{bmatrix} \end{array} \begin{array}{c} 1 \\ 2 \\ 3 \end{array}$$

$$\left[\bar{K}_{uu}\right]^{-1} = \frac{1}{E}\begin{bmatrix} 330.863 & -4.561 & -2.631 \\ -4.561 & 53.769 & -16.055 \\ -2.631 & -16.055 & 32.980 \end{bmatrix}$$

Now, $\left\{\bar{\delta}_u\right\} = \left[\bar{K}_{uu}\right]^{-1}\left\{\bar{J}_L\right\}_u$

$$\left\{\bar{\delta}_u\right\} = \frac{1}{E}\begin{Bmatrix} -26.307 \\ -160.545 \\ 329.804 \end{Bmatrix}$$

$$\left[\bar{\delta}\right]_{AB} = \frac{1}{E}\begin{Bmatrix} 0 \\ -26.307 \\ 0 \\ -160.545 \\ 0 \\ 329.804 \end{Bmatrix}, \left[\bar{\delta}\right]_{BC} = \frac{1}{E}\begin{Bmatrix} -26.307 \\ 0 \\ -160.545 \\ 0 \\ 329.804 \\ 0 \end{Bmatrix}$$

Step 7: Calculation of end moments and shear

$$\left\{\bar{M}\right\}_i = \left[\bar{K}\right]_i\left\{\bar{\delta}_i\right\} + \left\{F\bar{E}M\right\}_i$$

$$\left\{\bar{M}\right\}_{AB} = \begin{Bmatrix} \bar{M}_7 \\ \bar{M}_1 \\ \bar{V}_9 \\ \bar{V}_2 \\ \bar{H}_8 \\ \bar{H}_3 \end{Bmatrix} = \begin{Bmatrix} 0.1572 \\ 0.1384 \\ -0.0639 \\ 0.0639 \\ -0.1059 \\ 0.1059 \end{Bmatrix}$$

$$\left\{\bar{M}\right\}_{BC} = \begin{Bmatrix} \bar{M}_1 \\ \bar{M}_4 \\ \bar{V}_2 \\ \bar{V}_5 \\ \bar{H}_3 \\ \bar{H}_6 \end{Bmatrix} = \begin{Bmatrix} -0.1384 \\ -0.1174 \\ -0.0639 \\ 0.0639 \\ 9.8941 \\ -9.8941 \end{Bmatrix}$$

The member and final moments are shown below:

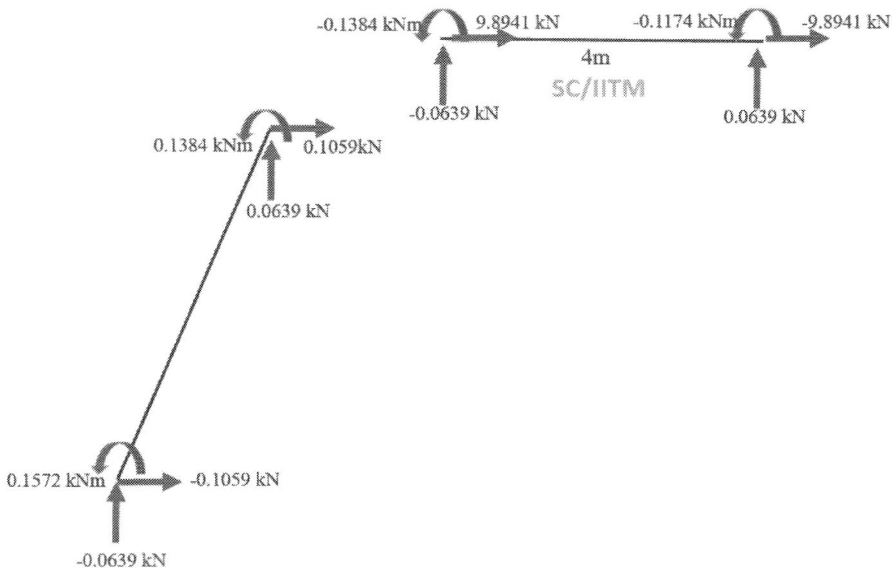

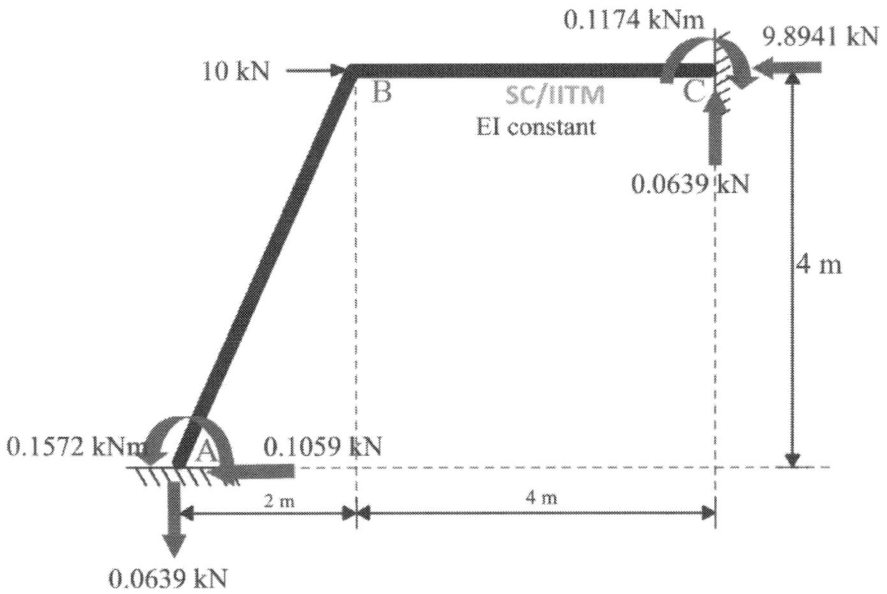

MATLAB program:

```
%% Non-orthogonal frame analysis using stiffness matrix method with
user inputs
clc; clear;

%% User Inputs
n = input('Enter the number of members: \n'); % Number of members
I = input('Enter the moments of inertia as a row vector [I1 I2 ...
In]: \n'); % Moment of inertia
L = input('Enter the lengths as a row vector [L1 L2 ... Ln]: \n');
% Lengths
A = input('Enter the areas as a row vector [A1 A2 ... An]: \n'); %
Area
theta = input('Enter the angles as a row vector [Theta1 Theta2 ...
ThetaN]: \n'); % Angles

uu = input('Enter the number of unrestrained degrees of
freedom: \n'); % Unrestrained DOF
ur = input('Enter the number of restrained degrees of freedom:
\n'); % Restrained DOF
uul = input('Enter the labels of unrestrained DOF as a row
vector [u1 u2 ...]: \n'); % Unrestrained labels
url = input('Enter the labels of restrained DOF as a row
vector [r1 r2 ...]: \n'); % Restrained labels

l = zeros(n, 6); % Global labels for each member
for i = 1:n
    fprintf('Enter global labels for Member %d as a row vector:
', i);
    l(i, :) = input('');
end

dof = uu + ur; % Total degrees of freedom
Ktotal = zeros(dof); % Initialize global stiffness matrix

%% Fixed End Moments Input
fem = zeros(6, n);
for i = 1:n
    fprintf('Enter fixed-end moments for Member %d as a column
vector : \n', i);
    fem(:, i) = input('');
end

%% Rotation coefficients for each member
rc1 = 4 . * I . / L;
rc2 = 2 . * I . / L;
rc3 = A . / L;
cx = cosd(theta);
cy = sind(theta);
```

```
%% Stiffness Matrix and Transformation Matrix Calculation
Tt = cell(1, n); % Transformation matrices for each member
Kg = cell(1, n); % Global stiffness matrices for each member
fembar = cell(1, n); % Transformed fixed-end moments

for i = 1:n
    % Local stiffness matrix
    k1 = [rc1(i); rc2(i); (rc1(i)+rc2(i))/L(i);
-(rc1(i)+rc2(i))/L(i); 0; 0];
    k2 = [rc2(i); rc1(i); (rc1(i)+rc2(i))/L(i);
-(rc1(i)+rc2(i))/L(i); 0; 0];
    k3 = [(rc1(i)+rc2(i))/L(i); (rc1(i)+rc2(i))/L(i);
2*(rc1(i)+rc2(i))/(L(i)^2); -2*(rc1(i)+rc2(i))/(L(i)^2); 0; 0];
    k4 = -k3;
    k5 = [0; 0; 0; 0; rc3(i); -rc3(i)];
    k6 = [0; 0; 0; 0; -rc3(i); rc3(i)];
    K = [k1 k2 k3 k4 k5 k6];

    % Transformation matrix
    T1 = [1; 0; 0; 0; 0; 0];
    T2 = [0; 1; 0; 0; 0; 0];
    T3 = [0; 0; cx(i); 0; cy(i); 0];
    T4 = [0; 0; 0; cx(i); 0; cy(i)];
    T5 = [0; 0; -cy(i); 0; cx(i); 0];
    T6 = [0; 0; 0; -cy(i); 0; cx(i)];
    T = [T1 T2 T3 T4 T5 T6];

    % Global stiffness matrix
    Ttr = T';
    Kg{i} = Ttr * K * T;

    % Transformation matrix and fixed-end moments storage
    Tt{i} = T;
    fembar{i} = Ttr * fem(:, i);

    % Assemble into global stiffness matrix
    Knew = zeros(dof);
    for p = 1:6
        for q = 1:6
            Knew(l(i, p), l(i, q)) = Kg{i}(p, q);
        end
    end
    Ktotal = Ktotal + Knew;
end

%% Unrestrained Stiffness Sub-matrix
Kunr = Ktotal(1:uu, 1:uu);
KuuInv = inv(Kunr);
```

```
%% Joint Load Vector
jl = input('Enter joint load vector as a column vector : ');
jlu = jl(1:uu);
delu = KuuInv * jlu;

%% Global Displacements
delr = zeros(ur, 1);
del = [delu; delr];

%% End Moments and Joint Forces
jf = zeros(dof, 1);
for i = 1:n
    deli = del(l(i, :));
    mbar = (Kg{i} * deli) + fembar{i};
    fprintf('Member %d Global End Moment Matrix [MBar]:\n', i);
    disp(mbar');

    % Assemble joint forces
    for p = 1:6
        jf(l(i, p)) = jf(l(i, p)) + mbar(p);
    end
end
```

MATLAB output:

```
Enter the number of members:
2
Enter the moments of inertia as a row vector [I1 I2 ... In]:
[0.0016 0.0016]
Enter the lengths as a row vector [L1 L2 ... Ln]:
[4.472 4]
Enter the areas as a row vector [A1 A2 ... An]:
[0.12 0.12]
Enter the angles as a row vector [Theta1 Theta2 ... ThetaN]:
[63.435 0]
Enter the number of unrestrained degrees of freedom:
3
Enter the number of restrained degrees of freedom:
6
Enter the labels of unrestrained DOF as a row vector [u1 u2
...]:
[1 2 3]
Enter the labels of restrained DOF as a row vector [r1 r2
...]:
[4 5 6 7 8 9]
Enter global labels for Member 1 as a row vector: [7 1 9 2 8
3]
Enter global labels for Member 2 as a row vector: [1 4 2 5 3
6]
```

```
Enter fixed-end moments for Member 1 as a column vector:
[0; 0; 0; 0; 0; 0]
Enter fixed-end moments for Member 2 as a column vector:
[0; 0; 0; 0; 0; 0]
Enter joint load vector as a column vector: [0; 0; 10; 0; 0;
0; 0; 0; 0]
Member Number =      1
Local Stiffness matrix of member, [K] =
    0.0014    0.0007    0.0005   -0.0005         0         0
    0.0007    0.0014    0.0005   -0.0005         0         0
    0.0005    0.0005    0.0002   -0.0002         0         0
   -0.0005   -0.0005   -0.0002    0.0002         0         0
         0         0         0         0    0.0268   -0.0268
         0         0         0         0   -0.0268    0.0268
Transformation matrix of member, [T] =
    1.0000         0         0         0         0         0
         0    1.0000         0         0         0         0
         0         0    0.4472         0   -0.8944         0
         0         0         0    0.4472         0   -0.8944
         0         0    0.8944         0    0.4472         0
         0         0         0    0.8944         0    0.4472
Transformation matrix Transpose, [T] =
    1.0000         0         0         0         0         0
         0    1.0000         0         0         0         0
         0         0    0.4472         0    0.8944         0
         0         0         0    0.4472         0    0.8944
         0         0   -0.8944         0    0.4472         0
         0         0         0   -0.8944         0    0.4472
Global Matrix, [K global] =
    0.0014    0.0007    0.0002   -0.0002   -0.0004    0.0004
    0.0007    0.0014    0.0002   -0.0002   -0.0004    0.0004
    0.0002    0.0002    0.0215   -0.0215    0.0106   -0.0106
   -0.0002   -0.0002   -0.0215    0.0215   -0.0106    0.0106
   -0.0004   -0.0004    0.0106   -0.0106    0.0055   -0.0055
    0.0004    0.0004   -0.0106    0.0106   -0.0055    0.0055

Member Number =      2
Local Stiffness matrix of member, [K] =
    0.0016    0.0008    0.0006   -0.0006         0         0
    0.0008    0.0016    0.0006   -0.0006         0         0
    0.0006    0.0006    0.0003   -0.0003         0         0
   -0.0006   -0.0006   -0.0003    0.0003         0         0
         0         0         0         0    0.0300   -0.0300
         0         0         0         0   -0.0300    0.0300
Transformation matrix of member, [T] =
     1     0     0     0     0     0
     0     1     0     0     0     0
     0     0     1     0     0     0
     0     0     0     1     0     0
     0     0     0     0     1     0
     0     0     0     0     0     1
```

```
Transformation matrix Transpose, [T] =
     1     0     0     0     0     0
     0     1     0     0     0     0
     0     0     1     0     0     0
     0     0     0     1     0     0
     0     0     0     0     1     0
     0     0     0     0     0     1
Global Matrix, [K global] =
    0.0016    0.0008    0.0006   -0.0006         0         0
    0.0008    0.0016    0.0006   -0.0006         0         0
    0.0006    0.0006    0.0003   -0.0003         0         0
   -0.0006   -0.0006   -0.0003    0.0003         0         0
         0         0         0         0    0.0300   -0.0300
         0         0         0         0   -0.0300    0.0300

Stiffness Matrix of complete structure, [Ktotal] =
 0.0030  0.0004  0.0004  0.0008 -0.0006       0  0.0007 -0.0004  0.0002
 0.0004  0.0218  0.0106  0.0006 -0.0003       0 -0.0002 -0.0106 -0.0215
 0.0004  0.0106  0.0355       0       0 -0.0300  0.0004 -0.0055 -0.0106
 0.0008  0.0006       0  0.0016 -0.0006       0       0       0       0
-0.0006 -0.0003       0 -0.0006  0.0003       0       0       0       0
      0       0 -0.0300       0       0  0.0300       0       0       0
 0.0007 -0.0002  0.0004       0       0       0  0.0014 -0.0004  0.0002
-0.0004 -0.0106 -0.0055       0       0       0 -0.0004  0.0055  0.0106
 0.0002 -0.0215 -0.0106       0       0       0  0.0002  0.0106  0.0215


Unrestrained Stiffness sub-matix, [Kuu] =
    0.0030    0.0004    0.0004
    0.0004    0.0218    0.0106
    0.0004    0.0106    0.0355
Inverse of Unrestrained Stiffness sub-matix, [KuuInverse] =
  330.8628   -4.5612   -2.6307
   -4.5612   53.7692  -16.0545
   -2.6307  -16.0545   32.9804

Joint Load vector, [Jl] =
     0     0    10     0     0     0     0     0     0
Unrestrained displacements, [DelU] =
  -26.3069 -160.5452  329.8036

Member Number =     1
Global displacement matrix [DeltaBar] =
         0  -26.3069         0 -160.5452         0  329.8036
Global End moment matrix [MBar] =
    0.1572    0.1384   -0.0639    0.0639   -0.1059    0.1059

Member Number =     2
Global displacement matrix [DeltaBar] =
  -26.3069         0 -160.5452         0  329.8036         0
```

```
Global End moment matrix [MBar] =
   -0.1384    -0.1174    -0.0639     0.0639     9.8941    -9.8941

Joint forces =
  0     0.0000    10.0000    -0.1174     0.0639    -9.8941
0.1572    -0.1059    -0.0639
```

Example 3.2:

Analyze the planar non-orthogonal structure shown below using the stiffness method.

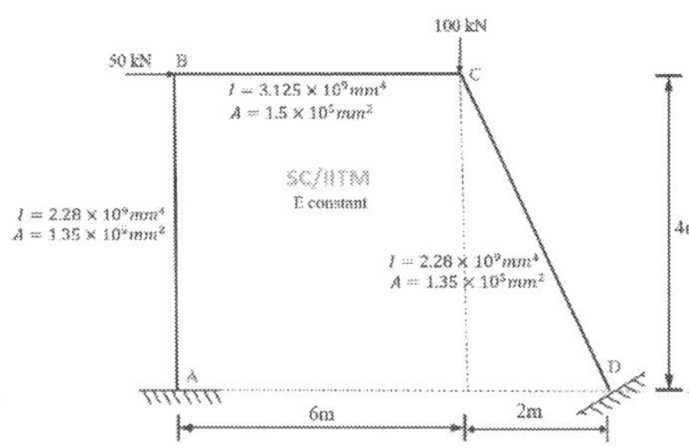

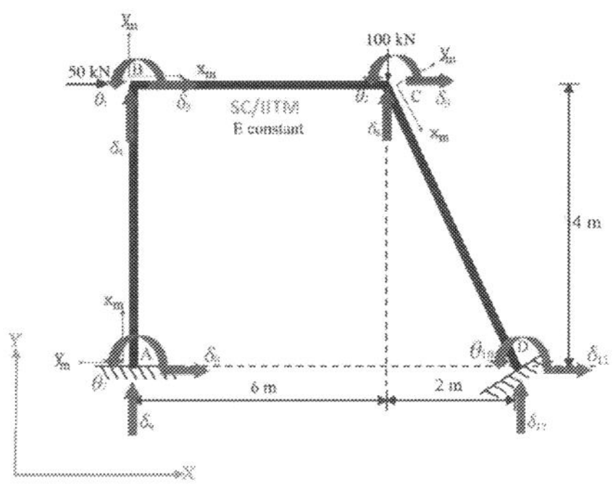

SOLUTION:

Step 1: Marking global axes, local axes of each member and degrees of freedom

Mark the local and global axes, and mark the unrestrained and restrained degrees of freedom, as shown above. Please check that all degrees of freedom are to be marked with reference to the global axes and not to the local axes of the member. Further, also ensure that there are three degrees of freedom in each node. The inclined (non-orthogonal) member (CD) has an inclination of θ (= $\tan^{-1}(4/2) = -63.435°$), which is negative. It is because x_m-axis of the member (AB) is at clockwise 63.435° with the global X-axis. In case of member (BC), this angle is zero (and not 180°) as the jth of the member (BC) is at B. Had the jth end of the member been assumed at C, then x_m-axis of the member BC would point toward B, making an angle of 180° with the X-axis. Similarly, in the case of member AB, the angle of inclination is +90°, as the x_m-axis of the member is normal to the X-axis of the reference frame. This is because jth end of the member AB is chosen at A; should it be chosen at B, the angle shall be –90°.

Step 2: Calculation of transformation matrix coefficients and global labels

The unrestrained degrees of freedom are 6 (θ_1, θ_2, δ_3, δ_4, δ_5, δ_6)

The restrained degrees of freedom: 6 (θ_7, δ_8, δ_9, θ_{10}, δ_{11}, δ_{12})

Thus, the size of the total stiffness matrix will be 12×12, in which the sub-matrix for the unrestrained degrees-of-freedom will be of size 6×6.

Member Number	Ends J	Ends k	Length (m)	θ (degrees)	C_x	C_y	Global Labels
1	A	B	4	90	0	1	(7,1,9,4,8,3)
2	B	C	6	0	1	0	(1,2,4,6,3,5)
3	C	D	4.472	–63.435	0.447	–0.894	(2,10,6,12,5,11)

Step 3: Calculation of local stiffness matrix

The stiffness matrix for the standard beam element, including the axial deformation, is given by,

$$[K]_i = \begin{bmatrix} \frac{4EI}{l} & \frac{2EI}{l} & \frac{6EI}{l^2} & -\frac{6EI}{l^2} & 0 & 0 \\ \frac{2EI}{l} & \frac{4EI}{l} & \frac{6EI}{l^2} & -\frac{6EI}{l^2} & 0 & 0 \\ \frac{6EI}{l^2} & \frac{6EI}{l^2} & \frac{12EI}{l^3} & -\frac{12EI}{l^3} & 0 & 0 \\ -\frac{6EI}{l^2} & -\frac{6EI}{l^2} & -\frac{12EI}{l^3} & \frac{12EI}{l^3} & 0 & 0 \\ 0 & 0 & 0 & 0 & \frac{AE}{l} & -\frac{AE}{l} \\ 0 & 0 & 0 & 0 & -\frac{AE}{l} & \frac{AE}{l} \end{bmatrix}$$

$$[K]_{AB} = E\times10^{-4} \begin{array}{c} \begin{array}{cccccc} 7 & 1 & 9 & 4 & 8 & 3 \end{array} \\ \begin{bmatrix} 23 & 11 & 9 & -9 & 0 & 0 \\ 11 & 23 & 6 & -9 & 0 & 0 \\ 9 & 9 & 4 & -4 & 0 & 0 \\ -9 & -9 & -4 & 4 & 0 & 0 \\ 0 & 0 & 0 & 0 & 338 & -338 \\ 0 & 0 & 0 & 0 & -338 & 338 \end{bmatrix} \end{array} \begin{array}{c} 7 \\ 1 \\ 9 \\ 4 \\ 8 \\ 3 \end{array}$$

$$[K]_{BC} = E\times10^{-4} \begin{array}{c} \begin{array}{cccccc} 1 & 2 & 4 & 6 & 3 & 5 \end{array} \\ \begin{bmatrix} 21 & 10 & 5 & -5 & 0 & 0 \\ 10 & 21 & 5 & -5 & 0 & 0 \\ 5 & 5 & 2 & -2 & 0 & 0 \\ -5 & -5 & -2 & 2 & 0 & 0 \\ 0 & 0 & 0 & 0 & 250 & -250 \\ 0 & 0 & 0 & 0 & -250 & 250 \end{bmatrix} \end{array} \begin{array}{c} 1 \\ 2 \\ 4 \\ 6 \\ 3 \\ 5 \end{array}$$

$$[K]_{CD} = E\times10^{-4}\begin{array}{c} \begin{matrix} 2 & \;\;10 & \;\;6 & \;\;12 & \;\;5 & \;\;11 \end{matrix} \\ \begin{bmatrix} 20 & 10 & 7 & -7 & 0 & 0 \\ 10 & 20 & 7 & -7 & 0 & 0 \\ 7 & 7 & 3 & -3 & 0 & 0 \\ -7 & -7 & -3 & 3 & 0 & 0 \\ 0 & 0 & 0 & 0 & 302 & -302 \\ 0 & 0 & 0 & 0 & -302 & 302 \end{bmatrix} \end{array} \begin{matrix} \\ 2 \\ 10 \\ 6 \\ 12 \\ 5 \\ 11 \end{matrix}$$

Step 4: Calculation of transformation matrix

The transformation matrix for any member 'i' is given by,

$$[T] = \begin{bmatrix} 1 & 0 & 0 & 0 & 0 & 0 \\ 0 & 1 & 0 & 0 & 0 & 0 \\ 0 & 0 & C_x & 0 & -C_y & 0 \\ 0 & 0 & 0 & C_x & 0 & -C_y \\ 0 & 0 & C_y & 0 & C_x & 0 \\ 0 & 0 & 0 & C_y & 0 & C_x \end{bmatrix}$$

$$[T]_{AB} = \begin{bmatrix} 1 & 0 & 0 & 0 & 0 & 0 \\ 0 & 1 & 0 & 0 & 0 & 0 \\ 0 & 0 & 0 & 0 & -1 & 0 \\ 0 & 0 & 0 & 0 & 0 & -1 \\ 0 & 0 & 1 & 0 & 0 & 0 \\ 0 & 0 & 0 & 1 & 0 & 0 \end{bmatrix}$$

$$[T]_{BC} = \begin{bmatrix} 1 & 0 & 0 & 0 & 0 & 0 \\ 0 & 1 & 0 & 0 & 0 & 0 \\ 0 & 0 & 1 & 0 & 0 & 0 \\ 0 & 0 & 0 & 1 & 0 & 0 \\ 0 & 0 & 0 & 0 & 1 & 0 \\ 0 & 0 & 0 & 0 & 0 & 1 \end{bmatrix}$$

$$[T]_{CD} = \begin{bmatrix} 1 & 0 & 0 & 0 & 0 & 0 \\ 0 & 1 & 0 & 0 & 0 & 0 \\ 0 & 0 & 0.4472 & 0 & 0.8944 & 0 \\ 0 & 0 & 0 & 0.4472 & 0 & 0.8944 \\ 0 & 0 & -0.8944 & 0 & 0.4472 & 0 \\ 0 & 0 & 0 & 0.8944 & 0 & 0.4472 \end{bmatrix}$$

Step 5: Estimation of joint load vector

As there is no member loading, {FEM} will be a null vector. Hence, the joint load vector can be written directly as follows:

$$\{J_L\}_{12x1} = \begin{Bmatrix} 0 \\ 0 \\ 50 \\ 0 \\ 0 \\ -100 \\ 0 \\ 0 \\ 0 \\ 0 \\ 0 \\ 0 \end{Bmatrix} \begin{matrix} 1 \\ 2 \\ 3 \\ 4 \\ 5 \\ 6 \\ 7 \\ 8 \\ 9 \\ 10 \\ 11 \\ 12 \end{matrix}$$

Thus, the joint load vector in unrestrained degrees of freedom is given by:

$$\{\overline{J}_{Lu}\} = \begin{Bmatrix} 0 \\ 0 \\ 50 \\ 0 \\ 0 \\ -100 \end{Bmatrix}_{6\times1}$$

Step 6: Calculation of global stiffness matrix

The stiffness matrix of every member with respect to the global axes system is obtained by the following equation:

$$\{\overline{K}\}_i = [T]_i^T [K]_i [T]_i$$

$$\begin{matrix} & 7 & 1 & 9 & 4 & 8 & 3 & \end{matrix}$$

$$\overline{K_{AB}} = E\times10^{-4} \begin{bmatrix} 23 & 11 & 0 & 0 & -9 & 9 \\ 11 & 23 & 0 & 0 & -9 & 9 \\ 0 & 0 & 338 & -338 & 0 & 0 \\ 0 & 0 & -338 & 338 & 0 & 0 \\ -9 & -9 & 0 & 0 & 4 & -4 \\ 9 & 9 & 0 & 0 & -4 & 4 \end{bmatrix} \begin{matrix} 7 \\ 1 \\ 9 \\ 4 \\ 8 \\ 3 \end{matrix}$$

$$\overline{K_{BC}} = E \times 10^{-4} \begin{array}{c} \begin{array}{cccccc} 1 & 2 & 3 & 4 & 5 & 6 \end{array} \\ \left[\begin{array}{cccccc} 21 & 10 & 5 & -5 & 0 & 9 \\ 10 & 21 & 5 & -5 & 0 & 9 \\ 5 & 5 & 2 & -2 & 0 & 0 \\ -5 & -5 & -2 & 2 & 0 & 0 \\ 0 & 0 & 0 & 0 & 250 & -250 \\ 0 & 0 & 0 & 0 & -250 & 250 \end{array}\right] \end{array} \begin{array}{c} 1 \\ 2 \\ 3 \\ 4 \\ 5 \\ 6 \end{array}$$

$$\bar{K}_{CD} = E \times 10^{-4} \begin{array}{c} \begin{array}{cccccc} 2 & 10 & 6 & 12 & 5 & 11 \end{array} \\ \left[\begin{array}{cccccc} 20 & 10 & 3 & -3 & 6 & -6 \\ 10 & 20 & 3 & -3 & 6 & -6 \\ 3 & 3 & 242 & -242 & -120 & 120 \\ -3 & -3 & -242 & 242 & 120 & -120 \\ 6 & 6 & -120 & 120 & 63 & -63 \\ -6 & -6 & 120 & 120 & -63 & 63 \end{array}\right] \end{array} \begin{array}{c} 2 \\ 10 \\ 6 \\ 12 \\ 5 \\ 11 \end{array}$$

$$\bar{K}_{TOTAL} = E \times 10^{-4}$$

$$\begin{array}{c} \begin{array}{cccccccccccc} 1 & 2 & 3 & 4 & 5 & 6 & 7 & 8 & 9 & 10 & 11 & 12 \end{array} \\ \left[\begin{array}{cccccccccccc} 44 & 10 & 9 & 5 & 0 & -5 & 11 & -9 & 0 & 0 & 0 & 0 \\ 10 & 41 & 0 & 5 & 6 & -2 & 0 & 0 & 0 & 10 & -6 & -3 \\ 9 & 0 & 254 & 0 & -250 & 0 & 9 & -4 & 0 & 0 & 0 & 0 \\ 5 & 5 & 0 & 339 & 0 & -2 & 0 & 0 & -338 & 0 & 0 & 0 \\ 0 & 6 & -250 & 0 & 313 & -120 & 0 & 0 & 0 & 6 & -63 & 120 \\ -5 & -2 & 0 & -2 & -120 & 244 & 0 & 0 & 0 & 3 & 120 & -242 \\ 0 & 0 & 9 & 0 & 0 & 0 & 23 & 23 & 0 & 0 & 0 & 0 \\ -9 & 0 & -4 & 0 & 0 & 0 & -9 & -9 & 0 & 0 & 0 & 0 \\ 0 & 0 & 0 & 338 & 0 & 0 & 0 & 0 & 338 & 0 & 0 & 0 \\ 0 & 10 & 0 & 0 & 6 & 3 & 0 & 0 & 0 & 20 & -6 & -3 \\ 0 & -6 & 0 & 0 & -63 & 120 & 0 & 0 & 0 & -6 & 63 & -120 \\ 0 & -3 & 0 & 0 & 120 & -242 & 0 & 0 & 0 & -3 & -120 & 242 \end{array}\right] \end{array} \begin{array}{c} 1 \\ 2 \\ 3 \\ 4 \\ 5 \\ 6 \\ 7 \\ 8 \\ 9 \\ 10 \\ 11 \\ 12 \end{array}$$

$$\left[\overline{K_{UU}}\right] = E \times 10^{-4} \begin{array}{c} \begin{array}{cccccc} 1 & 2 & 3 & 4 & 5 & 6 \end{array} \\ \left[\begin{array}{cccccc} 44 & 10 & 9 & 5 & 0 & -5 \\ 10 & 41 & 0 & 5 & 6 & -2 \\ 9 & 0 & 254 & 0 & -250 & 0 \\ 5 & 5 & 0 & 339 & 0 & -2 \\ 0 & 6 & -250 & 0 & 313 & -120 \\ -5 & -2 & 0 & -2 & -120 & 244 \end{array}\right] \end{array} \begin{array}{c} 1 \\ 2 \\ 3 \\ 4 \\ 5 \\ 6 \end{array}$$

$$\left[\bar{K}_{uu}\right]^{-1}=\frac{1}{E}\begin{bmatrix} 264 & -46.9 & -161.7 & -3.7 & -155.4 & -71 \\ -46.9 & 271 & -126.7 & -3.8 & -130.5 & -62.6 \\ -161.7 & -126.7 & 1376.3 & 7.8 & 1354.3 & 659.3 \\ -3.7 & -3.8 & 7.8 & 29.6 & 7.8 & 3.9 \\ -155.4 & -130.5 & 1354.3 & 7.8 & 1372.1 & 668.1 \\ -71 & -62.6 & 659.3 & 3.9 & 668.1 & 366.5 \end{bmatrix}$$

Now, $\left\{\bar{\delta}_u\right\}=\left[\bar{K}_{uu}\right]^{-1}\left\{\bar{J}_L\right\}_u$

$$\left\{\bar{\delta}_u\right\}=\frac{1}{E}\begin{Bmatrix} -986 \\ -75.8 \\ 2882.9 \\ -2.5 \\ 898.5 \\ -3682.2 \end{Bmatrix}$$

$$\left[\bar{\delta}\right]_{AB}=\frac{1}{E}\begin{Bmatrix} 0 \\ -986 \\ 0 \\ -2.5 \\ 0 \\ 2882.9 \end{Bmatrix},\left[\bar{\delta}\right]_{BC}=\frac{1}{E}\begin{Bmatrix} -986 \\ -75.8 \\ -2.5 \\ -3682.2 \\ 2882.9 \\ 898.5 \end{Bmatrix},\left[\bar{\delta}\right]_{CD}=\frac{1}{E}\begin{Bmatrix} -75.8 \\ 0 \\ -3682.2 \\ 0 \\ 898.5 \\ 0 \end{Bmatrix}$$

Step 7: Calculation of end moments and shear

$$\left\{\bar{M}\right\}_i=\left[\bar{K}\right]_i\left\{\bar{\delta}_i\right\}+\left\{F\bar{E}M\right\}_i$$

$$\left\{\bar{M}\right\}_{AB}=\begin{Bmatrix} \bar{M}_7 \\ \bar{M}_1 \\ \bar{V}_9 \\ \bar{V}_2 \\ \bar{H}_8 \\ \bar{H}_3 \end{Bmatrix}=\begin{Bmatrix} 1.3408 \\ 0.2167 \\ 0.0858 \\ -0.0858 \\ -0.3894 \\ 0.3894 \end{Bmatrix}$$

$$\overline{\{M\}}_{BC} = \begin{Bmatrix} \overline{M_1} \\ \overline{M_2} \\ \overline{V_4} \\ \overline{V_6} \\ \overline{H_3} \\ \overline{H_5} \end{Bmatrix} = \begin{Bmatrix} -0.2167 \\ 0.7314 \\ 0.0858 \\ -0.0858 \\ 49.6106 \\ -49.6106 \end{Bmatrix}$$

$$\overline{\{M\}}_{CD} = \begin{Bmatrix} \overline{M_2} \\ \overline{M_{10}} \\ \overline{V_6} \\ \overline{V_{12}} \\ \overline{H_5} \\ \overline{H_{11}} \end{Bmatrix} = \begin{Bmatrix} -0.7314 \\ -0.6541 \\ -99.9142 \\ 99.9142 \\ 49.6106 \\ -49.6106 \end{Bmatrix}$$

The member and final moments are shown below:

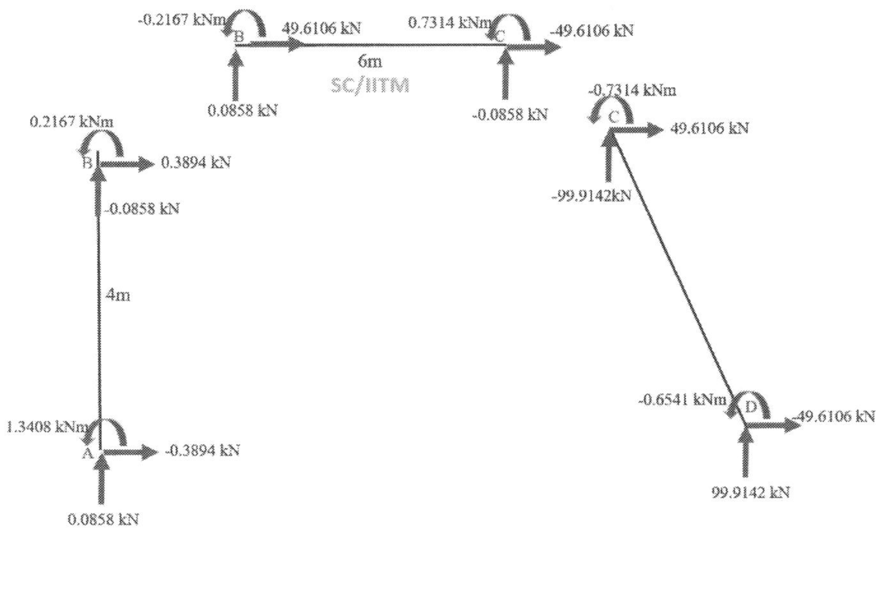

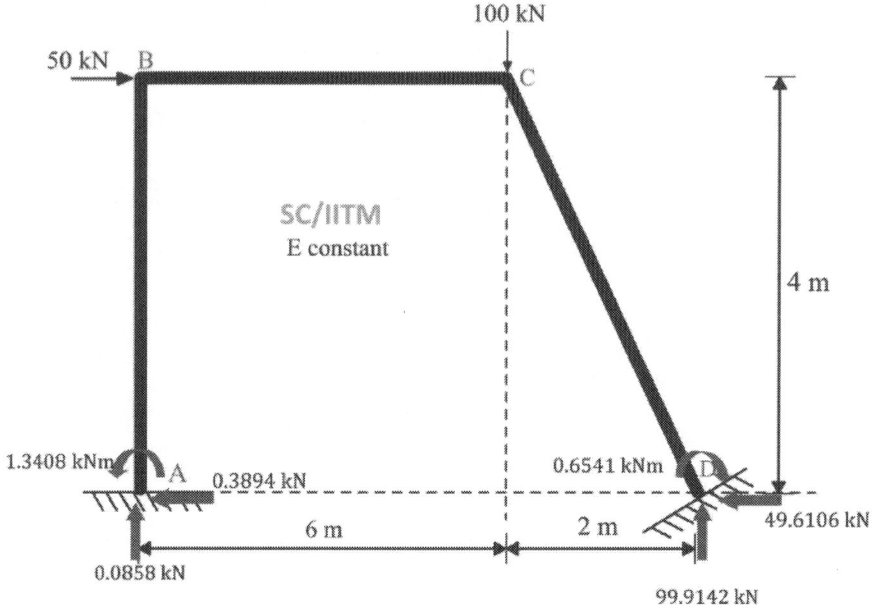

MATLAB output:

```
Enter the number of members:
3
Enter the moments of inertia as a row vector [I1 I2 ... In]:
[0.00228 0.003125 0.00228]
Enter the lengths as a row vector [L1 L2 ... Ln]:
[4 6 4.472]
Enter the areas as a row vector [A1 A2 ... An]:
[0.135 0.15 0.135]
Enter the angles as a row vector [Theta1 Theta2 ... ThetaN]:
[90 0 -63.435]
Enter the number of unrestrained degrees of freedom:
6
Enter the number of restrained degrees of freedom:
6
Enter the labels of unrestrained DOF as a row vector [u1 u2
...]:
[1 2 3 4 5 6]
Enter the labels of restrained DOF as a row vector [r1 r2
...]:
[7 8 9 10 11 12]
Enter global labels for Member 1 as a row vector: [7 1 9 4 8
3]
Enter global labels for Member 2 as a row vector: [1 2 4 6 3
5]
```

```
Enter global labels for Member 3 as a row vector: [2 10 6 12 5
11]
Enter fixed-end moments for Member 1 as a column vector :
[0; 0; 0; 0; 0; 0]
Enter fixed-end moments for Member 2 as a column vector:
[0; 0; 0; 0; 0; 0]
Enter fixed-end moments for Member 3 as a column vector:
[0; 0; 0; 0; 0; 0]
Enter joint load vector as a column vector: [0; 0; 50; 0; 0; -
100; 0; 0; 0; 0; 0; 0]
Member Number =     1
Local Stiffness matrix of member, [K] =
    0.0023    0.0011    0.0009   -0.0009         0         0
    0.0011    0.0023    0.0009   -0.0009         0         0
    0.0009    0.0009    0.0004   -0.0004         0         0
   -0.0009   -0.0009   -0.0004    0.0004         0         0
         0         0         0         0    0.0338   -0.0338
         0         0         0         0   -0.0338    0.0338
Transformation matrix of member, [T] =
     1     0     0     0     0     0
     0     1     0     0     0     0
     0     0     0     0    -1     0
     0     0     0     0     0    -1
     0     0     1     0     0     0
     0     0     0     1     0     0
Transformation matrix Transpose, [T] =
     1     0     0     0     0     0
     0     1     0     0     0     0
     0     0     0     0     1     0
     0     0     0     0     0     1
     0     0    -1     0     0     0
     0     0     0    -1     0     0
Global Matrix, [K global] =
    0.0023    0.0011         0         0   -0.0009    0.0009
    0.0011    0.0023         0         0   -0.0009    0.0009
         0         0    0.0338   -0.0338         0         0
         0         0   -0.0338    0.0338         0         0
   -0.0009   -0.0009         0         0    0.0004   -0.0004
    0.0009    0.0009         0         0   -0.0004    0.0004


Member Number =     2
Local Stiffness matrix of member, [K] =
    0.0021    0.0010    0.0005   -0.0005         0         0
    0.0010    0.0021    0.0005   -0.0005         0         0
    0.0005    0.0005    0.0002   -0.0002         0         0
   -0.0005   -0.0005   -0.0002    0.0002         0         0
         0         0         0         0    0.0250   -0.0250
         0         0         0         0   -0.0250    0.0250
```

```
Transformation matrix of member, [T] =
     1     0     0     0     0     0
     0     1     0     0     0     0
     0     0     1     0     0     0
     0     0     0     1     0     0
     0     0     0     0     1     0
     0     0     0     0     0     1
Transformation matrix Transpose, [T] =
     1     0     0     0     0     0
     0     1     0     0     0     0
     0     0     1     0     0     0
     0     0     0     1     0     0
     0     0     0     0     1     0
     0     0     0     0     0     1
Global Matrix, [K global] =
    0.0021    0.0010    0.0005   -0.0005         0         0
    0.0010    0.0021    0.0005   -0.0005         0         0
    0.0005    0.0005    0.0002   -0.0002         0         0
   -0.0005   -0.0005   -0.0002    0.0002         0         0
         0         0         0         0    0.0250   -0.0250
         0         0         0         0   -0.0250    0.0250


Member Number =      3
Local Stiffness matrix of member, [K] =
    0.0020    0.0010    0.0007   -0.0007         0         0
    0.0010    0.0020    0.0007   -0.0007         0         0
    0.0007    0.0007    0.0003   -0.0003         0         0
   -0.0007   -0.0007   -0.0003    0.0003         0         0
         0         0         0         0    0.0302   -0.0302
         0         0         0         0   -0.0302    0.0302
Transformation matrix of member, [T] =
    1.0000         0         0         0         0         0
         0    1.0000         0         0         0         0
         0         0    0.4472         0    0.8944         0
         0         0         0    0.4472         0    0.8944
         0         0   -0.8944         0    0.4472         0
         0         0         0   -0.8944         0    0.4472
Transformation matrix Transpose, [T] =
    1.0000         0         0         0         0         0
         0    1.0000         0         0         0         0
         0         0    0.4472         0   -0.8944         0
         0         0         0    0.4472         0   -0.8944
         0         0    0.8944         0    0.4472         0
         0         0         0    0.8944         0    0.4472
Global Matrix, [K global] =
    0.0020    0.0010    0.0003   -0.0003    0.0006   -0.0006
    0.0010    0.0020    0.0003   -0.0003    0.0006   -0.0006
    0.0003    0.0003    0.0242   -0.0242   -0.0120    0.0120
   -0.0003   -0.0003   -0.0242    0.0242    0.0120   -0.0120
```

```
    0.0006    0.0006   -0.0120    0.0120    0.0063   -0.0063
   -0.0006   -0.0006    0.0120   -0.0120   -0.0063    0.0063


Stiffness Matrix of complete structure, [Ktotal] =
  Columns 1 through 10
  0.0044   0.0010   0.0009   0.0005        0  -0.0005   0.0011  -0.0009        0        0
  0.0010   0.0041        0   0.0005   0.0006  -0.0002        0        0        0   0.0010
  0.0009        0   0.0254        0  -0.0250        0   0.0009  -0.0004        0        0
  0.0005   0.0005        0   0.0339        0  -0.0002        0        0  -0.0338        0
       0   0.0006  -0.0250        0   0.0313  -0.0120        0        0        0   0.0006
 -0.0005  -0.0002        0  -0.0002  -0.0120   0.0244        0        0        0   0.0003
  0.0011        0   0.0009        0        0        0   0.0023  -0.0009        0        0
 -0.0009        0  -0.0004        0        0        0  -0.0009   0.0004        0        0
       0        0        0  -0.0338        0        0        0        0   0.0338        0
       0   0.0010        0        0   0.0006   0.0003        0        0        0   0.0020
       0  -0.0006        0        0  -0.0063   0.0120        0        0        0  -0.0006
       0  -0.0003        0        0   0.0120  -0.0242        0        0        0  -0.0003
  Columns 11 through 12
         0         0
   -0.0006   -0.0003
         0         0
         0         0
   -0.0063    0.0120
    0.0120   -0.0242
         0         0
         0         0
         0         0
   -0.0006   -0.0003
    0.0063   -0.0120
   -0.0120    0.0242


Unrestrained Stiffness sub-matix, [Kuu] =
    0.0044    0.0010    0.0009    0.0005         0   -0.0005
    0.0010    0.0041         0    0.0005    0.0006   -0.0002
    0.0009         0    0.0254         0   -0.0250         0
    0.0005    0.0005         0    0.0339         0   -0.0002
         0    0.0006   -0.0250         0    0.0313   -0.0120
   -0.0005   -0.0002         0   -0.0002   -0.0120    0.0244
Inverse of Unrestrained Stiffness sub-matix, [KuuInverse] =
   1.0e+03 *
    0.2640   -0.0469   -0.1617   -0.0037   -0.1554   -0.0710
   -0.0469    0.2710   -0.1267   -0.0038   -0.1305   -0.0626
   -0.1617   -0.1267    1.3763    0.0078    1.3543    0.6593
   -0.0037   -0.0038    0.0078    0.0296    0.0078    0.0039
   -0.1554   -0.1305    1.3543    0.0078    1.3721    0.6681
   -0.0710   -0.0626    0.6593    0.0039    0.6681    0.3665
Joint Load vector, [Jl] =
     0     0    50     0     0  -100     0     0     0     0     0     0
Unrestrained displacements, [DelU] =
   1.0e+03 *
   -0.9860   -0.0758    2.8829   -0.0025    0.8985   -3.6822
```

```
Member Number =       1
Global displacement matrix [DeltaBar] =
   1.0e+03 *
         0    -0.9860          0    -0.0025          0     2.8829
Global End moment matrix [MBar] =
    1.3408     0.2167     0.0858    -0.0858    -0.3894     0.3894
Member Number =       2
Global displacement matrix [DeltaBar] =
   1.0e+03 *
   -0.9860    -0.0758    -0.0025    -3.6822     2.8829     0.8985
Global End moment matrix [MBar] =
   -0.2167     0.7314     0.0858    -0.0858    49.6106   -49.6106
Member Number =       3
Global displacement matrix [DeltaBar] =
   1.0e+03 *
   -0.0758          0    -3.6822          0     0.8985          0
Global End moment matrix [MBar] =
   -0.7314    -0.6541   -99.9142    99.9142    49.6106   -49.6106
Joint forces =
0.0000 -0.0000 50.0000 -0.0000 -0.0000 -100.0000 1.3408   -0.3894
0.0858    -0.6541   -49.6106    99.9142
```

Example 3.3:

Analyze the planar non-orthogonal structure shown below:

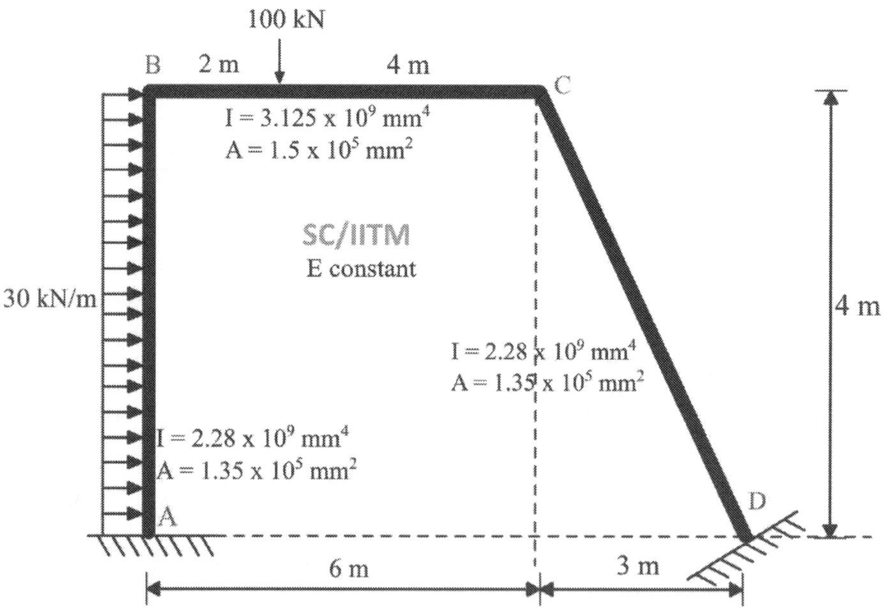

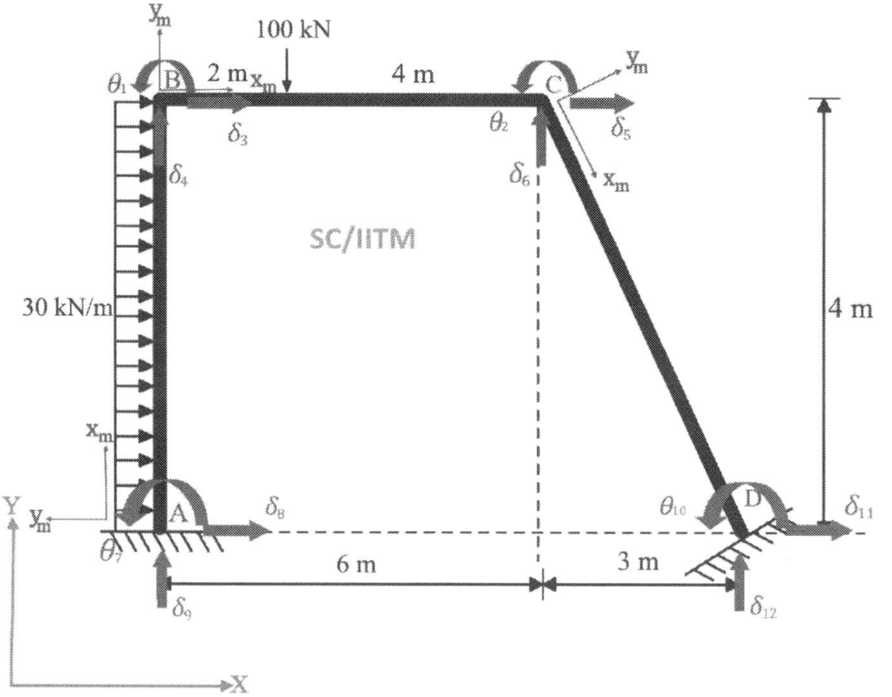

SOLUTION

Step 1: Marking global axes, local axes of each member and degrees of freedom

Mark the local and global axes, and mark the unrestrained and restrained degrees of freedom, as shown above. Please check that all degrees of freedom are to be marked with reference to the global axes and not to the local axes of the member. Further, also ensure that there are three degrees of freedom in each node. The inclined (non-orthogonal) member (CD) has an inclination of θ (= $\tan^{-1}(4/3) = -53.123°$), which is negative. It is because x_m-axis of the member (CD) is at clockwise 63.435° with the global X-axis. In case of member (BC), this angle is zero (and not 180°) as the jth of the member (BC) is at B. Had the jth end of the member been assumed at C, then x_m-axis of the member BC would point toward B, making an angle of 180° with the X-axis. Similarly, in the case of member AB, the angle of inclination is +90°, as the x_m-axis of the member is normal to the X-axis of the reference frame. This is because jth end of the member AB is chosen at A; should it be chosen at B, the angle shall be –90°.

Step 2: Calculation of transformation matrix coefficients and global labels

The unrestrained degrees of freedom are 6 (θ_1, θ_2, δ_3, δ_4, δ_5, δ_6)

The restrained degrees of freedom: 6 (θ_7, δ_8, δ_9, θ_{10}, δ_{11}, δ_{12})

Thus, the size of the total stiffness matrix will be 12×12, in which the sub-matrix for the unrestrained degrees-of-freedom will be of size 6×6.

Member Number	Ends j	Ends k	Length (m)	θ (degrees)	C_x	C_y	Global Labels
1	A	B	4	90	0	1	(7,1,9,4,8,3)
2	B	C	6	0	1	0	(1,2,4,6,3,5)
3	C	D	5	–53.123	0.6	–0.8	(2,10,6,12,5,11)

i. *Calculation of local stiffness matrix:*
   The stiffness matrix for the standard beam element, including the axial deformation, is given by,

$$[K]_i = \begin{bmatrix} \frac{4EI}{l} & \frac{2EI}{l} & \frac{6EI}{l^2} & -\frac{6EI}{l^2} & 0 & 0 \\ \frac{2EI}{l} & \frac{4EI}{l} & \frac{6EI}{l^2} & -\frac{6EI}{l^2} & 0 & 0 \\ \frac{6EI}{l^2} & \frac{6EI}{l^2} & \frac{12EI}{l^3} & -\frac{12EI}{l^3} & 0 & 0 \\ -\frac{6EI}{l^2} & -\frac{6EI}{l^2} & -\frac{12EI}{l^3} & \frac{12EI}{l^3} & 0 & 0 \\ 0 & 0 & 0 & 0 & \frac{AE}{l} & -\frac{AE}{l} \\ 0 & 0 & 0 & 0 & -\frac{AE}{l} & \frac{AE}{l} \end{bmatrix}$$

$$\begin{matrix} & 7 & 1 & 9 & 4 & 8 & 3 \end{matrix}$$

$$[K]_{AB} = E\times 10^{-4} \begin{bmatrix} 23 & 11 & 9 & -9 & 0 & 0 \\ 11 & 23 & 9 & -9 & 0 & 0 \\ 9 & 9 & 4 & -4 & 0 & 0 \\ -9 & -9 & -4 & 4 & 0 & 0 \\ 0 & 0 & 0 & 0 & 338 & -338 \\ 0 & 0 & 0 & 0 & -338 & 338 \end{bmatrix} \begin{matrix} 7 \\ 1 \\ 9 \\ 4 \\ 8 \\ 3 \end{matrix}$$

$$\begin{matrix} & 1 & 2 & 4 & 6 & 3 & 5 \end{matrix}$$

$$[K]_{BC} = E\times 10^{-4} \begin{bmatrix} 21 & 10 & 5 & -5 & 0 & 0 \\ 10 & 21 & 5 & -5 & 0 & 0 \\ 5 & 5 & 2 & -2 & 0 & 0 \\ -5 & -5 & -2 & 2 & 0 & 0 \\ 0 & 0 & 0 & 0 & 250 & -250 \\ 0 & 0 & 0 & 0 & -250 & 250 \end{bmatrix} \begin{matrix} 1 \\ 2 \\ 4 \\ 6 \\ 3 \\ 5 \end{matrix}$$

$$
[K]_{CD} = E \times 10^{-4} \begin{array}{c} \begin{array}{cccccc} 2 & 10 & 6 & 12 & 5 & 11 \end{array} \\ \begin{bmatrix} 18 & 9 & 5 & -5 & 0 & 0 \\ 9 & 18 & 5 & -5 & 0 & 0 \\ 5 & 5 & 2 & -2 & 0 & 0 \\ -5 & -5 & -2 & 2 & 0 & 0 \\ 0 & 0 & 0 & 0 & 270 & -270 \\ 0 & 0 & 0 & 0 & -270 & 270 \end{bmatrix} \end{array} \begin{array}{c} \\ 2 \\ 10 \\ 6 \\ 12 \\ 5 \\ 11 \end{array}
$$

ii. *Calculation of transformation matrix:*
The transformation matrix for any member 'i' is given by,

$$
[T] = \begin{bmatrix} 1 & 0 & 0 & 0 & 0 & 0 \\ 0 & 1 & 0 & 0 & 0 & 0 \\ 0 & 0 & C_x & 0 & -C_y & 0 \\ 0 & 0 & 0 & C_x & 0 & -C_y \\ 0 & 0 & C_y & 0 & C_x & 0 \\ 0 & 0 & 0 & C_y & 0 & C_x \end{bmatrix}
$$

$$
[T]_{AB} = \begin{bmatrix} 1 & 0 & 0 & 0 & 0 & 0 \\ 0 & 1 & 0 & 0 & 0 & 0 \\ 0 & 0 & 0 & 0 & -1 & 0 \\ 0 & 0 & 0 & 0 & 0 & -1 \\ 0 & 0 & 1 & 0 & 0 & 0 \\ 0 & 0 & 0 & 1 & 0 & 0 \end{bmatrix}
$$

$$
[T]_{BC} = \begin{bmatrix} 1 & 0 & 0 & 0 & 0 & 0 \\ 0 & 1 & 0 & 0 & 0 & 0 \\ 0 & 0 & 1 & 0 & 0 & 0 \\ 0 & 0 & 0 & 1 & 0 & 0 \\ 0 & 0 & 0 & 0 & 1 & 0 \\ 0 & 0 & 0 & 0 & 0 & 1 \end{bmatrix}
$$

$$
[T]_{CD} = \begin{bmatrix} 1 & 0 & 0 & 0 & 0 & 0 \\ 0 & 1 & 0 & 0 & 0 & 0 \\ 0 & 0 & 0.6001 & 0 & 0.7999 & 0 \\ 0 & 0 & 0 & 0.6001 & 0 & 0.7999 \\ 0 & 0 & -0.7999 & 0 & 0.6001 & 0 \\ 0 & 0 & 0 & -0.7999 & 0 & 0.6001 \end{bmatrix}
$$

iii. *Estimation of Joint load vector:*

The FEM on the members are calculated as follows:

Member AB:

$$\{FEM\}_{AB} = \begin{Bmatrix} 40 \\ -40 \\ 60 \\ 60 \\ 0 \\ 0 \end{Bmatrix}, \{F\bar{E}M\}_{AB} = \begin{Bmatrix} 40 \\ -40 \\ 0 \\ 0 \\ -60 \\ -60 \end{Bmatrix}$$

Member BC:

$$\{FEM\}_{BC} = \begin{Bmatrix} 89.89 \\ -44.44 \\ 66.67 \\ 33.333 \\ 0 \\ 0 \end{Bmatrix}, \{F\bar{E}M\}_{BC} = \begin{Bmatrix} 89.89 \\ -44.44 \\ 66.67 \\ 33.333 \\ 0 \\ 0 \end{Bmatrix}$$

The joint load vector is given by,

$$\{J_L\}_{12\times 1} = \begin{Bmatrix} -48.89 \\ 44.44 \\ 60 \\ -66.67 \\ 0 \\ -33.33 \\ -40 \\ 60 \\ 0 \\ 0 \\ 0 \\ 0 \end{Bmatrix} \begin{matrix} 1 \\ 2 \\ 3 \\ 4 \\ 5 \\ 6 \\ 7 \\ 8 \\ 9 \\ 10 \\ 11 \\ 12 \end{matrix}$$

Thus, the joint load vector in unrestrained degrees of freedom is given by,

$$\{J_{LU}\}_{6\times 1} = \begin{Bmatrix} -48.89 \\ 44.44 \\ 60 \\ -66.67 \\ 0 \\ -33.33 \end{Bmatrix} \begin{matrix} 1 \\ 2 \\ 3 \\ 4 \\ 5 \\ 6 \end{matrix}$$

iv. *Calculation of global stiffness matrix:*
The stiffness matrix of every member concerning the global axes system is obtained by the following equation:

$$\{\bar{K}\}_i = [T]_i^T [K]_i [T]_i$$

$$\bar{K}_{AB} = E\times10^{-4}\ \begin{array}{cccccc|c} 7 & 1 & 9 & 4 & 8 & 3 & \\ \hline 23 & 11 & 0 & 0 & -9 & 9 & 7 \\ 11 & 23 & 0 & 0 & -9 & 9 & 1 \\ 0 & 0 & 338 & -338 & 0 & 0 & 9 \\ 0 & 0 & -338 & 338 & 0 & 0 & 4 \\ -9 & -9 & 0 & 0 & 4 & -4 & 8 \\ 9 & 9 & 0 & 0 & -4 & 4 & 3 \end{array}$$

$$\bar{K}_{BC} = E\times10^{-4}\ \begin{array}{cccccc|c} 1 & 2 & 4 & 6 & 3 & 5 & \\ \hline 21 & 10 & 5 & -5 & 0 & 0 & 1 \\ 10 & 21 & 5 & -5 & 0 & 0 & 2 \\ 5 & 5 & 2 & -2 & 0 & 0 & 4 \\ -5 & -5 & -2 & 2 & 0 & 0 & 6 \\ 0 & 0 & 0 & 0 & 250 & -250 & 3 \\ 0 & 0 & 0 & 0 & -250 & 250 & 5 \end{array}$$

$$\bar{K}_{CD} = E\times10^{-4}\ \begin{array}{cccccc|c} 2 & 10 & 6 & 12 & 5 & 11 & \\ \hline 18 & 9 & 3 & -3 & 4 & -4 & 2 \\ 9 & 18 & 3 & -3 & 4 & -4 & 10 \\ 3 & 3 & 174 & -174 & -129 & 129 & 6 \\ -3 & -3 & -174 & 174 & 129 & -129 & 12 \\ 4 & 4 & -129 & 129 & 99 & -99 & 5 \\ -4 & -4 & 129 & -129 & -99 & 99 & 11 \end{array}$$

$$\bar{K}_{TOTAL} = E\times10^{-4}\ \begin{array}{cccccccccccc|c} 1 & 2 & 3 & 4 & 5 & 6 & 7 & 8 & 9 & 10 & 11 & 12 & \\ \hline 44 & 10 & 9 & 5 & 0 & -5 & 11 & -9 & 0 & 0 & 0 & 0 & 1 \\ 10 & 39 & 0 & 5 & 4 & -2 & 0 & 0 & 0 & 9 & -4 & -3 & 2 \\ 9 & 0 & 254 & 0 & -250 & 0 & 9 & -4 & 0 & 0 & 0 & 0 & 3 \\ 5 & 5 & 0 & 339 & 0 & -2 & 0 & 0 & -338 & 0 & 0 & 0 & 4 \\ 0 & 4 & -250 & 0 & 349 & -129 & 0 & 0 & 0 & 4 & -99 & 129 & 5 \\ -5 & -2 & 0 & -2 & -129 & 175 & 0 & 0 & 0 & 3 & 129 & 174 & 6 \\ 11 & 0 & 9 & 0 & 0 & 0 & 23 & -9 & 0 & 0 & 0 & 0 & 7 \\ -9 & 0 & -4 & 0 & 0 & 0 & -9 & 4 & 0 & 0 & 0 & 0 & 8 \\ 0 & 0 & 0 & -338 & 0 & 0 & 0 & 0 & 338 & 0 & 0 & 0 & 9 \\ 0 & 9 & 0 & 0 & 4 & 3 & 0 & 0 & 0 & 18 & -4 & -3 & 10 \\ 0 & -4 & 0 & 0 & -99 & 126 & 0 & 0 & 0 & -4 & 99 & -129 & 11 \\ 0 & -3 & 0 & 0 & 129 & -174 & 0 & 0 & 0 & -3 & -129 & 174 & 12 \end{array}$$

$$\left[\bar{K}_{UU}\right] = E\times10^{-4}\begin{array}{c}\begin{array}{cccccc}1 & 2 & 3 & 4 & 5 & 6\end{array}\\ \begin{bmatrix} 44 & 10 & 9 & 5 & 0 & -5 \\ 10 & 39 & 0 & 5 & 4 & -2 \\ 9 & 0 & 254 & 0 & -250 & 0 \\ 5 & 5 & 0 & 339 & 0 & -2 \\ 0 & 4 & -250 & 0 & 349 & -120 \\ -5 & -2 & 0 & -2 & -129 & 175 \end{bmatrix}\end{array}\begin{array}{c}1\\2\\3\\4\\5\\6\end{array}$$

$$\left[\bar{K}_{uu}\right]^{-1} = \frac{1}{E}\begin{bmatrix} 258.7 & -59.2 & -125.5 & -3.5 & -118.8 & -80.1 \\ -59.2 & 277.2 & -61.8 & -3.6 & -64.9 & -46.4 \\ -125.5 & -61.8 & 1262.3 & 7.5 & 1239.6 & 904.8 \\ -3.5 & -3.6 & 7.5 & 29.6 & 7.5 & 5.7 \\ -118.8 & -64.9 & 1239.6 & 7.5 & 1256.7 & 917.5 \\ -80.1 & -46.4 & 904.8 & 5.7 & 917.5 & 727.1 \end{bmatrix}$$

Now, $\left\{\bar{\delta}_u\right\} = \left[\bar{K}_{uu}\right]^{-1}\left\{\bar{J}_L\right\}_u$

$$\left\{\bar{\delta}_u\right\} = \frac{1}{E}\times10^4\begin{Bmatrix} -1.9906 \\ 1.3283 \\ 4.8465 \\ -0.1702 \\ 4.6213 \\ 3.1529 \end{Bmatrix}$$

$$\left[\bar{\delta}\right]_{AB} = \frac{1}{E}\times10^4\begin{Bmatrix} 0 \\ -1.9906 \\ 0 \\ -0.1702 \\ 0 \\ 4.8465 \end{Bmatrix}, \left[\bar{\delta}\right]_{BC} = \frac{1}{E}\times10^4\begin{Bmatrix} -1.9906 \\ 1.3283 \\ -0.1702 \\ 3.1529 \\ 4.8465 \\ 4.6213 \end{Bmatrix}, \left[\bar{\delta}\right]_{CD} = \frac{1}{E}\times10^4\begin{Bmatrix} 1.3283 \\ 0 \\ 3.1529 \\ 0 \\ 4.6213 \\ 0 \end{Bmatrix}$$

v. *Calculation of end moments and shear:*

$$\left\{\bar{M}\right\}_i = \left[\bar{K}\right]_i\left\{\bar{\delta}_i\right\} + \left\{F\bar{E}M\right\}_i$$

$$\overline{\{M\}}_{AB} = \begin{Bmatrix} \overline{M_7} \\ \overline{M_1} \\ \overline{V_9} \\ \overline{V_4} \\ \overline{H_8} \\ \overline{H_3} \end{Bmatrix} = \begin{Bmatrix} 58.745 \\ -43.9479 \\ 57.4513 \\ -57.4513 \\ -63.6993 \\ -56.3007 \end{Bmatrix}$$

$$\overline{\{M\}}_{BC} = \begin{Bmatrix} \overline{M_1} \\ \overline{M_2} \\ \overline{V_4} \\ \overline{V_6} \\ \overline{H_3} \\ \overline{H_5} \end{Bmatrix} = \begin{Bmatrix} 43.9479 \\ -54.8100 \\ 57.4513 \\ 42.5487 \\ 56.3007 \\ -56.3007 \end{Bmatrix}$$

$$\overline{\{M\}}_{CD} = \begin{Bmatrix} \overline{M_2} \\ \overline{M_{10}} \\ \overline{V_6} \\ \overline{V_{12}} \\ \overline{H_5} \\ \overline{H_{11}} \end{Bmatrix} = \begin{Bmatrix} 54.8100 \\ 42.6958 \\ -42.5517 \\ 42.5517 \\ 56.3007 \\ -56.3007 \end{Bmatrix}$$

The member and final moments are shown.

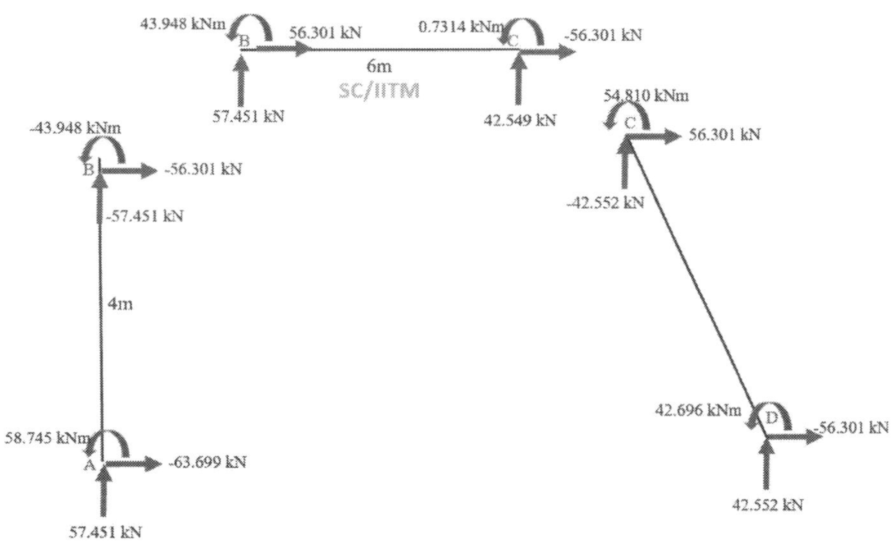

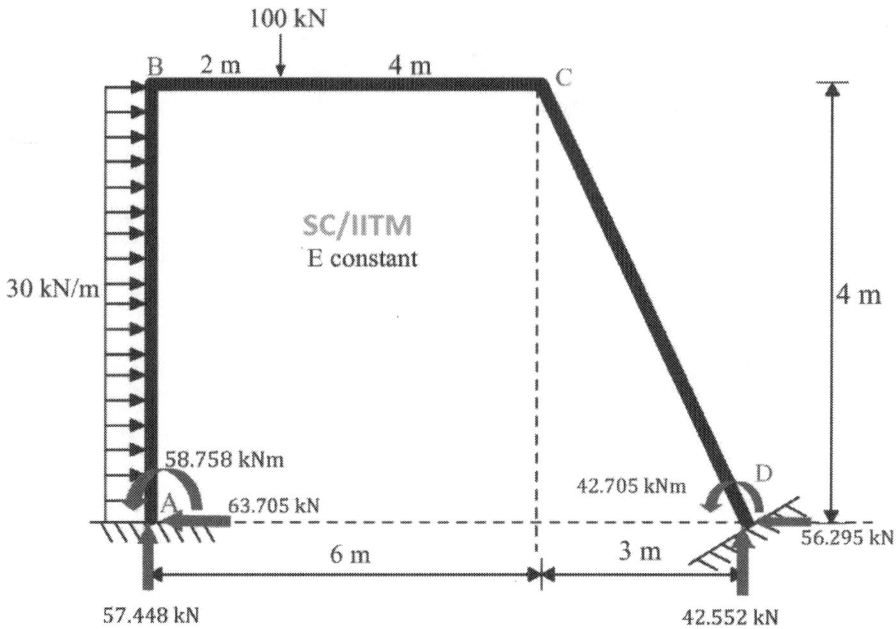


```
MATLAB output:
Enter the number of members:
3
Enter the moments of inertia as a row vector [I1 I2 ... In]:
[2.28E-3 3.125E-3 2.28E-3]
Enter the lengths as a row vector [L1 L2 ... Ln]:
[4 6 5]
Enter the areas as a row vector [A1 A2 ... An]:
[0.135 0.15 0.135]
Enter the angles as a row vector [Theta1 Theta2 ... ThetaN]:
[90 0 -53.123]
Enter the number of unrestrained degrees of freedom:
6
Enter the number of restrained degrees of freedom:
6
Enter the labels of unrestrained DOF as a row vector [u1 u2
...]:
[1 2 3 4 5 6]
Enter the labels of restrained DOF as a row vector [r1 r2
...]:
[7 8 9 10 11 12]
Enter global labels for Member 1 as a row vector: [7 1 9 4 8
3]
Enter global labels for Member 2 as a row vector: [1 2 4 6 3
5]
Enter global labels for Member 3 as a row vector: [2 10 6 12 5
11]
```

```
Enter fixed-end moments for Member 1 as a column vector :
[40; -40; 60; 60; 0; 0]
Enter fixed-end moments for Member 2 as a column vector :
[88.89; -44.44; 66.67; 33.33; 0; 0]
Enter fixed-end moments for Member 3 as a column vector :
[0; 0; 0; 0; 0; 0]
Enter joint load vector as a column vector : [-48.89; 44.44;
60; -66.67; 0; -33.333; -40; 60; 0; 0; 0; 0]
Member Number =      1
Local Stiffness matrix of member, [K] =
    0.0023     0.0011     0.0009    -0.0009          0          0
    0.0011     0.0023     0.0009    -0.0009          0          0
    0.0009     0.0009     0.0004    -0.0004          0          0
   -0.0009    -0.0009    -0.0004     0.0004          0          0
         0          0          0          0     0.0338    -0.0338
         0          0          0          0    -0.0338     0.0338
Transformation matrix of member, [T] =
     1     0     0     0     0     0
     0     1     0     0     0     0
     0     0     0     0    -1     0
     0     0     0     0     0    -1
     0     0     1     0     0     0
     0     0     0     1     0     0
Tranformation matrix Transpose, [T] =
     1     0     0     0     0     0
     0     1     0     0     0     0
     0     0     0     0     1     0
     0     0     0     0     0     1
     0     0    -1     0     0     0
     0     0     0    -1     0     0
Global Matrix, [K global] =
    0.0023     0.0011          0          0    -0.0009     0.0009
    0.0011     0.0023          0          0    -0.0009     0.0009
         0          0     0.0338    -0.0338          0          0
         0          0    -0.0338     0.0338          0          0
   -0.0009    -0.0009          0          0     0.0004    -0.0004
    0.0009     0.0009          0          0    -0.0004     0.0004

Member Number =      2
Local Stiffness matrix of member, [K] =
    0.0021     0.0010     0.0005    -0.0005          0          0
    0.0010     0.0021     0.0005    -0.0005          0          0
    0.0005     0.0005     0.0002    -0.0002          0          0
   -0.0005    -0.0005    -0.0002     0.0002          0          0
         0          0          0          0     0.0250    -0.0250
         0          0          0          0    -0.0250     0.0250
Transformation matrix of member, [T] =
     1     0     0     0     0     0
     0     1     0     0     0     0
     0     0     1     0     0     0
     0     0     0     1     0     0
```

```
     0      0      0      0      1      0
     0      0      0      0      0      1
Tranformation matrix Transpose, [T] =
     1      0      0      0      0      0
     0      1      0      0      0      0
     0      0      1      0      0      0
     0      0      0      1      0      0
     0      0      0      0      1      0
     0      0      0      0      0      1
Global Matrix, [K global] =
    0.0021    0.0010    0.0005   -0.0005         0         0
    0.0010    0.0021    0.0005   -0.0005         0         0
    0.0005    0.0005    0.0002   -0.0002         0         0
   -0.0005   -0.0005   -0.0002    0.0002         0         0
         0         0         0         0    0.0250   -0.0250
         0         0         0         0   -0.0250    0.0250

Member Number =      3
Local Stiffness matrix of member, [K] =
    0.0018    0.0009    0.0005   -0.0005         0         0
    0.0009    0.0018    0.0005   -0.0005         0         0
    0.0005    0.0005    0.0002   -0.0002         0         0
   -0.0005   -0.0005   -0.0002    0.0002         0         0
         0         0         0         0    0.0270   -0.0270
         0         0         0         0   -0.0270    0.0270
Transformation matrix of member, [T] =
    1.0000         0         0         0         0         0
         0    1.0000         0         0         0         0
         0         0    0.6001         0    0.7999         0
         0         0         0    0.6001         0    0.7999
         0         0   -0.7999         0    0.6001         0
         0         0         0   -0.7999         0    0.6001
Tranformation matrix Transpose, [T] =
    1.0000         0         0         0         0         0
         0    1.0000         0         0         0         0
         0         0    0.6001         0   -0.7999         0
         0         0         0    0.6001         0   -0.7999
         0         0    0.7999         0    0.6001         0
         0         0         0    0.7999         0    0.6001
Global Matrix, [K global] =
    0.0018    0.0009    0.0003   -0.0003    0.0004   -0.0004
    0.0009    0.0018    0.0003   -0.0003    0.0004   -0.0004
    0.0003    0.0003    0.0174   -0.0174   -0.0129    0.0129
   -0.0003   -0.0003   -0.0174    0.0174    0.0129   -0.0129
    0.0004    0.0004   -0.0129    0.0129    0.0099   -0.0099
   -0.0004   -0.0004    0.0129   -0.0129   -0.0099    0.0099
Stiffness Matrix of the complete structure, [Ktotal] =
  Columns 1 through 10
 0.0044  0.0010  0.0009  0.0005       0 -0.0005  0.0011 -0.0009       0       0
 0.0010  0.0039       0  0.0005  0.0004 -0.0002       0       0       0  0.0009
 0.0009       0  0.0254       0 -0.0250       0  0.0009 -0.0004       0       0
 0.0005  0.0005       0  0.0339       0 -0.0002       0       0 -0.0338       0
```

```
       0   0.0004  -0.0250        0   0.0349  -0.0129        0        0        0   0.0004
 -0.0005  -0.0002        0  -0.0002  -0.0129   0.0175        0        0        0   0.0003
  0.0011        0   0.0009        0        0        0   0.0023  -0.0009        0        0
 -0.0009        0  -0.0004        0        0        0  -0.0009   0.0004        0        0
       0        0        0  -0.0338        0        0        0        0   0.0338        0
       0   0.0009        0        0   0.0004   0.0003        0        0        0   0.0018
       0  -0.0004        0        0  -0.0099   0.0129        0        0        0  -0.0004
       0  -0.0003        0        0   0.0129  -0.0174        0        0        0  -0.0003
  Columns 11 through 12
         0         0
   -0.0004   -0.0003
         0         0
         0         0
   -0.0099    0.0129
    0.0129   -0.0174
         0         0
         0         0
         0         0
   -0.0004   -0.0003
    0.0099   -0.0129
   -0.0129    0.0174


Unrestrained Stiffness sub-matrix, [Kuu] =
    0.0044    0.0010    0.0009    0.0005         0   -0.0005
    0.0010    0.0039         0    0.0005    0.0004   -0.0002
    0.0009         0    0.0254         0   -0.0250         0
    0.0005    0.0005         0    0.0339         0   -0.0002
         0    0.0004   -0.0250         0    0.0349   -0.0129
   -0.0005   -0.0002         0   -0.0002   -0.0129    0.0175
The inverse of Unrestrained Stiffness sub-matrix, [KuuInverse] =
   1.0e+03 *
    0.2587   -0.0592   -0.1255   -0.0035   -0.1188   -0.0801
   -0.0592    0.2772   -0.0618   -0.0036   -0.0649   -0.0464
   -0.1255   -0.0618    1.2623    0.0075    1.2396    0.9048
   -0.0035   -0.0036    0.0075    0.0296    0.0075    0.0057
   -0.1188   -0.0649    1.2396    0.0075    1.2567    0.9175
   -0.0801   -0.0464    0.9048    0.0057    0.9175    0.7271

Joint Load vector, [Jl] =

 -48.8900 44.4400 60.0000 -66.6700       0  -33.3330  -40.0000   60.0000
0        0        0        0


Unrestrained displacements, [DelU] =
   1.0e+04 *
   -1.9906    1.3283    4.8465   -0.1702    4.6213    3.1529


Member Number =      1
Global displacement matrix [DeltaBar] =
   1.0e+04 *
         0   -1.9906         0   -0.1702         0    4.8465
```

```
Global End moment matrix [MBar] =
   58.7450  -43.9479   57.4513  -57.4513  -63.6993  -56.3007

Member Number =      2
Global displacement matrix [DeltaBar] =
   1.0e+04 *
   -1.9906    1.3283   -0.1702    3.1529    4.8465    4.6213
Global End moment matrix [MBar] =
   43.9479  -54.8100   57.4513   42.5487   56.3007  -56.3007

Member Number =      3
Global displacement matrix [DeltaBar] =
   1.0e+04 *
    1.3283         0    3.1529         0    4.6213         0
Global End moment matrix [MBar] =
   54.8100   42.6958  -42.5517   42.5517   56.3007  -56.3007

Joint forces =
  0.0000  0.0000  -0.0000  -0.0000  0.0000  -0.0030  58.7450  -63.6993
57.4513   42.6958  -56.3007   42.5517
```

4 Planar Truss Structures

4.1 PLANAR TRUSS

In a truss system, joints are assumed to be pinned, meaning there is no moment transfer. Therefore, they can only resist the axial force and axial deformation. Thus, at every node or joint in a truss system, there are only two possible independent components of joint translation for the reference axes system. Similarly, in the local axes system, each joint can have only two joint translations. Thus, there are no rotations at the ends.

4.2 DEGREE OF FREEDOM AND LABELING

Identifying the unrestrained and restrained degrees of freedom is the primary step in the analysis. Degrees of freedom are marked group-wise, considering the unrestrained set first and the restrained set later. Figure 4.1 shows the degrees of freedom of a planar truss.

4.3 TRANSFORMATION MATRIX

While the above figure shows the member axis for each member, it is to be noted that they are not parallel, necessitating the use of a transformation matrix similar to a planar non-orthogonal frame analysis. To proceed with the transformation, it is necessary to define two axes systems, namely the **local** and **global axes,** as shown in

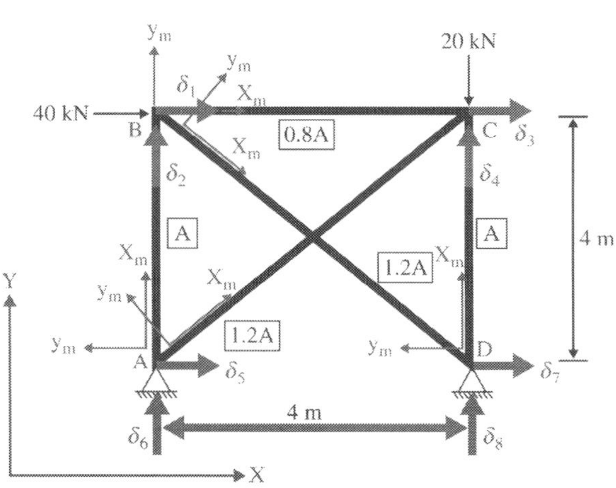

FIGURE 4.1 Degrees of freedom of a planar truss.

DOI: 10.1201/9781003639152-4

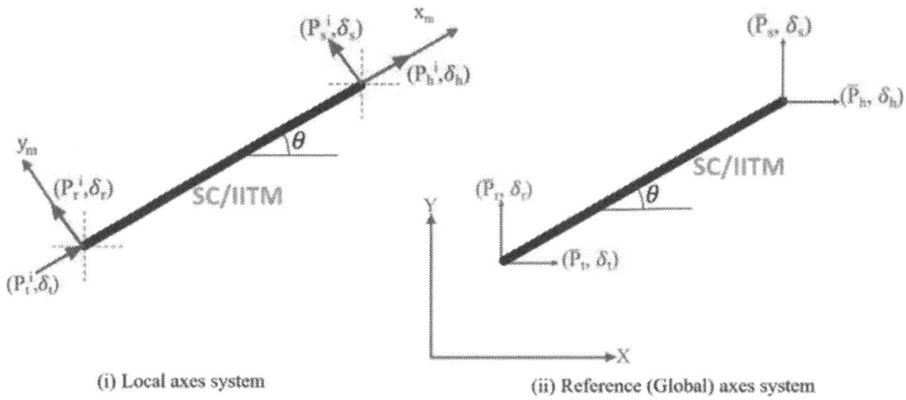

FIGURE 4.2 Transformation between local and global axes.

Figure 4.2. Consider a truss member inclined to an arbitrary angle of θ. This angle should always be measured from the X-axis toward the x_m-axis; if measured anti-clockwise, it is taken as positive.

To convert the local responses with respect to the reference axes responses, the transformation matrix can be written as follows:

$$\begin{Bmatrix} P_r \\ P_s \\ P_t \\ P_h \end{Bmatrix} = \begin{bmatrix} cos\theta & 0 & -sin\theta & 0 \\ 0 & cos\theta & 0 & -sin\theta \\ sin\theta & 0 & cos\theta & 0 \\ 0 & sin\theta & 0 & cos\theta \end{bmatrix} \begin{Bmatrix} \bar{P}_r \\ \bar{P}_s \\ \bar{P}_t \\ \bar{P}_h \end{Bmatrix} \tag{4.1}$$

$$\{P\}_i = [T]_i \{\bar{P}\}_i \tag{4.2}$$

$$\{\bar{P}\}_i = [T]^{-1}{}_i \{P\}_i = [T]^T{}_i \{P\}_i \tag{4.3}$$

$$\{\bar{\delta}_i\} = [T]^T{}_i \{\delta\}_i \tag{4.4}$$

Where,

$$\{\bar{\delta}_i\} = \begin{Bmatrix} \bar{\delta}_r \\ \bar{\delta}_s \\ \bar{\delta}_t \\ \bar{\delta}_h \end{Bmatrix} \tag{4.5a}$$

$$\{\delta\}_i = \begin{Bmatrix} \delta_r \\ \delta_s \\ \delta_t \\ \delta_h \end{Bmatrix} \tag{4.5b}$$

4.4 GLOBAL STIFFNESS MATRIX

Now, the global stiffness matrix and response of the truss member can be obtained using the following relationships:

$$\left[\bar{K}\right]_i = \left[T\right]^T_i \left[K\right]_i \left[T\right]_i \tag{4.6a}$$

$$\left[\bar{P}\right]_i = \left[\bar{K}\right]_i \left\{\bar{\delta}_i\right\} + \left[\bar{F}\right]_i \tag{4.6b}$$

Hence, solving a planar truss problem is simpler than the orthogonal and non-orthogonal structures problem, as the size of the matrix operation is limited to 4×4.

4.4.1 Axes Systems

To ensure consistency in analysis:

1. Each structural member's global and local axes systems should be marked.
2. The orientation of each axis is critical for determining the angle θ and performing transformations between the local and global coordinate systems.

Both the **unrestrained** and **restrained degrees of freedom** and the local and global axes systems are shown in Figure 4.1 for clarity.

4.5 ELEMENT STIFFNESS MATRIX

Now, the standard stiffness matrix of the truss element without end rotations is given as follows:

$$\left[K_T\right]_i = \begin{bmatrix} 0 & 0 & 0 & 0 \\ 0 & 0 & 0 & 0 \\ 0 & 0 & \frac{AE}{l} & -\frac{AE}{l} \\ 0 & 0 & -\frac{AE}{l} & \frac{AE}{l} \end{bmatrix} \tag{4.7}$$

4.6 LOAD VECTOR

When referring to a reference axis system for a truss structure, the load vector must be expressed in a specific format that aligns with the global coordinate system. This representation is crucial for accurately analyzing the structural behavior under applied loads. Here's how the load vector should be written for the truss shown in Figure 4.1:

$$[\bar{J}_L] = \begin{Bmatrix} \mathbf{40} \\ \mathbf{0} \\ \mathbf{0} \\ \mathbf{-20} \\ \mathbf{0} \\ \mathbf{0} \\ \mathbf{0} \\ \mathbf{0} \end{Bmatrix} \begin{matrix} 1 \\ 2 \\ 3 \\ 4 \\ 5 \\ 6 \\ 7 \\ 8 \end{matrix}$$

The unrestrained displacements are computed using the following relationship after portioning the joint load vector:

$$\{\bar{\delta}\} = \left[\bar{K}_{uu}\right]^{-1}\{\bar{J}_{Lu}\} \tag{4.8}$$

4.7 STEP-BY-STEP FRAMEWORK FOR COMPUTER-AIDED ANALYSIS

The step-by-step framework to analyze a truss using MATLAB is given in Figure 4.3.

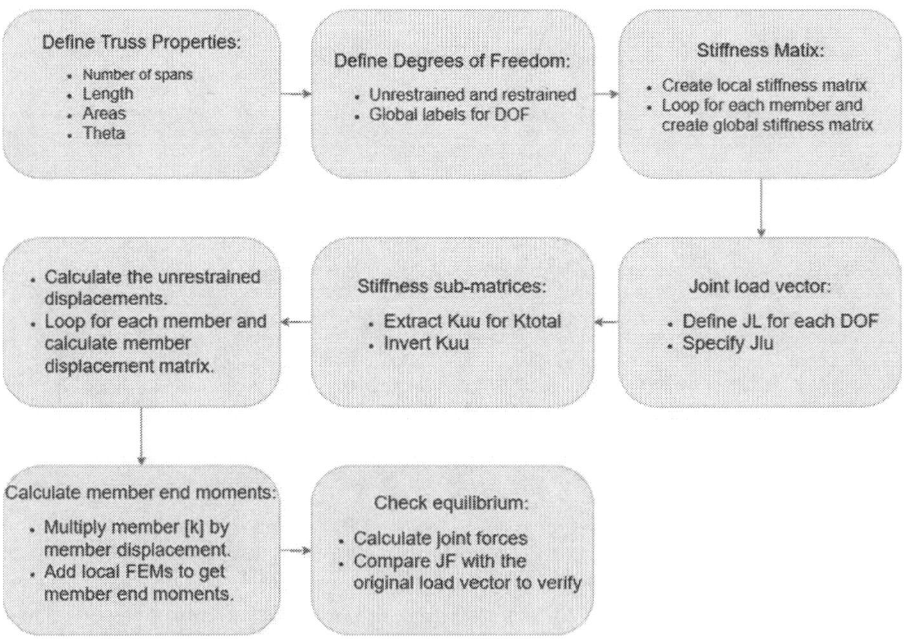

FIGURE 4.3 Steps to analyze a truss using MATLAB.

4.8 EXAMPLE PROBLEMS WITH A COMPUTER PROGRAM

Example 4.1:

Analyze the planar truss system shown below using the stiffness method. E is constant for all the members.

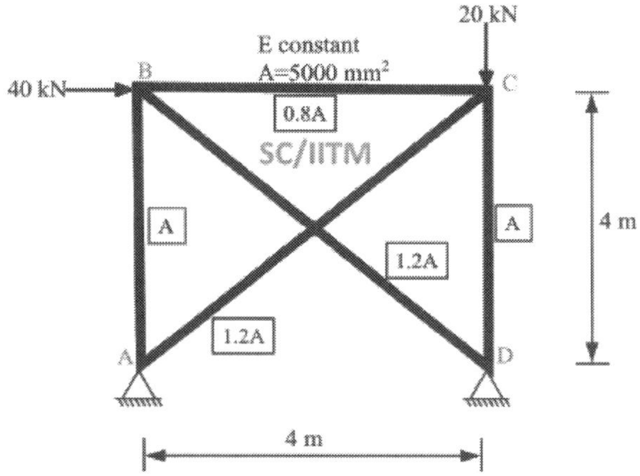

SOLUTION

i. *Marking restrained and unrestrained degrees of freedom:*

Unrestrained degrees-of-freedom = 4 (δ_1, δ_2, δ_3, δ_4) (marked in green color)
Restrained degrees-of-freedom = 4 (δ_5, δ_6, δ_7, δ_8) (marked in red color)

The global and local axes system for every member should be marked to find θ.

The unrestrained and restrained degrees of freedom are marked. The local and global axes systems are also shown in the following figure.

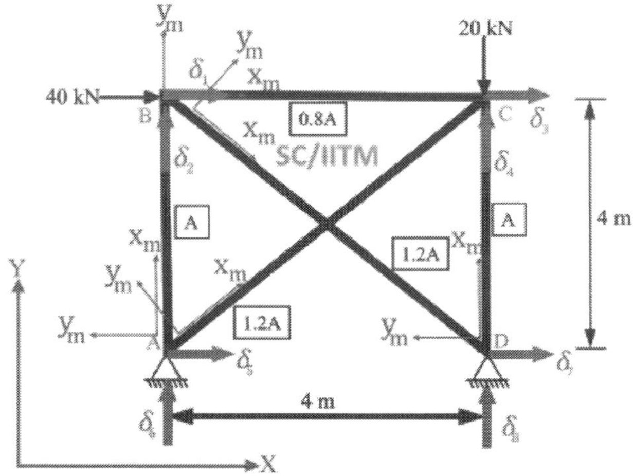

	Ends						
Member	**J**	**k**	**Length (m)**	**θ (degrees)**	**C_x**	**C_y**	**Global Labels**
AB	A	B	4	90	0	1	(6,2,5,1)
BC	B	C	4	0	1	0	(2,4,1,3)
CD	D	C	4	90	0	1	(8,4,7,3)
BD	B	D	5.656	–45	0.707	–0.707	(2,8,1,7)
AC	A	C	5.656	45	0.707	–0.707	(6,4,5,3)

ii. *Calculation of local stiffness matrix:*
The local stiffness matrix for the truss element is given by,

$$[K_T]_i = \begin{bmatrix} 0 & 0 & 0 & 0 \\ 0 & 0 & 0 & 0 \\ 0 & 0 & \frac{AE}{l} & -\frac{AE}{l} \\ 0 & 0 & -\frac{AE}{l} & \frac{AE}{l} \end{bmatrix}$$

$$\begin{array}{c} \quad 6 \quad 2 \quad 5 \quad 1 \end{array}$$

$$K_{AB} = E\times10^{-3}\begin{bmatrix} 0 & 0 & 0 & 0 \\ 0 & 0 & 0 & 0 \\ 0 & 0 & 1.25 & -1.25 \\ 0 & 0 & -1.25 & 1.25 \end{bmatrix}\begin{matrix} 6 \\ 2 \\ 5 \\ 1 \end{matrix}$$

$$\begin{array}{c} \quad 2 \quad 4 \quad 1 \quad 3 \end{array}$$

$$K_{BC} = E\times10^{-3}\begin{bmatrix} 0 & 0 & 0 & 0 \\ 0 & 0 & 0 & 0 \\ 0 & 0 & 1 & -1 \\ 0 & 0 & -1 & 1 \end{bmatrix}\begin{matrix} 2 \\ 4 \\ 1 \\ 3 \end{matrix}$$

$$\begin{array}{c} \quad 8 \quad 4 \quad 7 \quad 3 \end{array}$$

$$K_{CD} = E\times10^{-3}\begin{bmatrix} 0 & 0 & 0 & 0 \\ 0 & 0 & 0 & 0 \\ 0 & 0 & 1.25 & -1.25 \\ 0 & 0 & -1.25 & 1.25 \end{bmatrix}\begin{matrix} 8 \\ 4 \\ 7 \\ 3 \end{matrix}$$

$$\begin{array}{c} \quad 2 \quad 8 \quad 1 \quad 7 \end{array}$$

$$K_{BD} = E\times10^{-3}\begin{bmatrix} 0 & 0 & 0 & 0 \\ 0 & 0 & 0 & 0 \\ 0 & 0 & 1.061 & -1.061 \\ 0 & 0 & -1.061 & 1.061 \end{bmatrix}\begin{matrix} 2 \\ 8 \\ 1 \\ 7 \end{matrix}$$

$$K_{AC}=E\times10^{-3}\begin{array}{c}\begin{array}{cccc}6&4&5&3\end{array}\\\begin{bmatrix}0&0&0&0\\0&0&0&0\\0&0&1.061&-1.061\\0&0&-1.061&1.061\end{bmatrix}\end{array}\begin{array}{c}6\\4\\5\\3\end{array}$$

iii. *Calculation of transformation matrix:*

$$[T]_{AB}=\begin{bmatrix}0&0&-1&0\\0&0&0&-1\\1&0&0&0\\0&1&0&0\end{bmatrix},[T]_{BC}=\begin{bmatrix}1&0&0&0\\0&1&0&0\\0&0&1&0\\0&0&0&1\end{bmatrix},[T]_{CD}=\begin{bmatrix}0&0&-1&0\\0&0&0&-1\\1&0&0&0\\0&1&0&0\end{bmatrix}$$

$$[T]_{BD}=\begin{bmatrix}0.7071&0&0.7071&0\\0&0.7071&0&0.7071\\-0.7071&0&0.7071&0\\0&-0.7071&0&0.7071\end{bmatrix},$$

$$[T]_{AC}=\begin{bmatrix}0.7071&0&-0.7071&0\\0&0.7071&0&-0.7071\\0.7071&0&0.7071&0\\0&0.7071&0&0.7071\end{bmatrix}$$

iv. *Calculation of global stiffness matrix:*
The global stiffness matrix of all the members is calculated by using the following relationship:

$$\left[\bar{K}_T\right]_i=[T_T]_i^T\,[K_T]_i\,[T_T]_i$$

$$\bar{K}_{AB}=E\begin{array}{c}\begin{array}{cccc}6&2&5&1\end{array}\\\begin{bmatrix}0.0013&-0.0013&0&0\\-0.0013&0.0013&0&0\\0&0&0&0\\0&0&0&0\end{bmatrix}\end{array}\begin{array}{c}6\\2\\5\\1\end{array}$$

$$\bar{K}_{BC}=E\begin{array}{c}\begin{array}{cccc}2&4&1&3\end{array}\\\begin{bmatrix}0&0&0&0\\0&0&0&0\\0&0&1&-1\\0&0&-1&1\end{bmatrix}\end{array}\begin{array}{c}2\\4\\1\\3\end{array}$$

$$\bar{K}_{CD} = E\begin{bmatrix} 0.0013 & -0.0013 & 0 & 0 \\ -0.0013 & 0.0013 & 0 & 0 \\ 0 & 0 & 0 & 0 \\ 0 & 0 & 0 & 0 \end{bmatrix}\begin{matrix} 8 \\ 4 \\ 7 \\ 3 \end{matrix} \quad \text{(columns: 8, 4, 7, 3)}$$

$$\bar{K}_{BD} = E\times10^{-3}\begin{bmatrix} 0.5304 & -0.5304 & -0.5304 & 0.5304 \\ -0.5304 & 0.5304 & 0.5304 & -0.5304 \\ -0.5304 & 0.5304 & 0.5304 & -0.5304 \\ 0.5304 & -0.5304 & -0.5304 & 0.5304 \end{bmatrix}\begin{matrix} 2 \\ 8 \\ 1 \\ 7 \end{matrix} \quad \text{(columns: 2, 8, 1, 7)}$$

$$\bar{K}_{AC} = E\times10^{-3}\begin{bmatrix} 0.5304 & -0.5304 & 0.5304 & -0.5304 \\ -0.5304 & 0.5304 & -0.5304 & 0.5304 \\ 0.5304 & -0.5304 & 0.5304 & -0.5304 \\ -0.5304 & 0.5304 & -0.5304 & 0.5304 \end{bmatrix}\begin{matrix} 6 \\ 4 \\ 5 \\ 3 \end{matrix} \quad \text{(columns: 6, 4, 5, 3)}$$

The total stiffness matrix is obtained by assembling the global stiffness matrices of all members with respect to the global labels. From the total stiffness matrix, the stiffness matrix for the unrestrained degrees-of-freedom can be partitioned.

$$[K_{UU}] = E\begin{bmatrix} 0.0015 & -0.005 & -0.0010 & 0 \\ -0.005 & 0.0018 & 0 & 0 \\ -0.0010 & 0 & 0.0015 & 0.0005 \\ 0 & 0 & 0.0005 & 0.0018 \end{bmatrix}\begin{matrix} 1 \\ 2 \\ 3 \\ 4 \end{matrix} \quad \text{(columns: 1, 2, 3, 4)}$$

$$\left[K_{uu}\right]^{-1} = \frac{1}{E}\times10^{-3}\begin{bmatrix} 1.5534 & 0.4628 & 1.1319 & -0.3372 \\ 0.4628 & 0.6995 & 0.3372 & -0.1005 \\ 1.1319 & 0.3372 & 1.5534 & -0.4628 \\ -0.3372 & -0.1005 & -0.4628 & 0.6995 \end{bmatrix}$$

v. *Estimation of joint load vectors:*
The joint load vector can be written as follows:

$$\left[\bar{J}_L\right]=\begin{Bmatrix}+40\\0\\0\\-20\\0\\0\\0\\0\end{Bmatrix}\begin{matrix}1\\2\\3\\4\\5\\6\\7\\8\end{matrix}$$

The joint load vector in unrestrained degrees of freedom can be partitioned from the above matrix as follows:

$$\left[\bar{J}_{LU}\right]=\begin{Bmatrix}+40\\0\\0\\-20\end{Bmatrix}\begin{matrix}1\\2\\3\\4\end{matrix}$$

$$\left\{\bar{\delta}\right\}=\left[\bar{K}_{uu}\right]^{-1}\left\{\bar{J}_{Lu}\right\}$$

$$\left\{\bar{\delta}\right\}_u=\frac{1\times10^{-4}}{E}\begin{Bmatrix}6.881\\2.0521\\5.4532\\-2.7479\end{Bmatrix}$$

vi. *Calculation of member forces:*

$$\bar{M}_{AB}=\begin{Bmatrix}\bar{V}_6\\\bar{V}_2\\\bar{H}_5\\\bar{H}_1\end{Bmatrix}=\begin{Bmatrix}-25.6509\\25.6509\\0\\0\end{Bmatrix},\bar{M}_{BC}=\begin{Bmatrix}\bar{V}_2\\\bar{V}_4\\\bar{H}_1\\\bar{H}_5\end{Bmatrix}=\begin{Bmatrix}0\\0\\14.3491\\-14.3491\end{Bmatrix},\bar{M}_{CD}=\begin{Bmatrix}\bar{V}_8\\\bar{V}_4\\\bar{H}_7\\\bar{H}_3\end{Bmatrix}=\begin{Bmatrix}34.3491\\-34.3491\\0\\0\end{Bmatrix}$$

$$\bar{M}_{BD}=\begin{Bmatrix}\bar{V}_2\\\bar{V}_8\\\bar{H}_1\\\bar{H}_7\end{Bmatrix}=\begin{Bmatrix}-25.6509\\25.6509\\26.6509\\-25.6509\end{Bmatrix},\bar{M}_{AC}=\begin{Bmatrix}\bar{V}_6\\\bar{V}_4\\\bar{H}_5\\\bar{H}_3\end{Bmatrix}=\begin{Bmatrix}-14.3491\\14.3491\\-14.3491\\14.3491\end{Bmatrix}$$

The member and find end forces are shown.

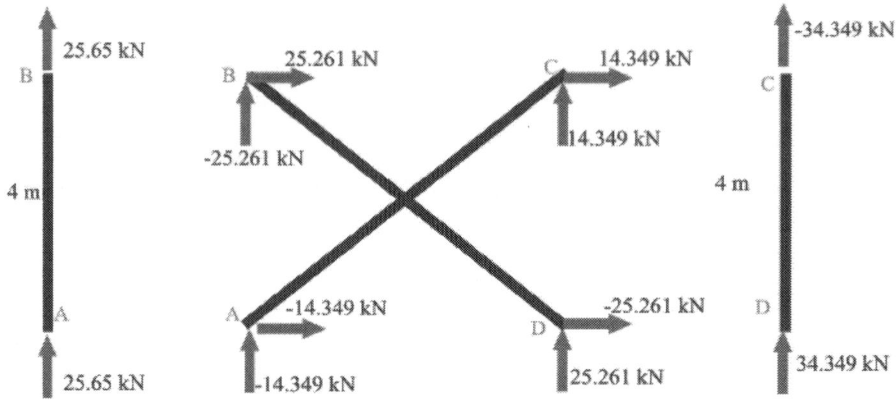

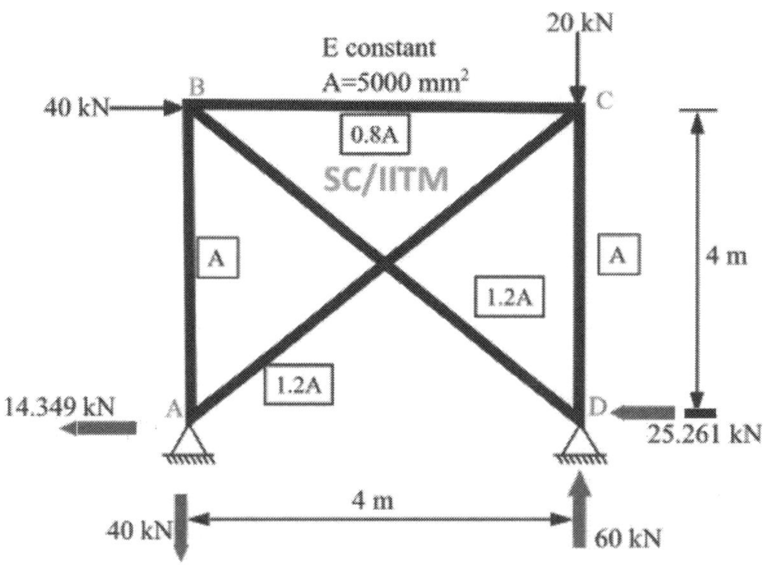

MATLAB program:

```
%% Stiffness matrix method for planar truss with user inputs
clc; clear;

%% User Inputs
n = input('Enter the number of members: \n'); % Number of members
L = input('Enter the lengths of each member as a row vector [L1 L2 ...
Ln]: \n'); % Lengths
A = input('Enter the areas for each member as a row vector [A1 A2 ...
An]: \n'); % Area
theta = input('Enter the angles for each member as a row
vector [Theta1 Theta2 ... ThetaN]: \n'); % Angles

uu = input('Enter the number of unrestrained degrees of freedom: \n');
% Unrestrained DOF
ur = input('Enter the number of restrained degrees of freedom: \n');
% Restrained DOF
uul = input('Enter the labels of unrestrained DOF as a row vector [u1
u2 ...]: \n'); % Unrestrained labels
url = input('Enter the labels of restrained DOF as a row vector [r1 r2
...]: \n'); % Restrained labels

l = zeros(n, 4); % Global labels for each member
for i = 1:n
    fprintf('Enter global labels for Member %d as a row vector: ', i);
    l(i, :) = input('');
end


dof = uu + ur; % Total degrees of freedom
Ktotal = zeros(dof); % Initialize global stiffness matrix


%% Fixed End Forces Input
fem = zeros(4, n);
for i = 1:n
    fprintf('Enter fixed-end forces for Member %d as a column
vector: \n', i);
    fem(:, i) = input('');
end

%% Rotation coefficients for each member
rc = A . / L;
cx = cosd(theta);
cy = sind(theta);

%% Stiffness Matrix and Transformation Matrix Calculation
Tt = cell(1, n); % Transformation matrices for each member
Kg = cell(1, n); % Global stiffness matrices for each member
fembar = cell(1, n); % Transformed fixed-end forces
```

```
for i = 1:n
    % Local stiffness matrix
    k1 = [0; 0; 0; 0];
    k2 = [0; 0; 0; 0];
    k3 = [0; 0; rc(i); -rc(i)];
    k4 = -k3;
    K = [k1 k2 k3 k4];

    % Transformation matrix
    T1 = [cx(i); 0; cy(i); 0];
    T2 = [0; cx(i); 0; cy(i)];
    T3 = [-cy(i); 0; cx(i); 0];
    T4 = [0; -cy(i); 0; cx(i)];
    T = [T1 T2 T3 T4];

    % Global stiffness matrix
    Ttr = T';
    Kg{i} = Ttr * K * T;

    % Transformation matrix and fixed-end forces storage
    Tt{i} = T;
    fembar{i} = Ttr * fem(:, i);

    % Assemble into global stiffness matrix
    Knew = zeros(dof);
    for p = 1:4
        for q = 1:4
            Knew(l(i, p), l(i, q)) = Kg{i}(p, q);
        end
    end
    Ktotal = Ktotal + Knew;
end

%% Unrestrained Stiffness Sub-matrix
Kunr = Ktotal(1:uu, 1:uu);
KuuInv = inv(Kunr);

%% Joint Load Vector
jl = input('Enter joint load vector as a column vector: \n');
jlu = jl(1:uu);
delu = KuuInv * jlu;

%% Global Displacements
delr = zeros(ur, 1);
del = [delu; delr];

%% End Forces and Joint Forces
jf = zeros(dof, 1);
for i = 1:n
    deli = del(l(i, :));
    mbar = (Kg{i} * deli) + fembar{i};
```

```
    fprintf('Member %d Global End Forces Matrix [MBar]:\n', i);
    disp(mbar');

    % Assemble joint forces
    for p = 1:4
        jf(l(i, p)) = jf(l(i, p)) + mbar(p);
    end
end

fprintf('Joint Forces = \n');
disp(jf');
```

MATLAB output:

```
Enter the number of members:
5
Enter the lengths of each member as a row vector [L1 L2 ...
Ln]:
[4 4 4 5.656 5.656]
Enter the areas for each member as a row vector [A1 A2 ...
An]:
[5E-3 4E-3 5E-3 6E-3 6E-3]
Enter the angles for each member as a row vector [Theta1
Theta2 ... ThetaN]:
[90 0 90 -45 45]
Enter the number of unrestrained degrees of freedom:
4
Enter the number of restrained degrees of freedom:
4
Enter the labels of unrestrained DOF as a row vector [u1 u2
...]:
[1 2 3 4]
Enter the labels of restrained DOF as a row vector [r1 r2
...]:
[5 6 7 8]
Enter global labels for Member 1 as a row vector: [6 2 5 1]
Enter global labels for Member 2 as a row vector: [2 4 1 3]
Enter global labels for Member 3 as a row vector: [8 4 7 3]
Enter global labels for Member 4 as a row vector: [2 8 1 7]
Enter global labels for Member 5 as a row vector: [6 4 5 3]
Enter fixed-end forces for Member 1 as a column vector:
[0; 0; 0; 0]
Enter fixed-end forces for Member 2 as a column vector:
[0; 0; 0; 0]
Enter fixed-end forces for Member 3 as a column vector:
[0; 0; 0; 0]
Enter fixed-end forces for Member 4 as a column vector:
[0; 0; 0; 0]
Enter fixed-end forces for Member 5 as a column vector:
[0; 0; 0; 0]
```

```
Enter joint load vector as a column vector:
[40; 0; 0; -20; 0; 0; 0; 0]
Member Number =      1
Local Stiffness matrix of member, [K] =
         0         0         0         0
         0         0         0         0
         0         0    0.0013   -0.0013
         0         0   -0.0013    0.0013
Transformation matrix of member, [T] =
     0     0    -1     0
     0     0     0    -1
     1     0     0     0
     0     1     0     0
Transformation matrix Transpose, [T] =
     0     0     1     0
     0     0     0     1
    -1     0     0     0
     0    -1     0     0
Global Matrix, [K global] =
    0.0013   -0.0013         0         0
   -0.0013    0.0013         0         0
         0         0         0         0
         0         0         0         0

Member Number =      2
Local Stiffness matrix of member, [K] =
   1.0e-03 *
         0         0         0         0
         0         0         0         0
         0         0    1.0000   -1.0000
         0         0   -1.0000    1.0000
Transformation matrix of member, [T] =
     1     0     0     0
     0     1     0     0
     0     0     1     0
     0     0     0     1
Transformation matrix Transpose, [T] =
     1     0     0     0
     0     1     0     0
     0     0     1     0
     0     0     0     1
Global Matrix, [K global] =
   1.0e-03 *
         0         0         0         0
         0         0         0         0
         0         0    1.0000   -1.0000
         0         0   -1.0000    1.0000

Member Number =      3
Local Stiffness matrix of member, [K] =
         0         0         0         0
```

```
         0            0            0            0
         0            0       0.0013      -0.0013
         0            0      -0.0013       0.0013
Transformation matrix of member, [T] =
     0     0    -1     0
     0     0     0    -1
     1     0     0     0
     0     1     0     0
Transformation matrix Transpose, [T] =
     0     0     1     0
     0     0     0     1
    -1     0     0     0
     0    -1     0     0
Global Matrix, [K global] =
    0.0013      -0.0013            0            0
   -0.0013       0.0013            0            0
         0            0            0            0
         0            0            0            0

Member Number =      4
Local Stiffness matrix of member, [K] =
         0            0            0            0
         0            0            0            0
         0            0       0.0011      -0.0011
         0            0      -0.0011       0.0011
Transformation matrix of member, [T] =
    0.7071            0       0.7071            0
         0       0.7071            0       0.7071
   -0.7071            0       0.7071            0
         0      -0.7071            0       0.7071
Transformation matrix Transpose, [T] =
    0.7071            0      -0.7071            0
         0       0.7071            0      -0.7071
    0.7071            0       0.7071            0
         0       0.7071            0       0.7071
Global Matrix, [K global] =
   1.0e-03 *
    0.5313      -0.5313      -0.5313       0.5313
   -0.5313       0.5313       0.5313      -0.5313
   -0.5313       0.5313       0.5313      -0.5313
    0.5313      -0.5313      -0.5313       0.5313

Member Number =      5
Local Stiffness matrix of member, [K] =
         0            0            0            0
         0            0            0            0
         0            0       0.0011      -0.0011
         0            0      -0.0011       0.0011
Transformation matrix of member, [T] =
    0.7071            0      -0.7071            0
```

```
         0    0.7071         0   -0.7071
    0.7071         0    0.7071         0
         0    0.7071         0    0.7071
Transformation matrix Transpose, [T] =
    0.7071         0    0.7071         0
         0    0.7071         0    0.7071
   -0.7071         0    0.7071         0
         0   -0.7071         0    0.7071
Global Matrix, [K global] =
   1.0e-03 *
    0.5313   -0.5313    0.5313   -0.5313
   -0.5313    0.5313   -0.5313    0.5313
    0.5313   -0.5313    0.5313   -0.5313
   -0.5313    0.5313   -0.5313    0.5313


Stiffness Matrix of complete structure, [Ktotal] =
    0.0015   -0.0005   -0.0010         0         0         0   -0.0005    0.0005
   -0.0005    0.0018         0         0         0   -0.0013    0.0005   -0.0005
   -0.0010         0    0.0015    0.0005   -0.0005   -0.0005         0         0
         0         0    0.0005    0.0018   -0.0005   -0.0005         0   -0.0013
         0         0   -0.0005   -0.0005    0.0005    0.0005         0         0
         0   -0.0013   -0.0005   -0.0005    0.0005    0.0018         0         0
   -0.0005    0.0005         0         0         0         0    0.0005   -0.0005
    0.0005   -0.0005         0   -0.0013         0         0   -0.0005    0.0018
Unrestrained Stiffness sub-matrix, [Kuu] =
    0.0015   -0.0005   -0.0010         0
   -0.0005    0.0018         0         0
   -0.0010         0    0.0015    0.0005
         0         0    0.0005    0.0018
Inverse of Unrestrained Stiffness sub-matrix, [KuuInverse] =
   1.0e+03 *
    1.5517    0.4629    1.1303   -0.3371
    0.4629    0.6994    0.3371   -0.1006
    1.1303    0.3371    1.5517   -0.4629
   -0.3371   -0.1006   -0.4629    0.6994
Joint Load vector, [Jl] =
    40     0     0   -20     0     0     0     0
Unrestrained displacements, [DelU] =
   1.0e+04 *
    6.8812    2.0525    5.4468   -2.7475

Member Number =      1
Global displacement matrix [DeltaBar] =
   1.0e+04 *
         0    2.0525         0    6.8812
Global End moment matrix [MBar] =
  -25.6568   25.6568         0         0

Member Number =      2
Global displacement matrix [DeltaBar] =
   1.0e+04 *
    2.0525   -2.7475    6.8812    5.4468
```

```
Global End moment matrix [MBar] =
         0          0   14.3432  -14.3432

Member Number =      3
Global displacement matrix [DeltaBar] =
   1.0e+04 *
         0   -2.7475         0    5.4468
Global End moment matrix [MBar] =
   34.3432  -34.3432         0         0


Member Number =      4
Global displacement matrix [DeltaBar] =
   1.0e+04 *
    2.0525         0    6.8812         0
Global End moment matrix [MBar] =
  -25.6568   25.6568   25.6568  -25.6568

Member Number =      5
Global displacement matrix [DeltaBar] =
   1.0e+04 *
         0   -2.7475         0    5.4468
Global End moment matrix [MBar] =
  -14.3432   14.3432  -14.3432   14.3432
Joint forces =
 40.0000  0.0000 -0.0000 -20.0000 -14.3432 -40.0000 -25.6568   60.0000
```

Example 4.2:

Analyze the planar truss system shown below using the stiffness method. E is constant for all the members.

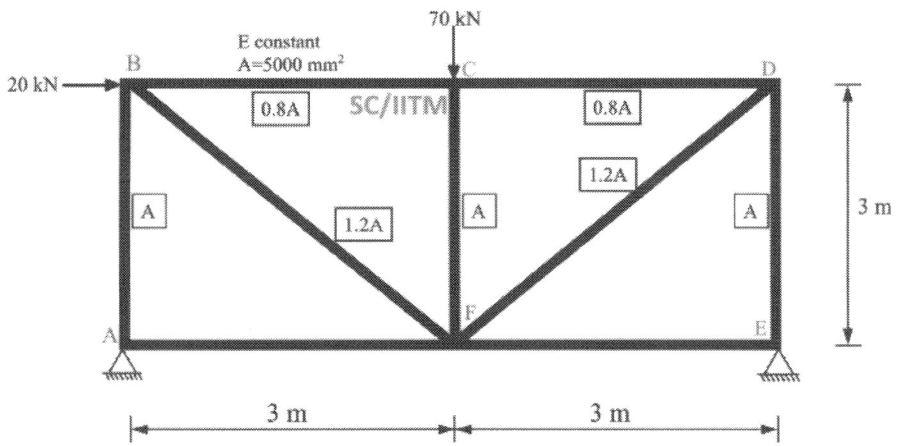

SOLUTION

i. *Marking restrained and unrestrained degrees of freedom:*

Unrestrained degrees-of-freedom = 4 (δ_1, δ_2, δ_3, δ_4, δ_5, δ_6, δ_7, δ_8) (marked in green color)
Restrained degrees-of-freedom = 4 (δ_9, δ_{10}, δ_{11}, δ_{12}) (marked in red color)

The unrestrained and restrained degrees-of-freedom are marked in the truss as shown.

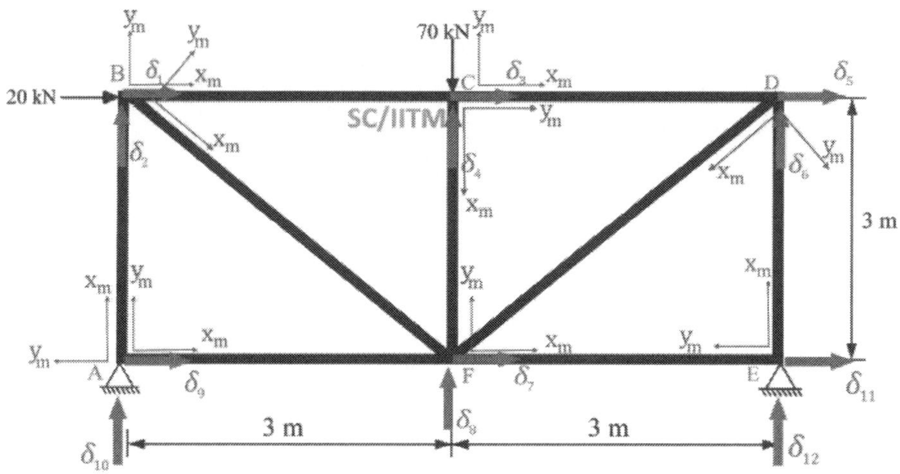

Member	Ends j	k	Length (m)	θ (degrees)	C_x	C_y	Global Labels
AB	A	B	3	90	0	1	(10,2,9,1)
BC	B	C	3	0	1	0	(2,4,1,3)
CD	C	D	3	0	1	0	(4,6,3,5)
DE	E	D	3	90	0	1	(12,6,11,5)
AF	A	F	3	0	1	0	(10,8,9,7)
FE	F	E	3	0	1	0	(8,12,7,11)
BF	B	F	4.242	–45	0.707	–0.707	(2,8,1,7)
CF	F	C	3	+90	0	1	(8,4,7,3)
DF	D	F	4.242	–135	–0.707	0.707	(6,8,5,7)

ii. *Calculation of local stiffness matrix:*
The local stiffness matrix for the truss element is given by,

$$[K_T]_i = \begin{bmatrix} 0 & 0 & 0 & 0 \\ 0 & 0 & 0 & 0 \\ 0 & 0 & \frac{AE}{l} & -\frac{AE}{l} \\ 0 & 0 & -\frac{AE}{l} & \frac{AE}{l} \end{bmatrix}$$

$$K_{AB} = E \times 10^{-3} \begin{bmatrix} 0 & 0 & 0 & 0 \\ 0 & 0 & 0 & 0 \\ 0 & 0 & 1.7 & -1.7 \\ 0 & 0 & -1.7 & 1.7 \end{bmatrix}, K_{BC} = E \times 10^{-3} \begin{bmatrix} 0 & 0 & 0 & 0 \\ 0 & 0 & 0 & 0 \\ 0 & 0 & 1.3 & -1.3 \\ 0 & 0 & -1.3 & 1.3 \end{bmatrix}$$

$$K_{CD} = E \times 10^{-3} \begin{bmatrix} 0 & 0 & 0 & 0 \\ 0 & 0 & 0 & 0 \\ 0 & 0 & 1.3 & -1.3 \\ 0 & 0 & -1.3 & 1.3 \end{bmatrix}, K_{DE} = E \times 10^{-3} \begin{bmatrix} 0 & 0 & 0 & 0 \\ 0 & 0 & 0 & 0 \\ 0 & 0 & 1.7 & -1.7 \\ 0 & 0 & -1.7 & 1.7 \end{bmatrix}$$

$$K_{AF} = E \times 10^{-3} \begin{bmatrix} 0 & 0 & 0 & 0 \\ 0 & 0 & 0 & 0 \\ 0 & 0 & 1.3 & -1.3 \\ 0 & 0 & -1.3 & 1.3 \end{bmatrix}, K_{FE} = E \times 10^{-3} \begin{bmatrix} 0 & 0 & 0 & 0 \\ 0 & 0 & 0 & 0 \\ 0 & 0 & 1.3 & -1.3 \\ 0 & 0 & -1.3 & 1.3 \end{bmatrix}$$

$$K_{BF} = E \times 10^{-3} \begin{bmatrix} 0 & 0 & 0 & 0 \\ 0 & 0 & 0 & 0 \\ 0 & 0 & 1.4 & -1.4 \\ 0 & 0 & -1.4 & 1.4 \end{bmatrix}, K_{CF} = E \times 10^{-3} \begin{bmatrix} 0 & 0 & 0 & 0 \\ 0 & 0 & 0 & 0 \\ 0 & 0 & 1.7 & -1.7 \\ 0 & 0 & -1.7 & 1.7 \end{bmatrix}$$

$$K_{DF} = E \times 10^{-3} \begin{bmatrix} 0 & 0 & 0 & 0 \\ 0 & 0 & 0 & 0 \\ 0 & 0 & 1.4 & -1.4 \\ 0 & 0 & -1.4 & 1.4 \end{bmatrix}$$

iii. *Calculation of transformation matrix:*

$$[T]_{AB} = \begin{bmatrix} 0 & 0 & -1 & 0 \\ 0 & 0 & 0 & -1 \\ 1 & 0 & 0 & 0 \\ 0 & 1 & 0 & 0 \end{bmatrix}, [T]_{BC} = \begin{bmatrix} 1 & 0 & 0 & 0 \\ 0 & 1 & 0 & 0 \\ 0 & 0 & 1 & 0 \\ 0 & 0 & 0 & 1 \end{bmatrix}, [T]_{CD} = \begin{bmatrix} 1 & 0 & 0 & 0 \\ 0 & 1 & 0 & 0 \\ 0 & 0 & 1 & 0 \\ 0 & 0 & 0 & 1 \end{bmatrix}$$

$$[T]_{DE}=\begin{bmatrix}0&0&-1&0\\0&0&0&-1\\1&0&0&0\\0&1&0&0\end{bmatrix},[T]_{AF}=\begin{bmatrix}1&0&0&0\\0&1&0&0\\0&0&1&0\\0&0&0&1\end{bmatrix},[T]_{FE}=\begin{bmatrix}1&0&0&0\\0&1&0&0\\0&0&1&0\\0&0&0&1\end{bmatrix}$$

$$[T]_{BF}=\begin{bmatrix}0.7071&0&0.7071&0\\0&0.7071&0&0.7071\\-0.7071&0&0.7071&0\\0&-0.7071&0&0.7071\end{bmatrix},[T]_{CF}=\begin{bmatrix}0&0&-1&0\\0&0&0&-1\\1&0&0&0\\0&1&0&0\end{bmatrix}$$

$$[T]_{DF}=\begin{bmatrix}-0.7071&0&0.7071&0\\0&-0.7071&0&0.7071\\-0.7071&0&-0.7071&0\\0&-0.7071&0&-0.7071\end{bmatrix}$$

iv. *Calculation of global stiffness matrix:*
The global stiffness matrix of all the members is calculated by using the following relationship:

$$\left[\bar{K}_T\right]_i=[T_T]_i^T\,[K_T]_i\,[T_T]_i$$

$$\bar{K}_{AB}=E\begin{array}{c}\begin{array}{cccc}10&2&9&1\end{array}\\\begin{bmatrix}0.0017&-0.0017&0&0\\-0.0017&0.0017&0&0\\0&0&0&0\\0&0&0&0\end{bmatrix}\end{array}\begin{array}{c}\\10\\2\\9\\1\end{array}$$

$$\bar{K}_{BC}=E\begin{array}{c}\begin{array}{cccc}2&4&1&3\end{array}\\\begin{bmatrix}0&0&0&0\\0&0&0&0\\0&0&0.0013&-0.0013\\0&0&-0.0013&0.0013\end{bmatrix}\end{array}\begin{array}{c}\\2\\4\\1\\3\end{array}$$

$$\bar{K}_{CD}=E\begin{array}{c}\begin{array}{cccc}4&6&3&5\end{array}\\\begin{bmatrix}0&0&0&0\\0&0&0&0\\0&0&0.0013&-0.0013\\0&0&-0.0013&0.0013\end{bmatrix}\end{array}\begin{array}{c}\\4\\6\\3\\5\end{array}$$

$$\bar{K}_{DE} = E\begin{array}{c}\begin{array}{cccc} 12 & 6 & 11 & 5 \end{array}\\ \begin{bmatrix} 0.0017 & -0.0017 & 0 & 0 \\ -0.0017 & 0.0017 & 0 & 0 \\ 0 & 0 & 0 & 0 \\ 0 & 0 & 0 & 0 \end{bmatrix}\end{array}\begin{array}{c} \\ 12 \\ 6 \\ 11 \\ 5 \end{array}$$

$$\bar{K}_{AF} = E\begin{array}{c}\begin{array}{cccc} 10 & 8 & 9 & 7 \end{array}\\ \begin{bmatrix} 0 & 0 & 0 & 0 \\ 0 & 0 & 0 & 0 \\ 0 & 0 & 0.0013 & -0.0013 \\ 0 & 0 & -0.0013 & 0.0013 \end{bmatrix}\end{array}\begin{array}{c} \\ 10 \\ 8 \\ 9 \\ 7 \end{array}$$

$$\bar{K}_{FE} = E\begin{array}{c}\begin{array}{cccc} 8 & 12 & 7 & 11 \end{array}\\ \begin{bmatrix} 0 & 0 & 0 & 0 \\ 0 & 0 & 0 & 0 \\ 0 & 0 & 0.0013 & -0.0013 \\ 0 & 0 & -0.0013 & 0.0013 \end{bmatrix}\end{array}\begin{array}{c} \\ 8 \\ 12 \\ 7 \\ 11 \end{array}$$

$$\bar{K}_{BF} = E\times 10^{-3}\begin{array}{c}\begin{array}{cccc} 2 & 8 & 1 & 7 \end{array}\\ \begin{bmatrix} 0.7072 & -0.7072 & -0.7072 & 0.7072 \\ -0.7072 & 0.7072 & 0.7072 & -0.7072 \\ -0.7072 & 0.7072 & 0.7072 & -0.7072 \\ 0.7072 & -0.7072 & -0.7072 & 0.7072 \end{bmatrix}\end{array}\begin{array}{c} \\ 2 \\ 8 \\ 1 \\ 7 \end{array}$$

$$\bar{K}_{CF} = E\begin{array}{c}\begin{array}{cccc} 8 & 4 & 7 & 3 \end{array}\\ \begin{bmatrix} 0.0017 & -0.0017 & 0 & 0 \\ -0.0017 & 0.0017 & 0 & 0 \\ 0 & 0 & 0 & 0 \\ 0 & 0 & 0 & 0 \end{bmatrix}\end{array}\begin{array}{c} \\ 8 \\ 4 \\ 7 \\ 3 \end{array}$$

$$\bar{K}_{DF} = E\times 10^{-3}\begin{array}{c}\begin{array}{cccc} 6 & 8 & 5 & 7 \end{array}\\ \begin{bmatrix} 0.7072 & -0.7072 & 0.7072 & -0.7072 \\ -0.7072 & 0.7072 & -0.7072 & 0.7072 \\ 0.7072 & -0.7072 & 0.7072 & -0.7072 \\ -0.7072 & 0.7072 & -0.7072 & 0.7072 \end{bmatrix}\end{array}\begin{array}{c} \\ 6 \\ 8 \\ 5 \\ 7 \end{array}$$

The total stiffness matrix is obtained by assembling the global stiffness matrices of all members for the global labels. From the total stiffness matrix, the stiffness matrix for the unrestrained degrees of freedom can be partitioned.

$$
\begin{array}{cccccccc} 1 & 2 & 3 & 4 & 5 & 6 & 7 & 8 \end{array}
$$

$$
\left[K_{UU}\right] = E \times 10^{-3} \begin{bmatrix} 2 & 0.7 & -1.3 & 0 & 0 & 0 & -0.7 & 0.7 \\ 0.7 & 2.4 & 0 & 0 & 0 & 0 & 0.7 & -0.7 \\ 1.3 & 0 & 2.7 & 0 & -1.3 & 0 & 0 & 0 \\ 0 & 0 & 0 & 1.7 & 0 & 0 & 0 & -1.7 \\ 0 & 0 & -1.3 & 0 & 1.2 & 0.7 & -0.7 & -0.7 \\ 0 & 0 & 0 & 0 & 0.7 & 2.4 & -0.7 & -0.7 \\ -0.7 & 0.7 & 0 & 0 & -0.7 & -0.7 & 4.1 & 0 \\ 0.7 & -0.7 & 0 & -1.7 & -0.7 & -0.7 & 0 & 3.1 \end{bmatrix} \begin{array}{c} 1 \\ 2 \\ 3 \\ 4 \\ 5 \\ 6 \\ 7 \\ 8 \end{array}
$$

$$
\left[K_{UU}\right]^{-1} = \frac{1}{E} \times 10^{-3} \begin{bmatrix} 1.757 & 0.3 & 1.382 & -0.375 & 1.007 & -0.3 & 0.375 & -0.375 \\ 0.3 & 0.6 & 0.3 & 0.3 & 1.3 & 0 & 0 & 0.3 \\ 1.382 & 0.3 & 1.757 & 0 & 1.382 & -0.3 & 0.375 & 0 \\ -0.375 & 0.3 & 0 & 1.982 & 0.375 & 0.3 & 0 & 1.382 \\ 1.007 & 0.3 & 1.382 & 0.375 & 1.757 & -0.3 & 0.375 & 0.375 \\ -0.3 & 0 & -0.3 & 0.3 & -0.3 & 0.6 & 0 & 0.3 \\ 0.375 & 0 & 0.375 & 0 & 0.375 & 0 & 0.375 & 0 \\ -0.375 & 0.3 & 0 & 1.382 & 0.375 & 0.3 & 0 & 1.382 \end{bmatrix}
$$

v. *Estimation of joint load vectors:*
The joint load vector can be written as follows:

$$
\left[J_L\right] = \begin{Bmatrix} \mathbf{20} \\ \mathbf{0} \\ \mathbf{0} \\ \mathbf{-70} \\ \mathbf{0} \\ \mathbf{0} \\ \mathbf{0} \\ \mathbf{0} \\ \mathbf{0} \\ \mathbf{0} \\ \mathbf{0} \\ \mathbf{0} \end{Bmatrix} \begin{array}{c} 1 \\ 2 \\ 3 \\ 4 \\ 5 \\ 6 \\ 7 \\ 8 \\ 9 \\ 10 \\ 11 \\ 12 \end{array}
$$

The joint load vector in unrestrained degrees of freedom can be partitioned from the above matrix as follows:

$$\left[\bar{J}_{Lu}\right]=\begin{Bmatrix}20\\0\\0\\-70\\0\\0\\0\\0\end{Bmatrix}$$

$$\left\{\bar{\delta}\right\}=\left[\bar{K}_{uu}\right]^{-1}\left\{\bar{J}_{Lu}\right\}$$

$$\left\{\bar{\delta}\right\}_U=\frac{1\times10^{-5}}{E}\begin{Bmatrix}0.6139\\-0.15\\0.2764\\-1.4624\\-0.0611\\-0.27\\0.075\\-1.0424\end{Bmatrix}$$

vi. *Calculation of member forces:*

$$\bar{M}_{AB}=\begin{Bmatrix}\bar{V}_{10}\\\bar{V}_2\\\bar{H}_9\\\bar{H}_1\end{Bmatrix}=\begin{Bmatrix}25\\-25\\0\\0\end{Bmatrix},\bar{M}_{BC}=\begin{Bmatrix}\bar{V}_2\\\bar{V}_4\\\bar{H}_1\\\bar{H}_3\end{Bmatrix}=\begin{Bmatrix}0\\0\\45\\-45\end{Bmatrix},\bar{M}_{CD}=\begin{Bmatrix}\bar{V}_4\\\bar{V}_6\\\bar{H}_3\\\bar{H}_5\end{Bmatrix}=\begin{Bmatrix}0\\0\\45\\-45\end{Bmatrix}$$

$$\bar{M}_{DE}=\begin{Bmatrix}\bar{V}_{12}\\\bar{V}_6\\\bar{H}_{11}\\\bar{H}_5\end{Bmatrix}=\begin{Bmatrix}45\\-45\\0\\0\end{Bmatrix},\bar{M}_{AF}=\begin{Bmatrix}\bar{V}_{10}\\\bar{V}_8\\\bar{H}_9\\\bar{H}_7\end{Bmatrix}=\begin{Bmatrix}0\\0\\-10\\10\end{Bmatrix},\bar{M}_{FE}=\begin{Bmatrix}\bar{V}_8\\\bar{V}_{12}\\\bar{H}_7\\\bar{H}_{11}\end{Bmatrix}=\begin{Bmatrix}0\\0\\10\\-10\end{Bmatrix}$$

$$\bar{M}_{BF}=\begin{Bmatrix}\bar{V}_2\\\bar{V}_8\\\bar{H}_1\\\bar{H}_7\end{Bmatrix}=\begin{Bmatrix}25\\-25\\-25\\25\end{Bmatrix},\bar{M}_{CF}=\begin{Bmatrix}\bar{V}_8\\\bar{V}_4\\\bar{H}_7\\\bar{H}_3\end{Bmatrix}=\begin{Bmatrix}70\\-70\\0\\0\end{Bmatrix},\bar{M}_{DF}=\begin{Bmatrix}\bar{V}_6\\\bar{V}_8\\\bar{H}_5\\\bar{H}_7\end{Bmatrix}=\begin{Bmatrix}45\\-45\\45\\-45\end{Bmatrix}$$

The member end forces and final end forces are given below:

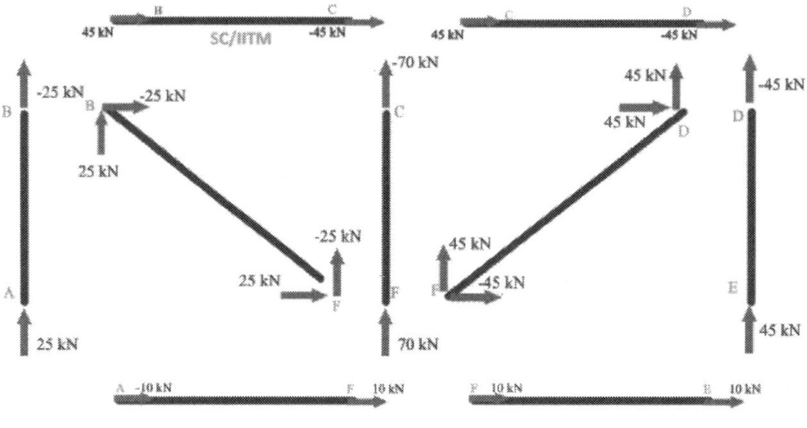

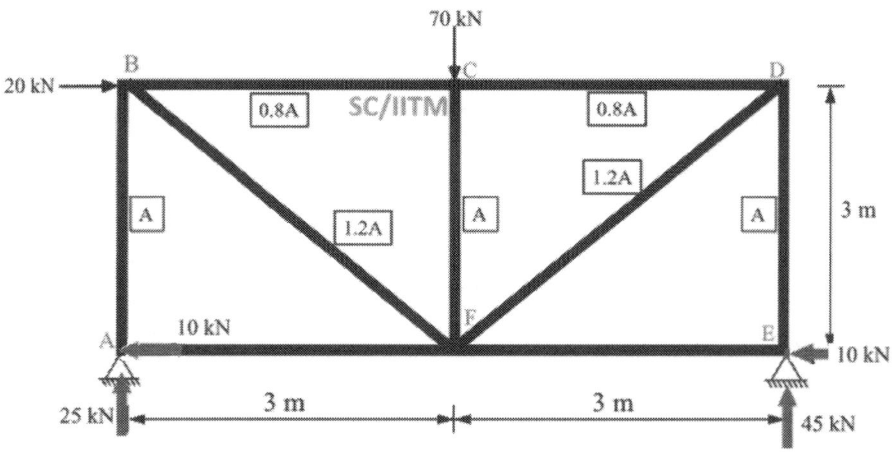

MATLAB output:

```
Enter the number of members:
9
Enter the lengths of each member as a row vector [L1 L2 ...
Ln]:
[3 3 3 3 3 3 4.242 3 4.242]
Enter the areas for each member as a row vector [A1 A2 ...
An]:
[5E-3 4E-3 4E-3 5E-3 4E-3 4E-3 6E-3 5E-3 6E-3]
Enter the angles for each member as a row vector [Theta1
Theta2 ... ThetaN]:
```

```
[90 0 0 90 0 0 -45 90 -135]
Enter the number of unrestrained degrees of freedom:
8
Enter the number of restrained degrees of freedom:
4
Enter the labels of unrestrained DOF as a row vector [u1 u2
...]:
[1 2 3 4 5 6 7 8]
Enter the labels of restrained DOF as a row vector [r1 r2
...]:
[9 10 11 12]
Enter global labels for Member 1 as a row vector: [10 2 9 1]
Enter global labels for Member 2 as a row vector: [2 4 1 3]
Enter global labels for Member 3 as a row vector: [4 6 3 5]
Enter global labels for Member 4 as a row vector: [12 6 11 5]
Enter global labels for Member 5 as a row vector: [10 8 9 7]
Enter global labels for Member 6 as a row vector: [8 12 7 11]
Enter global labels for Member 7 as a row vector: [2 8 1 7]
Enter global labels for Member 8 as a row vector: [8 4 7 3]
Enter global labels for Member 9 as a row vector: [6 8 5 7]
Enter fixed-end forces for Member 1 as a column vector:
[0; 0; 0; 0]
Enter fixed-end forces for Member 2 as a column vector:
[0; 0; 0; 0]
Enter fixed-end forces for Member 3 as a column vector:
[0; 0; 0; 0]
Enter fixed-end forces for Member 4 as a column vector:
[0; 0; 0; 0]
Enter fixed-end forces for Member 5 as a column vector:
[0; 0; 0; 0]
Enter fixed-end forces for Member 6 as a column vector:
[0; 0; 0; 0]
Enter fixed-end forces for Member 7 as a column vector:
[0; 0; 0; 0]
Enter fixed-end forces for Member 8 as a column vector:
[0; 0; 0; 0]
Enter fixed-end forces for Member 9 as a column vector:
[0; 0; 0; 0]
Enter joint load vector as a column vector:
[20; 0; 0; -70; 0; 0; 0; 0; 0; 0; 0; 0]
Member Number =      1
Local Stiffness matrix of member, [K] =
         0          0          0          0
         0          0          0          0
         0          0     0.0017    -0.0017
         0          0    -0.0017     0.0017
Transformation matrix of member, [T] =
     0     0    -1     0
     0     0     0    -1
     1     0     0     0
     0     1     0     0
```

```
Transformation matrix Transpose, [T] =
     0      0      1      0
     0      0      0      1
    -1      0      0      0
     0     -1      0      0
Global Matrix, [K global] =
    0.0017   -0.0017          0          0
   -0.0017    0.0017          0          0
         0         0          0          0
         0         0          0          0

Member Number =      2
Local Stiffness matrix of member, [K] =
         0         0          0          0
         0         0          0          0
         0         0     0.0013    -0.0013
         0         0    -0.0013     0.0013
Transformation matrix of member, [T] =
     1      0      0      0
     0      1      0      0
     0      0      1      0
     0      0      0      1
Transformation matrix Transpose, [T] =
     1      0      0      0
     0      1      0      0
     0      0      1      0
     0      0      0      1
Global Matrix, [K global] =
         0         0          0          0
         0         0          0          0
         0         0     0.0013    -0.0013
         0         0    -0.0013     0.0013

Member Number =      3
Local Stiffness matrix of member, [K] =
         0         0          0          0
         0         0          0          0
         0         0     0.0013    -0.0013
         0         0    -0.0013     0.0013
Transformation matrix of member, [T] =
     1      0      0      0
     0      1      0      0
     0      0      1      0
     0      0      0      1
Transformation matrix Transpose, [T] =
     1      0      0      0
     0      1      0      0
     0      0      1      0
     0      0      0      1
Global Matrix, [K global] =
         0         0          0          0
```

```
         0         0         0         0
         0         0    0.0013   -0.0013
         0         0   -0.0013    0.0013

Member Number =      4
Local Stiffness matrix of member, [K] =
         0         0         0         0
         0         0         0         0
         0         0    0.0017   -0.0017
         0         0   -0.0017    0.0017
Transformation matrix of member, [T] =
     0     0    -1     0
     0     0     0    -1
     1     0     0     0
     0     1     0     0
Transformation matrix Transpose, [T] =
     0     0     1     0
     0     0     0     1
    -1     0     0     0
     0    -1     0     0
Global Matrix, [K global] =
    0.0017   -0.0017         0         0
   -0.0017    0.0017         0         0
         0         0         0         0
         0         0         0         0

Member Number =      5
Local Stiffness matrix of member, [K] =
         0         0         0         0
         0         0         0         0
         0         0    0.0013   -0.0013
         0         0   -0.0013    0.0013
Transformation matrix of member, [T] =
     1     0     0     0
     0     1     0     0
     0     0     1     0
     0     0     0     1
Transformation matrix Transpose, [T] =
     1     0     0     0
     0     1     0     0
     0     0     1     0
     0     0     0     1
Global Matrix, [K global] =
         0         0         0         0
         0         0         0         0
         0         0    0.0013   -0.0013
         0         0   -0.0013    0.0013

Member Number =      6
Local Stiffness matrix of member, [K] =
         0         0         0         0
```

```
         0         0         0         0
         0         0    0.0013   -0.0013
         0         0   -0.0013    0.0013
Transformation matrix of member, [T] =
     1     0     0     0
     0     1     0     0
     0     0     1     0
     0     0     0     1
Transformation matrix Transpose, [T] =
     1     0     0     0
     0     1     0     0
     0     0     1     0
     0     0     0     1
Global Matrix, [K global] =
         0         0         0         0
         0         0         0         0
         0         0    0.0013   -0.0013
         0         0   -0.0013    0.0013


Member Number =      7
Local Stiffness matrix of member, [K] =
         0         0         0         0
         0         0         0         0
         0         0    0.0014   -0.0014
         0         0   -0.0014    0.0014
Transformation matrix of member, [T] =
    0.7071         0    0.7071         0
         0    0.7071         0    0.7071
   -0.7071         0    0.7071         0
         0   -0.7071         0    0.7071
Transformation matrix Transpose, [T] =
    0.7071         0   -0.7071         0
         0    0.7071         0   -0.7071
    0.7071         0    0.7071         0
         0    0.7071         0    0.7071
Global Matrix, [K global] =
   1.0e-03 *
    0.7072   -0.7072   -0.7072    0.7072
   -0.7072    0.7072    0.7072   -0.7072
   -0.7072    0.7072    0.7072   -0.7072
    0.7072   -0.7072   -0.7072    0.7072


Member Number =      8
Local Stiffness matrix of member, [K] =
         0         0         0         0
         0         0         0         0
         0         0    0.0017   -0.0017
         0         0   -0.0017    0.0017
Transformation matrix of member, [T] =
     0     0    -1     0
```

```
     0      0      0     -1
     1      0      0      0
     0      1      0      0
Transformation matrix Transpose, [T] =
     0      0      1      0
     0      0      0      1
    -1      0      0      0
     0     -1      0      0
Global Matrix, [K global] =
    0.0017   -0.0017         0         0
   -0.0017    0.0017         0         0
         0         0         0         0
         0         0         0         0

Member Number =      9
Local Stiffness matrix of member, [K] =
         0         0         0         0
         0         0         0         0
         0         0    0.0014   -0.0014
         0         0   -0.0014    0.0014
Transformation matrix of member, [T] =
   -0.7071         0    0.7071         0
         0   -0.7071         0    0.7071
   -0.7071         0   -0.7071         0
         0   -0.7071         0   -0.7071
Transformation matrix Transpose, [T] =
   -0.7071         0   -0.7071         0
         0   -0.7071         0   -0.7071
    0.7071         0   -0.7071         0
         0    0.7071         0   -0.7071
Global Matrix, [K global] =
   1.0e-03 *
    0.7072   -0.7072    0.7072   -0.7072
   -0.7072    0.7072   -0.7072    0.7072
    0.7072   -0.7072    0.7072   -0.7072
   -0.7072    0.7072   -0.7072    0.7072

Stiffness Matrix of the complete structure, [Ktotal] =
  Columns 1 through 9
    0.0020   -0.0007   -0.0013         0         0         0   -0.0007    0.0007         0
   -0.0007    0.0024         0         0         0         0    0.0007   -0.0007         0
   -0.0013         0    0.0027         0   -0.0013         0         0         0         0
         0         0         0    0.0017         0         0         0   -0.0017         0
         0         0   -0.0013         0    0.0020    0.0007   -0.0007   -0.0007         0
         0         0         0         0    0.0007    0.0024   -0.0007   -0.0007         0
   -0.0007    0.0007         0         0   -0.0007   -0.0007    0.0041         0   -0.0013
    0.0007   -0.0007         0   -0.0017   -0.0007   -0.0007         0    0.0031         0
         0         0         0         0         0         0   -0.0013         0    0.0013
         0   -0.0017         0         0         0         0         0         0         0
         0         0         0         0         0         0   -0.0013         0         0
         0         0         0         0         0   -0.0017         0         0         0
  Columns 10 through 12
         0         0         0
```

```
  -0.0017         0         0
        0         0         0
        0         0         0
        0         0         0
        0         0   -0.0017
        0   -0.0013         0
        0         0         0
        0         0         0
   0.0017         0         0
        0    0.0013         0
        0         0    0.0017
Unrestrained Stiffness sub-matrix, [Kuu] =
    0.0020   -0.0007   -0.0013         0         0         0   -0.0007    0.0007
   -0.0007    0.0024         0         0         0         0    0.0007   -0.0007
   -0.0013         0    0.0027         0   -0.0013         0         0         0
         0         0         0    0.0017         0         0         0   -0.0017
         0         0   -0.0013         0    0.0020    0.0007   -0.0007   -0.0007
         0         0         0         0    0.0007    0.0024   -0.0007   -0.0007
   -0.0007    0.0007         0         0   -0.0007   -0.0007    0.0041         0
    0.0007   -0.0007         0   -0.0017   -0.0007   -0.0007         0    0.0031
Inverse of Unrestrained Stiffness sub-matrix, [KuuInverse] =
   1.0e+03 *
    1.7570    0.3000    1.3820   -0.3750    1.0070   -0.3000    0.3750   -0.3750
    0.3000    0.6000    0.3000    0.3000    0.3000         0    0.0000    0.3000
    1.3820    0.3000    1.7570   -0.0000    1.3820   -0.3000    0.3750   -0.0000
   -0.3750    0.3000   -0.0000    1.9820    0.3750    0.3000   -0.0000    1.3820
    1.0070    0.3000    1.3820    0.3750    1.7570   -0.3000    0.3750    0.3750
   -0.3000         0   -0.3000    0.3000   -0.3000    0.6000   -0.0000    0.3000
    0.3750    0.0000    0.3750   -0.0000    0.3750   -0.0000    0.3750   -0.0000
   -0.3750    0.3000   -0.0000    1.3820    0.3750    0.3000   -0.0000    1.3820
Joint Load vector, [Jl] =
    20     0     0   -70     0     0     0     0     0     0     0     0
Unrestrained displacements, [DelU] =
   1.0e+05 *
    0.6139   -0.1500    0.2764   -1.4624   -0.0611   -0.2700    0.0750   -1.0424


Member Number =      1
Global displacement matrix [DeltaBar] =
   1.0e+04 *
        0   -1.5000         0    6.1390
Global End moment matrix [MBar] =
   25.0000  -25.0000         0         0
Member Number =      2
Global displacement matrix [DeltaBar] =
   1.0e+05 *
   -0.1500   -1.4624    0.6139    0.2764
Global End moment matrix [MBar] =
        0         0   45.0000  -45.0000
Member Number =      3
Global displacement matrix [DeltaBar] =
   1.0e+05 *
   -1.4624   -0.2700    0.2764   -0.0611
Global End moment matrix [MBar] =
        0         0   45.0000  -45.0000
```

```
Member Number =     4
Global displacement matrix [DeltaBar] =
   1.0e+04 *
         0    -2.7000         0    -0.6110
Global End moment matrix [MBar] =
   45.0000  -45.0000         0         0
Member Number =     5
Global displacement matrix [DeltaBar] =
   1.0e+05 *
         0    -1.0424         0     0.0750
Global End moment matrix [MBar] =
         0         0  -10.0000   10.0000
Member Number =     6
Global displacement matrix [DeltaBar] =
   1.0e+05 *
   -1.0424         0    0.0750         0
Global End moment matrix [MBar] =
         0         0   10.0000  -10.0000
Member Number =     7
Global displacement matrix [DeltaBar] =
   1.0e+05 *
   -0.1500   -1.0424    0.6139    0.0750
Global End moment matrix [MBar] =
   25.0000  -25.0000  -25.0000   25.0000
Member Number =     8
Global displacement matrix [DeltaBar] =
   1.0e+05 *
   -1.0424   -1.4624    0.0750    0.2764
Global End moment matrix [MBar] =
    70   -70     0     0
Member Number =     9
Global displacement matrix [DeltaBar] =
   1.0e+05 *
   -0.2700   -1.0424   -0.0611    0.0750
Global End moment matrix [MBar] =
   45.0000  -45.0000   45.0000  -45.0000
Joint forces =
20.0000    0.0000    0.0000  -70.0000         0    0.0000   -0.0000   -0.0000
-10.0000    25.0000  -10.0000   45.0000
```

5 Three-Dimensional Analysis of Space Frames

5.1 THREE-DIMENSIONAL ANALYSIS OF STRUCTURES

In the previous section, a detailed explanation of the stiffness method of analysis of planar orthogonal and non-orthogonal structures using beam elements and planar truss elements was given. The same algorithm and sign convention are now extended to solve the three-dimensional structures to make the analysis very simple and easier.

Let us extend the basics of beam elements to three-dimensional structures. We know that the equation for the joint equilibrium of the planar structure is given below:

$$[K]_{complte}\{\Delta\}_{complte} = \{J_L\}_{complete} + \{R\}_{complete} \tag{5.1}$$

The above equation is also expandable to solve three-dimensional structures. Similarly, the matrix equation describing the equilibrium of the beam element is given by:

$$\{M\}_i = [K_i][T_i]\{\delta_i\} + \{FEM\}_i \tag{5.2}$$

This equation can also be extended to analyze three-dimensional structures consisting of beam elements arbitrarily oriented in space. The first task in three-dimensional analysis is to develop the stiffness matrix of the complete structure, which can be simply done by the summation of member stiffness matrices of individual elements. It is very important to note that the complete stiffness matrix of the space system will be established in the reference axes system.

Sign Convention:

Consider three axes as shown in Figure 5.1, where the vector representing the translation or forces is marked. The right-hand thumb rule is used to mark the direction of the moment on every axis. If the thumb points toward the arrows, then the direction of the remaining four fingers will indicate the direction of the moment. This indicates the direction of rotation or moment. All these are considered to be positively established using the right-hand thumb rule of orthogonal coordinate axes. Also, please note that the X-axis is fixed as horizontal; the Y-axis is 90° anticlockwise to the X-axis, and the Z-axis is 90° clockwise to the X-axis.

DOI: 10.1201/9781003639152-5

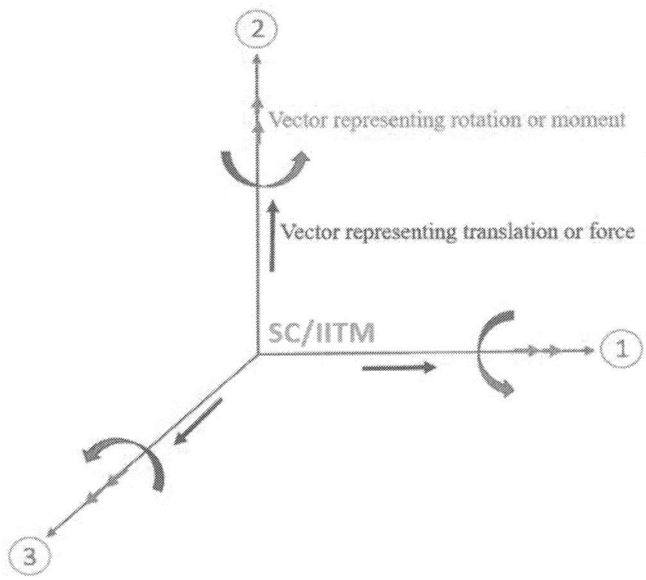

FIGURE 5.1 Reference axes and force representation.

5.2 BEAM ELEMENT

Consider a member with two ends (j,k) as shown in Figure 5.2. The local axes system is represented by (x_m-y_m-z_m), which are orthogonal to each other. The reference axes system is represented by (X-Y-Z). At each end, there are three translations (along

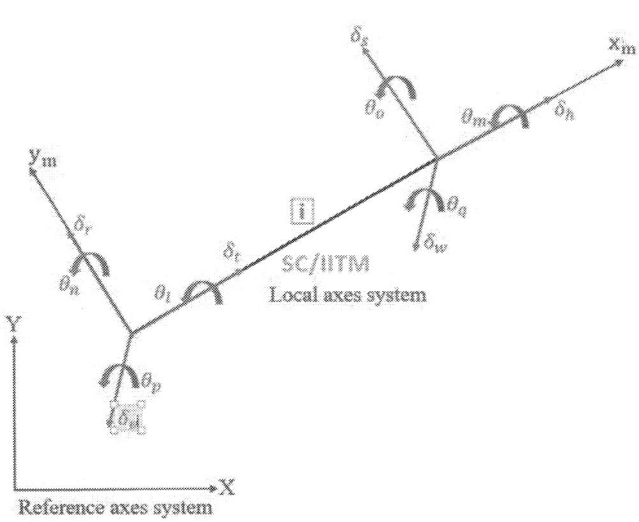

FIGURE 5.2 Degree of freedom in local axes system.

x-y-z) and three rotations about (x-y-z) axes, as shown. This makes the modular beam element possess 12 degrees of freedom.

5.3 STIFFNESS MATRIX

a. Translation along (x_m-y_m-z_m) axes

Let us consider unit displacement along (x_m-y_m-z_m) axes at the jth end as shown in Figure 5.3. The corresponding stiffness coefficients are marked in the figure.

Let us apply unit rotation at the jth end about (x_m-y_m-z_m) axes, as shown in Figure 5.4. The respective stiffness coefficients are marked in the figure.

The same procedure can be extended for the kth end to derive the rest of the stiffness coefficients. The stiffness matrix is shown in Figure 5.5.

Now, the values of these stiffness coefficients are determined from the first principles to obtain the complete stiffness matrix:

$$[K]=$$

$$\begin{bmatrix}
\frac{EA_x}{l} & 0 & 0 & 0 & 0 & 0 & -\frac{EA_x}{l} & 0 & 0 & 0 & 0 & 0 \\
0 & \frac{12EI_z}{l^3} & 0 & 0 & 0 & \frac{6EI_z}{l^2} & 0 & -\frac{12EI_z}{l^3} & 0 & 0 & 0 & \frac{6EI_z}{l^2} \\
0 & 0 & \frac{12EI_y}{l^3} & 0 & -\frac{6EI_y}{l^2} & 0 & 0 & 0 & -\frac{12EI_y}{l^3} & 0 & -\frac{6EI_y}{l^2} & 0 \\
0 & 0 & 0 & \frac{GI_x}{l} & 0 & 0 & 0 & 0 & 0 & -\frac{GI_x}{l} & 0 & 0 \\
0 & 0 & -\frac{6EI_y}{l^2} & 0 & \frac{4EI_y}{l^2} & 0 & 0 & 0 & \frac{6EI_y}{l^2} & 0 & \frac{2EI_y}{l} & 0 \\
0 & -\frac{6EI_z}{l^2} & 0 & 0 & 0 & \frac{4EI_z}{l} & 0 & -\frac{6EI_z}{l^2} & 0 & 0 & 0 & \frac{2EI_z}{l} \\
-\frac{EA_x}{l} & 0 & 0 & 0 & 0 & 0 & \frac{EA_x}{l} & 0 & 0 & 0 & 0 & 0 \\
0 & -\frac{12EI_z}{l^3} & 0 & 0 & 0 & \frac{6EI_z}{l^2} & 0 & \frac{12EI_z}{l^3} & 0 & 0 & 0 & -\frac{6EI_z}{l^2} \\
0 & 0 & -\frac{12EI_y}{l^3} & 0 & \frac{6EI_y}{l^2} & 0 & 0 & 0 & \frac{12EI_y}{l^3} & 0 & \frac{6EI_y}{l^2} & 0 \\
0 & 0 & 0 & -\frac{GI_x}{l} & 0 & 0 & 0 & 0 & 0 & \frac{GI_x}{l} & 0 & 0 \\
0 & 0 & -\frac{6EI_y}{l^2} & 0 & -\frac{2EI_y}{l} & 0 & 0 & 0 & \frac{6EI_y}{l^2} & 0 & \frac{4EI_y}{l} & 0 \\
0 & -\frac{6EI_z}{l^2} & 0 & 0 & 0 & \frac{2EI_z}{l} & 0 & -\frac{6EI_z}{l^2} & 0 & 0 & 0 & \frac{4EI_z}{1}
\end{bmatrix}$$

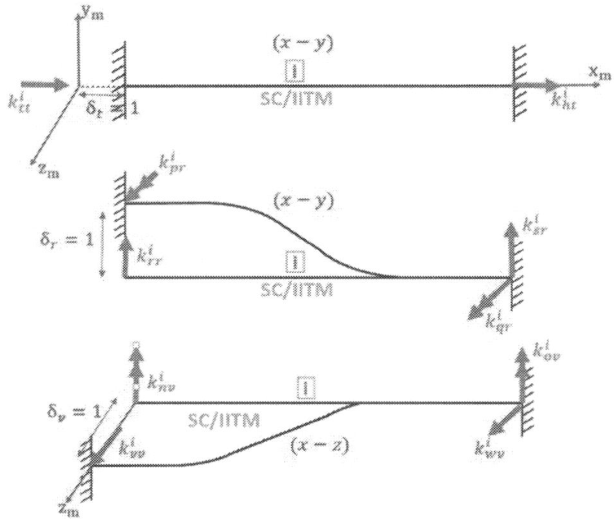

FIGURE 5.3 Unit translation at jth end in x_m-y_m-z_m directions.

I_x is called a torsional constant. For the beam element with a rectangular cross-section, I_x is given by:

$$I_x = \frac{ht^3}{12}\left[1 - 0.63\left(\frac{t}{h}\right) + 0.052\left(\frac{t}{h}\right)^3\right] \text{for h} \gg \text{t} \tag{5.3}$$

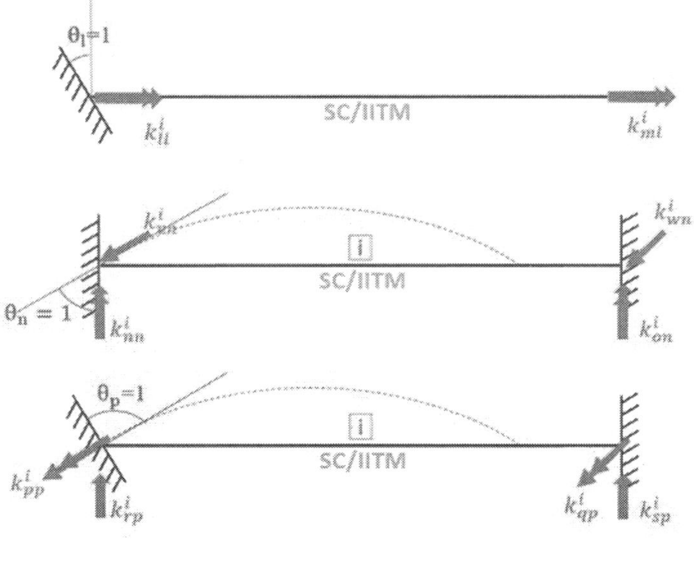

FIGURE 5.4 Unit rotation at jth end in x_m-y_m-z_m directions.

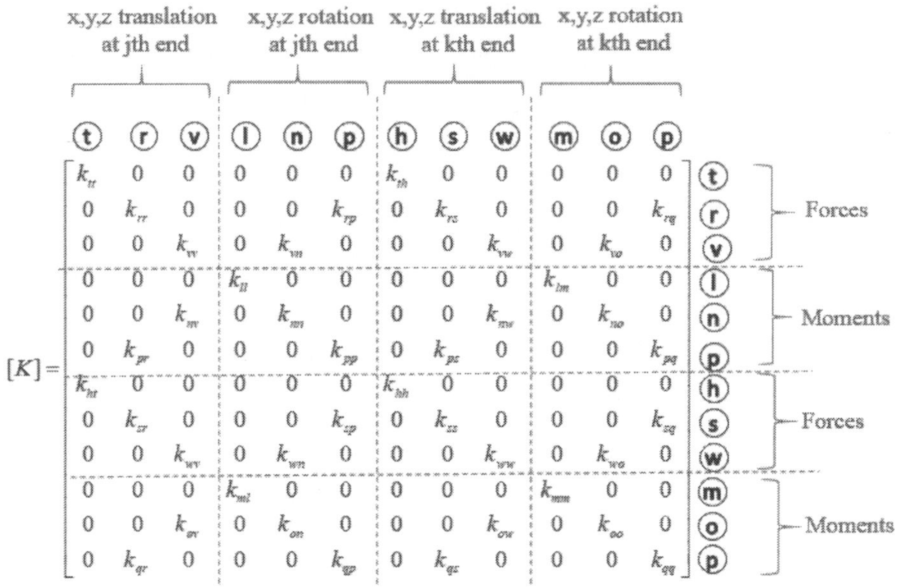

FIGURE 5.5 Stiffness matrix of 3D beam element.

where h is the depth of the rectangular section and t is the width of the section. For rectangular cross-sections with a very large value of h/t,

$$I_x = \frac{ht^3}{12} \tag{5.4}$$

In the case of I-sections, as shown, the torsional constant is given by:

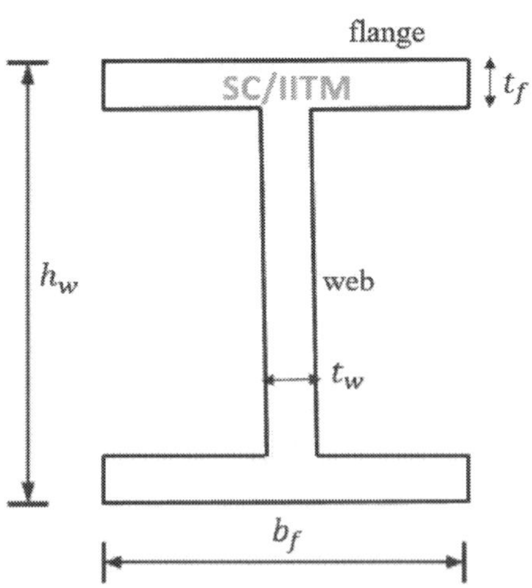

$$I_x = \frac{1}{3}\sum ht^3 \tag{5.5}$$

where h is the longer dimension and t is the shorter dimension. Hence,

$$I_x = \frac{1}{3}\left\{\left(b_f t_f^3\right) + \left(h_w t_w^3\right) + \left(b_f t_f^3\right)\right\} \tag{5.6}$$

5.4 TRANSFORMATION MATRIX

In a space frame, members can be oriented in any fashion. The member's local axes may not coincide with the reference axes system. In such a situation, the stiffness matrix needs to be transformed concerning the reference axes system. In addition, the load applied on the local axes also needs to be transformed concerning the reference axes system. Most importantly, forces, end moments, and reactions are to be computed concerning the reference axes system.

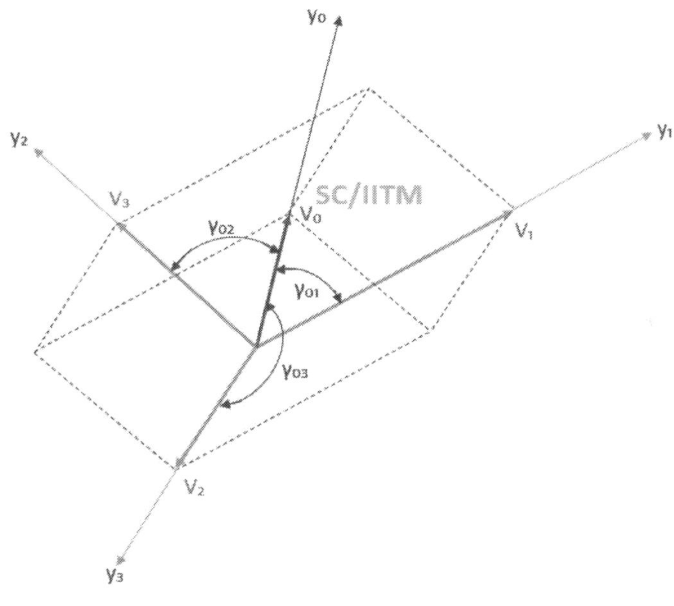

Let V_0 be a vector that is arbitrarily oriented along the axis Y_0. This has to be transformed into the reference axes system (Y_1-Y_2-Y_3). The vector V_0 has its components along the Y_1, Y_2, and Y_3 axes. To find the components, the inclination or position of the vector concerning the three axes should be known. As seen in the above figure, γ_{01} is the angle between the axes Y_0 and Y_1. Similarly, γ_{02} is the angle between the axes Y_0, Y_2, and γ_{03} is the angle between the axes Y_0 and Y_3. The corresponding components are V_1, V_2, and V_3. Thus, the following relationship will be valid.

$$
\begin{aligned}
V_1 &= V_0 \, cos\,\gamma_{01} \\
V_2 &= V_0 \, cos\,\gamma_{02} \\
V_3 &= V_0 \, cos\,\gamma_{03}
\end{aligned}
\tag{5.7}
$$

where γ_{01}, γ_{02}, and γ_{03} are defined as angles between the vector V_0 or the axis Y_0 to which the vector is aligned and the coordinate axes (Y_1-Y_2-Y_3), respectively. In the above equation, the terms cos γ_{01}, cos γ_{02}, and cos γ_{03} are called direction cosines. Now, the vector V_0 along (Y_1-Y_2-Y_3) is a set of coordinate axes. Let us resolve or transform this to the standard reference axes system (X-Y-Z).

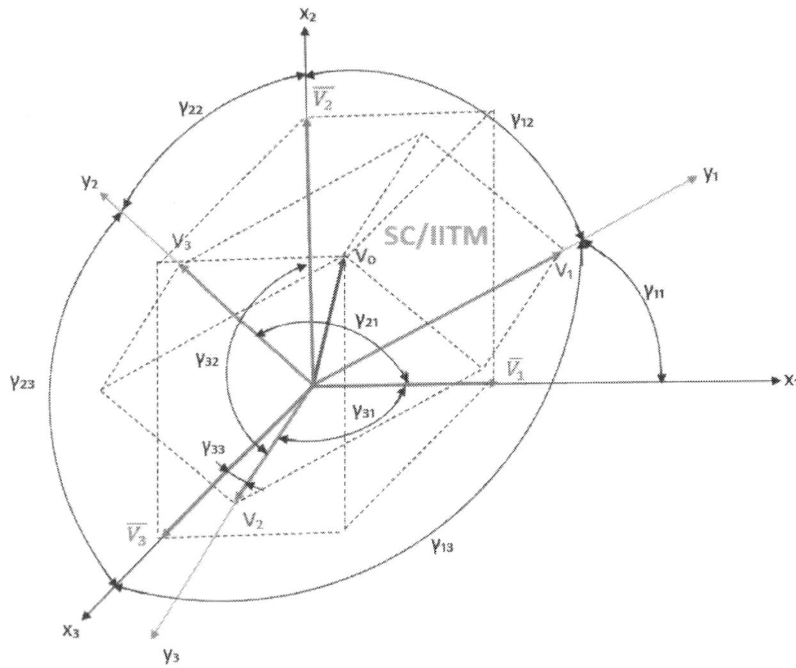

Let, X_1-X_2-X_3 be the reference axes systems. Let us define the angles of X_1-X_2-X_3 and vectors V_1, V_2, V_3. V_1, V_2, and V_3 are already resolved along (Y_1-Y_2-Y_3). Let γ_{11} be the angle between the V_1 axis and X_1 axis. By this logic, the angle between the vector and the reference axes system can be defined. Thus, the V_1 vector makes an angle (γ_{11}, γ_{12}, γ_{13}) with the reference axes system; the V_2 vector makes an angle (γ_{11}, γ_{22}, γ_{23}) with the reference axes, and the V_3 vector makes an angle (γ_{31}, γ_{32}, γ_{33}) with the reference axes, as shown in the above figure. The set of equations that connects the transformed components of these vectors with the known components of vectors is given by:

$$
\begin{aligned}
\overline{V_1} &= V_1 \, cos\,\gamma_{11} + V_2 \, cos\,\gamma_{21} + V_1 \, cos\,\gamma_{31} \\
\overline{V_2} &= V_1 \, cos\,\gamma_{12} + V_2 \, cos\,\gamma_{22} + V_1 \, cos\,\gamma_{32} \\
\overline{V_3} &= V_1 \, cos\,\gamma_{13} + V_2 \, cos\,\gamma_{23} + V_1 \, cos\,\gamma_{33}
\end{aligned}
\tag{5.8}
$$

Let $C_{ij} = \cos \gamma_{ij}$, where I represents the (Y_1-Y_2-Y_3) axes system and j represents the (X_1-X_2-X_3) axes system. Thus, the above equation can be written in matrix form as follows:

$$\begin{Bmatrix} \bar{V}_1 \\ \bar{V}_2 \\ \bar{V}_3 \end{Bmatrix} = \begin{bmatrix} C_{11} & C_{21} & C_{31} \\ C_{12} & C_{22} & C_{32} \\ C_{13} & C_{23} & C_{33} \end{bmatrix} \tag{5.9}$$

$$\{\bar{V}\} = [C_s]^T \{V\} \tag{5.10}$$

where, $[C_s] = \begin{bmatrix} C_{11} & C_{12} & C_{13} \\ C_{21} & C_{22} & C_{23} \\ C_{31} & C_{32} & C_{33} \end{bmatrix}$

It also verifies that $[C_s]^T = [C_s]^{-1}$. $\bar{V}$ Refer to the reference axes system, and V refers to the coordinate axes system. Please note that they do not refer to the axis of the original vector V_0.

5.5 MEMBER ROTATION MATRIX

Consider a beam element with length L_i. The member has two nodes, 'j' and 'k'. The member is oriented along its local axes system (x_m-y_m-z_m). This local axes system is placed arbitrarily in space concerning the standard reference axes system (X-Y-Z).

γ_x, γ_y, and γ_z are the angles of the x_m axis with respect to X-Y-Z axes, respectively, as shown in Figure 5.6.

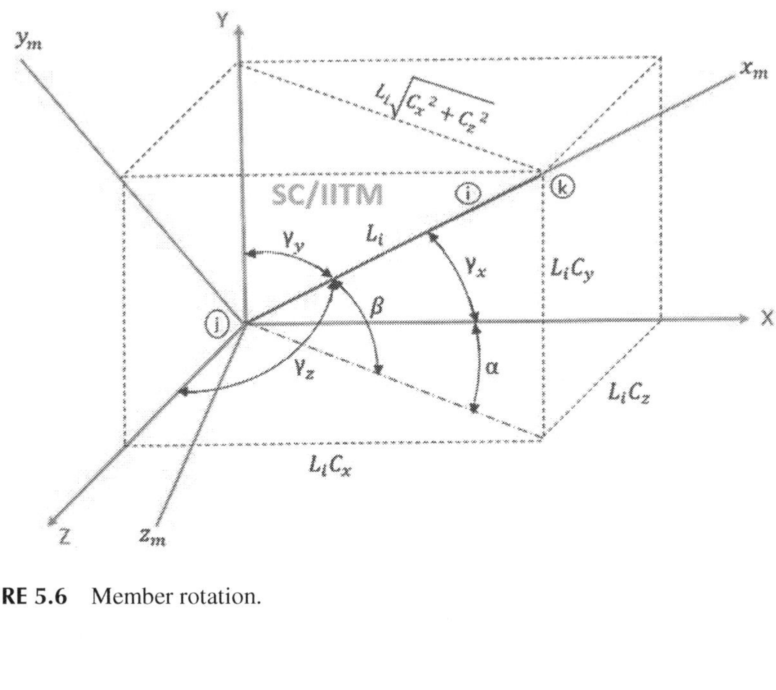

FIGURE 5.6 Member rotation.

Let,

$$C_x = \cos\gamma_x$$

$$C_y = \cos\gamma_y$$

$$C_z = \cos\gamma_z$$

As the j and k coordinates of the member positioned in space are known, the direction cosines can be written as follows:

$$C_x = \frac{X_k - X_j}{L_i} \tag{5.11}$$

$$C_y = \frac{Y_k - Y_j}{L_i} \tag{5.12}$$

$$C_z = \frac{Z_k - Z_j}{L_i} \tag{5.13}$$

$$L_i = \sqrt{\left(X_k - X_j\right)^2 + \left(Y_k - Y_j\right)^2 + \left(Z_k - Z_j\right)^2} \tag{5.14}$$

where (X_k,Y_k,Z_k) and (X_j,Y_j,Z_j) are the coordinates of the beam element placed in space. It is now important to know that the direction cosines give the components of beam element only along the reference axes system, but important information on the orientation of the local axes system concerning the reference axes system is not known. Therefore, the beam element is oriented along the local axes $(x_m\text{-}y_m\text{-}z_m)$. Hence, the orientation of $(x_m\text{-}y_m\text{-}z_m)$, axes or the local axes system concerning the reference axes system, is called ψ angle, which has to be estimated. Thus, to find the orientation of the local axes system concerning the reference axes system, we need to know the direction cosines and ψ angle.

5.6 Y-Z-X TRANSFORMATION

The direction cosines give the components of the member along the reference axes system. Further, ψ angle provides information about the orientation of the local axes system concerning the reference axes system. It is very important to note that the procedure of aligning the reference axes system to that of the local axes system is called transformation. There are various schemes involved in this transformation. One such scheme is the Y-Z-X transformation, as shown in Figure 5.7, which means rotating the reference axes about the Y-axis, then the Z-axis, and finally about the X-axis. Y-Z-X highlights the order or sequence of rotation to be carried out. The reference axes system (X-Y-Z) will be rotated about the Y-axis first, then about the Z-axis, and then about the X-axis last. The amount of rotation happening in the Y and Z axes are α and β, respectively, and

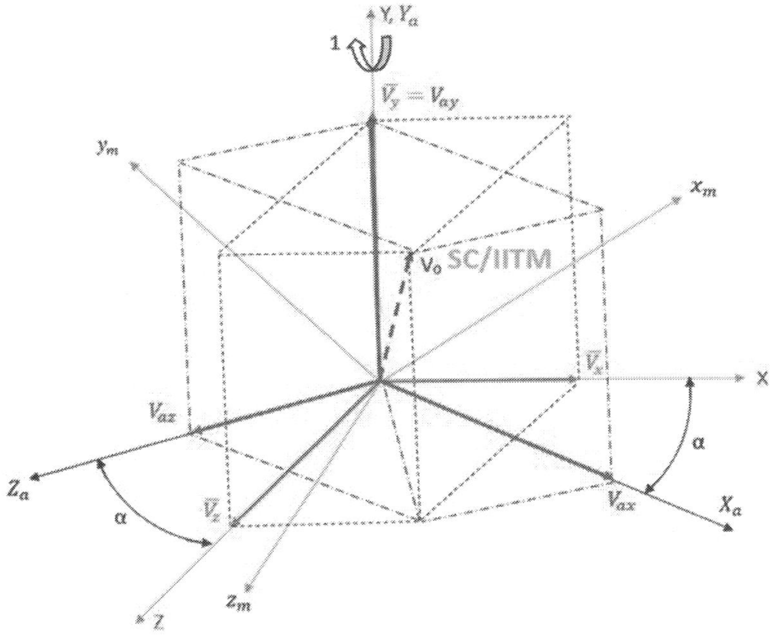

FIGURE 5.7 Y-Z-X transformation.

ultimately, we will get ψ angle after the rotation about the X-axis. The procedure is to hold the orthogonal axes system (X-Y-Z) and rotate this axes system about the Y-axis by α degrees.

Let V_0 be the vector placed arbitrarily. Now, the reference axes system is rotated about the Y-axis, where Y and Y_α remain the same. X-axis and Z-axis will move to X_α and Z_α, respectively, since the angle of rotation is α degree. $\bar{V}_x, \bar{V}_y, \bar{V}_z$ are the components of the vector concerning the reference axes (X-Y-Z).

From the figure,

$$\sin\alpha = \frac{C_z}{\sqrt{C_x^2 + C_z^2}} \tag{5.15a}$$

$$\cos\alpha = \frac{C_x}{\sqrt{C_x^2 + C_z^2}} \tag{5.15b}$$

Thus,

$$\begin{Bmatrix} V_{\alpha x} \\ V_{\alpha y} \\ V_{\alpha z} \end{Bmatrix} = \begin{bmatrix} \cos\alpha & 0 & \sin\alpha \\ 0 & 1 & 0 \\ -\sin\alpha & 0 & \cos\alpha \end{bmatrix} \begin{Bmatrix} \bar{V}_x \\ \bar{V}_y \\ \bar{V}_z \end{Bmatrix} \tag{5.15c}$$

By substituting the values of *cos* α and *sin* α in the above matrix equation,

$$\begin{Bmatrix} V_{\alpha x} \\ V_{\alpha y} \\ V_{\alpha z} \end{Bmatrix} = \begin{bmatrix} \frac{C_x}{\sqrt{C_x^2 + C_z^2}} & 0 & \frac{C_z}{\sqrt{C_x^2 + C_z^2}} \\ 0 & 1 & 0 \\ -\frac{C_z}{\sqrt{C_x^2 + C_z^2}} & 0 & \frac{C_x}{\sqrt{C_x^2 + C_z^2}} \end{bmatrix} \begin{Bmatrix} \bar{V}_x \\ \bar{V}_y \\ \bar{V}_z \end{Bmatrix} \tag{5.15d}$$

$$\{V_\alpha\} = [C_\alpha]\{\bar{V}\} \tag{5.16}$$

The next step is to rotate about the Z-axis by β degrees. Since the rotation takes place about the Z-axis, Z_α and Z_β remain the same. X_α and Y_α will shift to X_β and Y_β, respectively. Now, the vector components are $V_{\beta x}$, $V_{\beta y}$, and $V_{\beta z}$ as shown in Figure 5.8.

Now,

$$\begin{Bmatrix} V_{\beta x} \\ V_{\beta y} \\ V_{\beta z} \end{Bmatrix} = \begin{bmatrix} \cos\beta & \sin\beta & 0 \\ -\sin\beta & \cos\beta & 0 \\ 0 & 0 & 1 \end{bmatrix} \begin{Bmatrix} V_{\alpha x} \\ V_{\alpha y} \\ V_{\alpha z} \end{Bmatrix} \tag{5.17a}$$

$$\sin\beta = C_y \tag{5.17b}$$

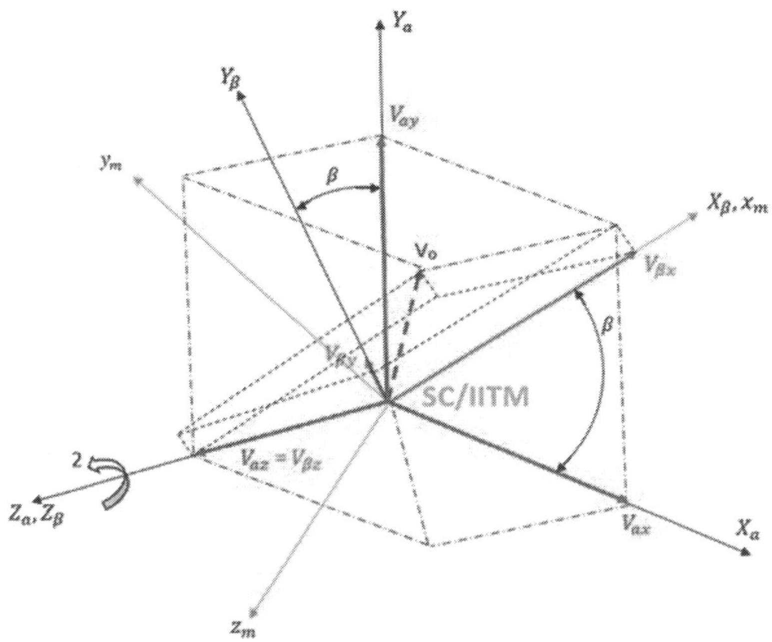

FIGURE 5.8 Z-axis rotation.

$$cos\beta = \sqrt{C_x^2 + C_z^2} \tag{5.17c}$$

By substituting these values, we get:

$$\begin{Bmatrix} V_{\beta x} \\ V_{\beta y} \\ V_{\beta z} \end{Bmatrix} = \begin{bmatrix} \sqrt{C_x^2 + C_z^2} & C_y & 0 \\ -C_y & \sqrt{C_x^2 + C_z^2} & 0 \\ 0 & 0 & 1 \end{bmatrix} \begin{Bmatrix} V_{\alpha x} \\ V_{\alpha y} \\ V_{\alpha z} \end{Bmatrix} \tag{5.17d}$$

$$\{V_\beta\} = [C_\beta]\{V_\alpha\} \tag{5.17e}$$

Finally, let us rotate about X-axis by an angle ψ_y as shown in Figure 5.9. Here X_β will remain the same as x_m. Y_β and Z_β will shift to y_m and z_m, respectively. Thus, ψ_y is the angle between Y_β and y_m measured from Y_β toward y_m or angle between Z_β and z_m measured Z_β toward z_m. The idea is to bring y_m and Y_β aligned and x_m and X_β aligned. x_m and X_β are already aligned, since the rotation is about X-axis.

Thus,

$$\begin{Bmatrix} V_x \\ V_y \\ V_z \end{Bmatrix} = \begin{bmatrix} 1 & 0 & 0 \\ 0 & cos\psi_y & sin\psi_y \\ 0 & -sin\psi_y & cos\psi_y \end{bmatrix} \begin{Bmatrix} V_{\beta x} \\ V_{\beta y} \\ V_{\beta z} \end{Bmatrix} \tag{5.18a}$$

$$\{V\} = [C_{\psi_y}]\{V_\beta\} \tag{5.18b}$$

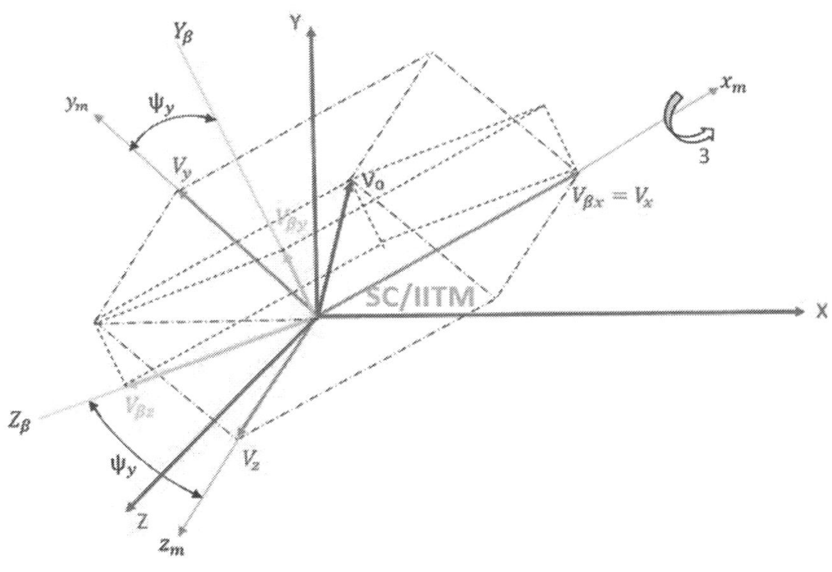

FIGURE 5.9 Rotation about X-axis.

Hence,

$$\{V\}=\left[C_{\psi_y}\right]\{V_\beta\}=\left[C_{\psi_y}\right]\left[C_\beta\right]\{V_\alpha\}=\left[C_{\psi_y}\right]\left[C_\beta\right]\left[C_\alpha\right]\{\bar{V}\} \tag{5.18c}$$

$$\{V\}=\left[C_{\psi_y}\right]\left[C_\beta\right]\left[C_\alpha\right]\{\bar{V}\} \tag{5.19a}$$

$$\text{Let,}\left[C_y\right]=\left[C_{\psi_y}\right]\left[C_\beta\right]\left[C_\alpha\right] \tag{5.19b}$$

By substituting the values,

$$\left[C_y\right]=\begin{bmatrix}1 & 0 & 0\\ 0 & cos\psi_y & sin\psi_y\\ 0 & -sin\psi_y & cos\psi_y\end{bmatrix}\begin{bmatrix}\sqrt{C_x^2+C_z^2} & C_y & 0\\ -C_y & \sqrt{C_x^2+C_z^2} & 0\\ 0 & 0 & 1\end{bmatrix}\begin{bmatrix}\frac{C_x}{\sqrt{C_x^2+C_z^2}} & 0 & \frac{C_z}{\sqrt{C_x^2+C_z^2}}\\ 0 & 1 & 0\\ -\frac{C_z}{\sqrt{C_x^2+C_z^2}} & 0 & \frac{C_x}{\sqrt{C_x^2+C_z^2}}\end{bmatrix} \tag{5.19c}$$

$$\left[C_y\right]=\begin{bmatrix}C_x & C_y & C_z\\ \frac{-C_xC_ycos\psi_y-C_zsin\psi_y}{\sqrt{C_x^2+C_z^2}} & cos\psi_y\sqrt{C_x^2+C_z^2} & \frac{-C_zC_ycos\psi_y+C_xsin\psi_y}{\sqrt{C_x^2+C_z^2}}\\ \frac{C_xC_ysin\psi_y-C_zsin\psi_y}{\sqrt{C_x^2+C_y^2}} & -sin\psi_y\sqrt{C_x^2+C_z^2} & \frac{C_yC_zsin\psi_y-C_xcos\psi_y}{\sqrt{C_x^2+C_z^2}}\end{bmatrix} \tag{5.19d}$$

$[C_y]$ is called the rotation matrix, which is expressed in terms if direction cosines. It also defines the position of x_m or the longitudinal axis of the member in the local axes system with θ, in relation to the reference axes system and through the ψ angle.

5.7 Z-Y-X TRANSFORMATION

In this transformation, rotate about Z-axis first, then about Y-axis and finally about X-axis. Here, one can get ψ_z angle. As shown in Figure 5.10, the local axes system is (x_m, y_m, z_m), and the global axes system is (X-Y-Z).

When rotated about Z-axis by an angle of θ degrees, Z and Z_θ axes will remain the same. X and Y will shift to X_θ and Y_θ, respectively. In the second step, rotation about Y-axis by an angle α degrees, will leave Y_θ and Y_λ axes the same. X_θ and Z_θ

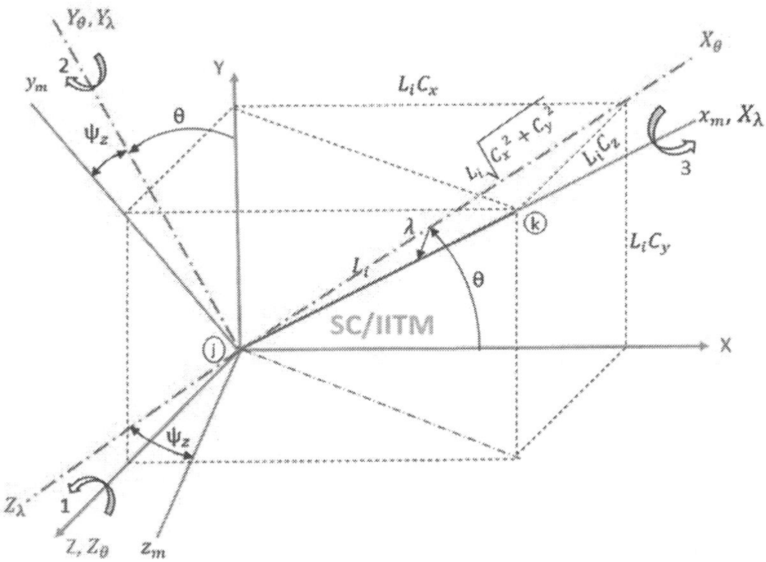

FIGURE 5.10 Z-Y-X transformation.

will shift to X_λ and Z_λ, respectively. In the final step, by rotating about X-axis by an angle ψ_z, X_λ and x_m will remain the same. But, Y_λ and Z_λ will shift to y_m and z_m, respectively. Thus, ψ_z is measured from Y_λ to y_m or Z_λ to z_m.

Hence, the rotation matrix can be directly written as follows:

$$\left[C_{\psi_z}\right]=\begin{bmatrix}1 & 0 & 0\\ 0 & cos\psi_z & sin\psi_z\\ 0 & -sin\psi_z & cos\psi_z\end{bmatrix} \tag{5.20a}$$

We can also write,

$$\left[C_z\right]=\left[C_{\psi_z}\right]\left[C_\lambda\right]\left[C_\theta\right] \tag{5.20b}$$

$$\left[C_z\right]=$$

$$\begin{bmatrix}cos\theta cos\lambda & sin\theta cos\lambda & sin\lambda\\ (-sin\theta cos\psi_z)-(cos\theta sin\lambda sin\psi_z) & (cos\theta cos\psi_z)-(sin\theta sin\lambda sin\psi_z) & cos\lambda sin\psi_z\\ (sin\theta sin\psi_z)-(cos\theta sin\lambda cos\psi_z) & (-cos\theta sin\psi_z)-(sin\theta sin\lambda cos\psi_z) & cos\lambda\cos\psi_z\end{bmatrix} \tag{5.20c}$$

From Figure 5.10, we know the following:

$$sin\theta = \frac{C_y}{\sqrt{C_x^2 + C_y^2}} \tag{5.20d}$$

$$cos\theta = \frac{C_x}{\sqrt{C_x^2 + C_y^2}} \tag{5.20e}$$

$$sin\lambda = C_z \tag{5.20f}$$

$$cos\lambda = \sqrt{C_x^2 + C_y^2} \tag{5.20g}$$

By substituting the above values in C_z,

$$\left[C_z\right] = \begin{bmatrix} C_x & C_y & C_z \\ \frac{(-C_xC_z sin\psi_z) - (C_y cos\psi_z)}{\sqrt{C_x^2 + C_y^2}} & \frac{(-C_yC_z sin\psi_z) - (C_x cos\psi_z)}{\sqrt{C_x^2 + C_y^2}} & \frac{sin\psi_z}{\sqrt{C_x^2 + C_y^2}} \\ \frac{(-C_xC_z cos\psi_z) - (C_y sin\psi_z)}{\sqrt{C_x^2 + C_y^2}} & \frac{(-C_yC_z cos\psi_z) - (C_x sin\psi_z)}{\sqrt{C_x^2 + C_y^2}} & \frac{cos\psi_z}{\sqrt{C_x^2 + C_y^2}} \end{bmatrix} \tag{5.20h}$$

In general, one can use either Y-Z-X transformation or Z-Y-X transformation to transform the reference axes system to the local axes system. But choice of transformation order can make the evaluation of ψ angle simpler. In both the transformations, the rotation in the last step takes place about X-axis only. Most importantly, in both the transformations, ψ angles are calculated. But ψ_y and ψ_z are completely different. Thus, for a given member, which transformation order is to be followed? If the member is positioned in the frame of reference axes such that the longitudinal axes of the member correspond to Y-axis of the reference system, then use Z-Y-X transformation. Similarly, if a member is placed in the frame of reference axes system such that the longitudinal axes of the member correspond to Z-axis of the reference system, then use Y-Z-X transformation.

5.8 Ψ ANGLE

Let us consider a three-axes system (X-Y-Z) as shown in Figure 5.11. The local axes system is (x_m-y_m-z_m). The transformed axes system is (X_β, Y_β, Z_β). The line of sight is along the X-axis. The longitudinal axis of the member coincides with the X-axis. Thus, this is the Y-Z-X transformation. The axes are then marked on

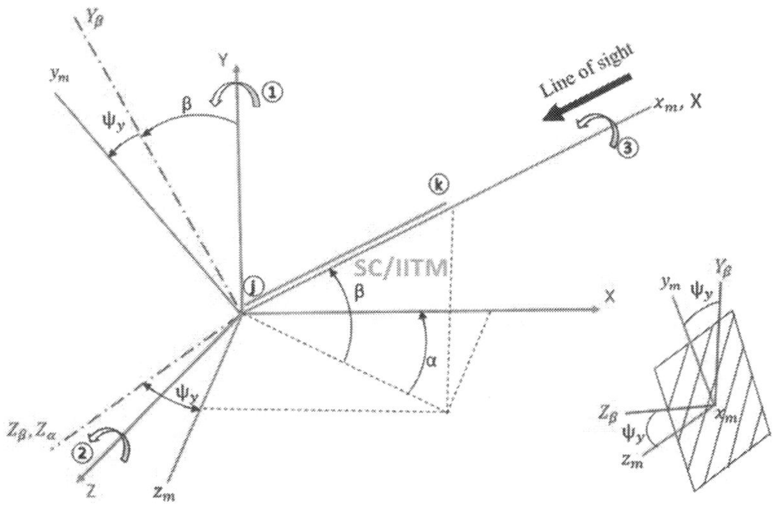

FIGURE 5.11 Y-Z-X transformation.

the cross section of the member seen from the line of sight. The angle ψ_y is measured anticlockwise when viewing the cross section of the member toward the jth end from the kth end. So, the direction cosines define the location of the x_m axis. ψ_y defines the location of minor principal axis. All parameters are geometrically dependent.

Similarly, for Z-Y-X transformation, as shown in Figure 5.12, the line of sight is along the x_m axis. The longitudinal axis of the member coincides with the x_m axis.

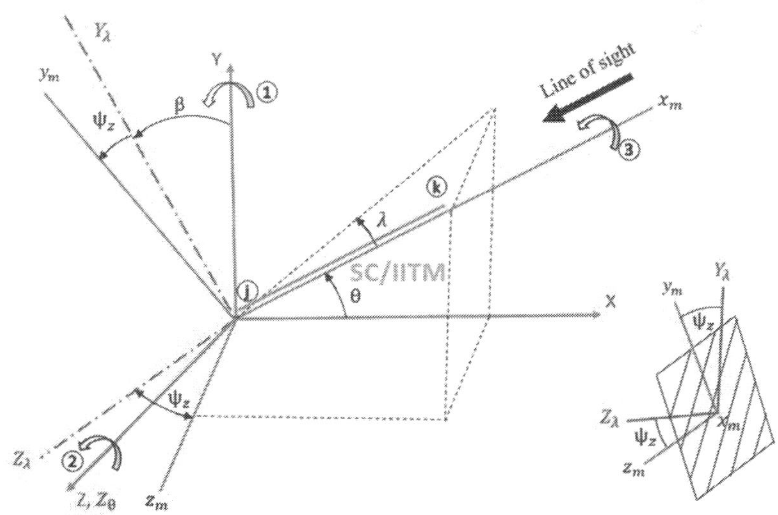

FIGURE 5.12 Z-Y-X transformation.

Here, ψ_z is measured anticlockwise when viewing the cross section of the member toward the jth end from the kth end.

5.9 ANALYSIS OF SPACE FRAME

For any member, which is arbitrarily oriented in space with respect to the reference axes system, one needs to estimate two parameters:

i. Direction cosines
ii. ψ angle

The important points to be noted in the analysis of space frame are as follows:

1. Three direction cosines define the location of the longitudinal axis of the member (x_m) axis with respect to the reference axes system.
2. ψ angle defines the location of minor principal axis.
3. All parameters of direction cosines and ψ angle are geometrically dependent. It actually depends on the position or orientation of the member with respect to the reference axes system.
4. Direction cosines of each member can be readily computed, but ψ angle needs to be carefully estimated.
5. One can do two types of transformations or rotations: Y-Z-X and Z-Y-X transformations. ψ angle computed from both the transformations will be different. Obviously, the rotation process enables alignment of X-Y-Z axes to x-y-z axes of the member.

The primary objective is to extend the knowledge of two-dimensional analysis to three-dimensional analysis. Consider a fixed beam element arbitrarily oriented with two nodes 'j' and 'k'. The member is designated as 'i' as shown in Figure 5.13. Each node will have three translational and three rotational degrees-of-freedom. The

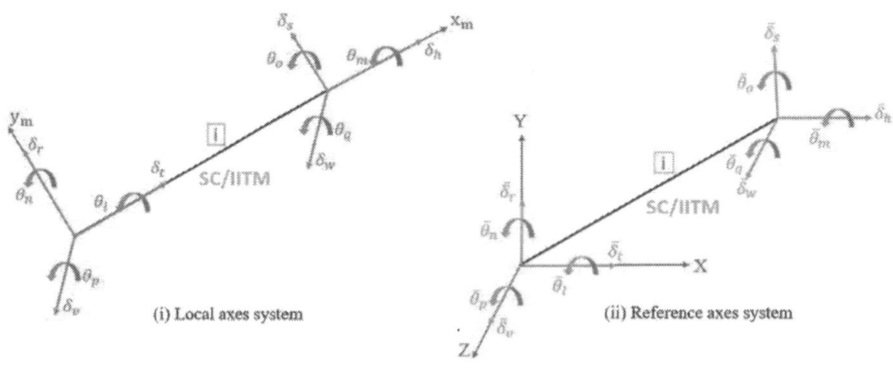

FIGURE 5.13 Local and reference axes system.

beam element will have twelve degrees-of-freedom, making the stiffness matrix of size 12×12, unlike two-dimensional analysis.

It is assumed that the orientation of the local axes system with respect to the reference axes system can be described in terms of the direction cosines C_x,C_y,C_z and ψ angle. We already know that,

$$C_x = \frac{X_k - X_j}{L_i}$$

$$C_y = \frac{Y_k - Y_j}{L_i}$$

$$C_z = \frac{Z_k - Z_j}{L_i}$$

$$L_i = \sqrt{\left(X_k - X_j\right)^2 + \left(Y_k - Y_j\right)^2 + \left(Z_k - Z_j\right)^2}$$

The set of translations and rotations at jth end measured in local axes system can be connected to the reference axes system as below:

$$\begin{Bmatrix} \delta_t \\ \delta_r \\ \delta_v \end{Bmatrix}_i = \begin{bmatrix} C_{11} & C_{12} & C_{13} \\ C_{21} & C_{22} & C_{23} \\ C_{31} & C_{32} & C_{33} \end{bmatrix} \begin{Bmatrix} \bar{\delta}_t \\ \bar{\delta}_r \\ \bar{\delta}_v \end{Bmatrix}_i$$

$$\begin{Bmatrix} \theta_l \\ \theta_n \\ \theta_p \end{Bmatrix}_i = \begin{bmatrix} C_{11} & C_{12} & C_{13} \\ C_{21} & C_{22} & C_{23} \\ C_{31} & C_{32} & C_{33} \end{bmatrix} \begin{Bmatrix} \bar{\theta}_l \\ \bar{\theta}_n \\ \bar{\theta}_p \end{Bmatrix}_i$$

Similarly, for the kth end,

$$\begin{Bmatrix} \delta_h \\ \delta_s \\ \delta_w \end{Bmatrix}_i = \begin{bmatrix} C_{11} & C_{12} & C_{13} \\ C_{21} & C_{22} & C_{23} \\ C_{31} & C_{32} & C_{33} \end{bmatrix} \begin{Bmatrix} \bar{\delta}_h \\ \bar{\delta}_s \\ \bar{\delta}_w \end{Bmatrix}_i$$

$$\begin{Bmatrix} \theta_m \\ \theta_o \\ \theta_q \end{Bmatrix}_i = \begin{bmatrix} C_{11} & C_{12} & C_{13} \\ C_{21} & C_{22} & C_{23} \\ C_{31} & C_{32} & C_{33} \end{bmatrix} \begin{Bmatrix} \bar{\theta}_m \\ \bar{\theta}_o \\ \bar{\theta}_q \end{Bmatrix}_i$$

By combining all twelve degrees of freedom, we get:

$$\begin{Bmatrix} \delta_t \\ \delta_r \\ \delta_v \\ \theta_l \\ \theta_n \\ \theta_p \\ \delta_h \\ \delta_s \\ \delta_w \\ \theta_m \\ \theta_o \\ \theta_q \end{Bmatrix} = \begin{bmatrix} C_{11} & C_{12} & C_{13} & 0 & 0 & 0 & 0 & 0 & 0 & 0 & 0 & 0 \\ C_{21} & C_{22} & C_{23} & 0 & 0 & 0 & 0 & 0 & 0 & 0 & 0 & 0 \\ C_{31} & C_{32} & C_{33} & 0 & 0 & 0 & 0 & 0 & 0 & 0 & 0 & 0 \\ 0 & 0 & 0 & C_{11} & C_{12} & C_{13} & 0 & 0 & 0 & 0 & 0 & 0 \\ 0 & 0 & 0 & C_{21} & C_{22} & C_{23} & 0 & 0 & 0 & 0 & 0 & 0 \\ 0 & 0 & 0 & C_{31} & C_{32} & C_{33} & 0 & 0 & 0 & 0 & 0 & 0 \\ 0 & 0 & 0 & 0 & 0 & 0 & C_{11} & C_{12} & C_{13} & 0 & 0 & 0 \\ 0 & 0 & 0 & 0 & 0 & 0 & C_{21} & C_{22} & C_{23} & 0 & 0 & 0 \\ 0 & 0 & 0 & 0 & 0 & 0 & C_{31} & C_{32} & C_{33} & 0 & 0 & 0 \\ 0 & 0 & 0 & 0 & 0 & 0 & 0 & 0 & 0 & C_{11} & C_{12} & C_{13} \\ 0 & 0 & 0 & 0 & 0 & 0 & 0 & 0 & 0 & C_{21} & C_{22} & C_{23} \\ 0 & 0 & 0 & 0 & 0 & 0 & 0 & 0 & 0 & C_{31} & C_{32} & C_{33} \end{bmatrix} \begin{Bmatrix} \bar{\delta}_t \\ \bar{\delta}_r \\ \bar{\delta}_v \\ \bar{\theta}_l \\ \bar{\theta}_n \\ \bar{\theta}_p \\ \bar{\delta}_h \\ \bar{\delta}_s \\ \bar{\delta}_w \\ \bar{\theta}_m \\ \bar{\theta}_o \\ \bar{\theta}_q \end{Bmatrix}$$

$$\text{Thus, } \{\delta^{(s)}\}_i = [T_i^{(s)}]\{\bar{\delta}^{(s)}\}_i \tag{5.21}$$

where,

$$[T_i] = \begin{bmatrix} [C_i] & [0] & [0] & [0] \\ [0] & [C_i] & [0] & [0] \\ [0] & [0] & [C_i] & [0] \\ [0] & [0] & [0] & [C_i] \end{bmatrix}_{12\times12}$$

This is the transformation matrix in three-dimensional space, which is orthogonal. Hence, the following condition holds good:

$$\left[T^{(s)}\right]_i^T = \left[T^{(s)}\right]_i^{-1} \tag{5.22}$$

$$\left[C_i\right]^T = \left[C_i\right]^{-1} \tag{5.23}$$

Therefore,

$$\left[\bar{\delta}_i\right] = \left[T\right]_i^T \{\delta\}_i \tag{5.24}$$

Where [C] is the rotation matrix, whose elements are the direction cosines. Thus, C_y can be used for Y-Z-X transformation and C_z for Z-Y-X transformation, which contains the direction cosines and ψ angle. In addition, the following equations used in two-dimensional analysis are also valid for three-dimensional analysis.

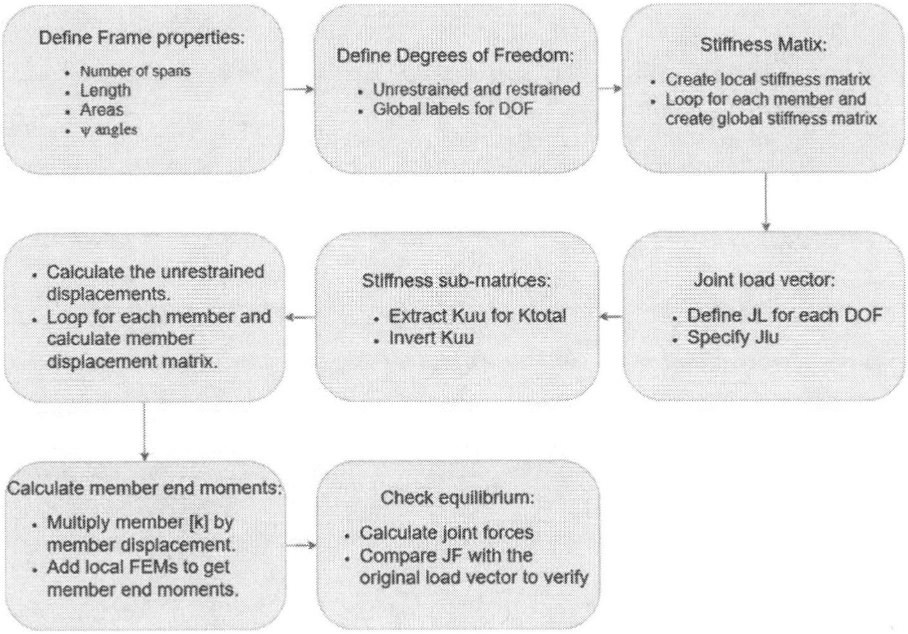

FIGURE 5.14 Steps to analyze a space frame using MATLAB.

$$
\begin{gathered}
\{m\}_i = [T]_i \{\bar{m}\}_i \\
[\bar{K}]_i = \left[T^{(s)}\right]_i^T \left[K^{(s)}\right]_i \left[T^{(s)}\right]_i \\
[K_{uu}]\{\Delta_u\} = \{J_L\}_u \\
\{M\}_i = [K]_i [T]\{\bar{\delta}\} + \{FEM\}_i
\end{gathered}
\tag{5.25}
$$

5.10 STEP-BY-STEP FRAMEWORK OF ANALYSIS OF SPACE FRAME

The step-by-step framework to analyze a space frame using MATLAB is given in Figure 5.14.

5.11 EXERCISE PROBLEMS SOLVED USING MATLAB PROGRAM

Example 5.1:

Determine the ψ angle for the members of space frame shown below:

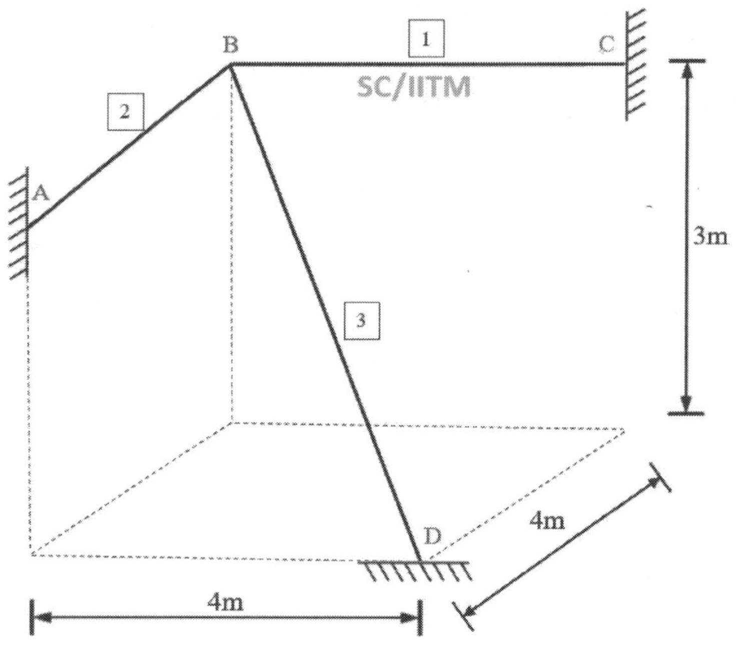

SOLUTION

Mark the global axis and local axes for all the members.

Member	j	k	Length (m)
1	1	2	4
2	3	2	4
3	2	4	6.402

Member 1: x_m axis of member 1 is toward Y-axis. Therefore, we should use Z-Y-X transformation.

Here, ψ angle = ψ_z = angle between y_m and Y-axis = 90 degrees

Member 2: x_m axis of member 2 is toward X-axis. Therefore, we can use either Y-Z-X or Z-Y-X transformation.

Let us use Y-Z-X transformation.

Here, ψ angle = ψ_y = angle between y_m and Y-axis = 90 degrees

Member 3: x_m axis of member 2 is neither aligned with X, Y, nor Z, but it is inclined. Let us try Y-Z-X transformation. To understand this transformation, let us consider a simple example. Consider a vector that is placed arbitrarily in the space as shown in Figure 5.15. Let us consider a point 'p' in x_m-y_m plane in the frame of X-Y-Z, but the point should not be located along the x_m axis. The point is projected on other planes. The coordinates of this point will describe the position of $\overline{X}_p$, $\overline{Y}_p$, $\overline{Z}_p$ with respect to the reference axes.

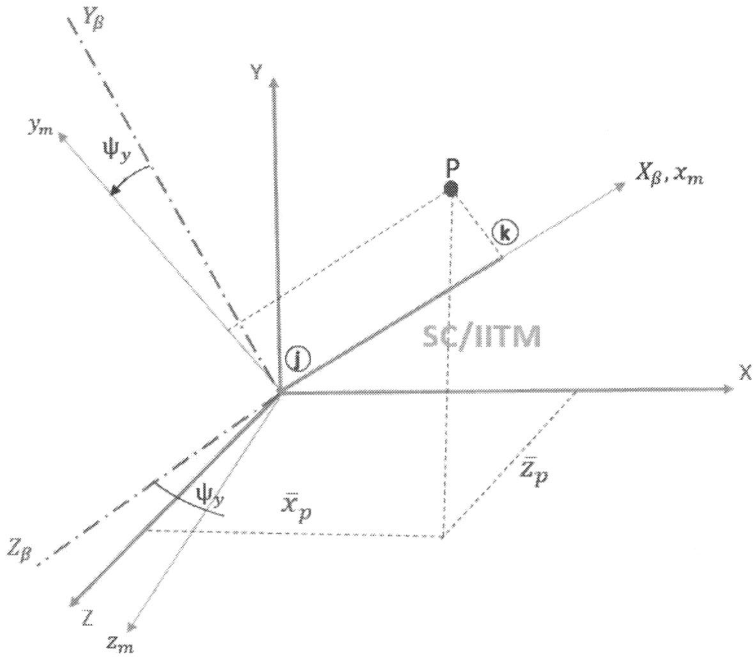

FIGURE 5.15 Transformation.

For Y-Z-X transformation, we already know that,

$$\begin{Bmatrix} X_{\beta p} \\ Y_{\beta p} \\ Z_{\beta p} \end{Bmatrix} = \begin{bmatrix} \sqrt{C_x^2+C_z^2} & C_y & 0 \\ -C_y & \sqrt{C_x^2+C_z^2} & 0 \\ 0 & 0 & 1 \end{bmatrix} \begin{bmatrix} \dfrac{C_x}{\sqrt{C_x^2+C_z^2}} & 0 & \dfrac{C_z}{\sqrt{C_x^2+C_z^2}} \\ 0 & 1 & 0 \\ -\dfrac{C_z}{\sqrt{C_x^2+C_z^2}} & 0 & \dfrac{C_x}{\sqrt{C_x^2+C_z^2}} \end{bmatrix} \begin{Bmatrix} \bar{X}_p \\ \bar{Y}_p \\ \bar{Z}_p \end{Bmatrix}$$

$$\begin{Bmatrix} X_{\beta p} \\ Y_{\beta p} \\ Z_{\beta p} \end{Bmatrix} = \begin{bmatrix} C_x & C_y & C_z \\ \dfrac{-C_x C_y}{\sqrt{C_x^2+C_z^2}} & \sqrt{C_x^2+C_z^2} & \dfrac{-C_y C_z}{\sqrt{C_x^2+C_z^2}} \\ -\dfrac{C_z}{\sqrt{C_x^2+C_z^2}} & 0 & \dfrac{C_x}{\sqrt{C_x^2+C_z^2}} \end{bmatrix} \begin{Bmatrix} \bar{X}_p \\ \bar{Y}_p \\ \bar{Z}_p \end{Bmatrix}$$

The coordinates of the point 'p' on the (x_m,y_m) plane can be written as $Y_{\beta p}$ and $Z_{\beta p}$.

Thus,

$$\sin \psi_y = \frac{Z_{\beta p}}{\sqrt{Z_{\beta p}^2 + Y_{\beta p}^2}}$$

$$\cos \psi_y = \frac{Y_{\beta p}}{\sqrt{Z_{\beta p}^2 + Y_{\beta p}^2}}$$

So, the coordinates of point 'p' with respect to the reference axes system should be P(4,4,0). Similarly, the coordinates of point 'p' with respect to jth end of the member of 3 are P(0,0,–3). Thus, the position vector is given by,

$$\begin{Bmatrix} \overline{X}_p \\ \overline{Y}_p \\ \overline{Z}_p \end{Bmatrix} = \begin{Bmatrix} 0 \\ 0 \\ -3 \end{Bmatrix}$$

Now, the coordinates of the jth end with respect to X-Y-Z axes system are (4,4,3). Similarly, the coordinates of kth end with respect to X-Y-Z system are (0,0,0).

i. *Direction cosines:*

$$C_x = \frac{X_k - X_j}{L_i} = \frac{0-4}{6.402} = -0.625$$

$$C_y = \frac{Y_k - Y_j}{L_i} = \frac{0-4}{6.402} = -0.625$$

$$C_z = \frac{Z_k - Z_j}{L_i} = \frac{0-3}{6.402} = -0.469$$

ii. *ψ angle:*
Now,

$$X_{\beta p} = C_x\overline{X}_p + C_y\overline{Y}_p + C_z\overline{Z}_p = 0 + 0 + (0.469 \times 3) = 1.407$$

$$Y_{\beta p} = \frac{-C_xC_y}{\sqrt{C_x^2 + C_z^2}}\overline{X}_p + \sqrt{C_x^2 + C_z^2}\,\overline{Y}_p - \frac{C_yC_z}{\sqrt{C_x^2 + C_z^2}}\overline{Z}_p$$

$$= 0 + 0 - \frac{-(-0.625)(-0.469)(-3)}{\sqrt{0.625^2 + 0.429^2}} = +1.125$$

$$Z_{\beta p} = -\frac{C_z}{\sqrt{C_x^2 + C_z^2}}\bar{X}_p + \frac{C_x}{\sqrt{C_x^2 + C_z^2}}\bar{Z}_p = 0 + \frac{(-0.625)(-3)}{\sqrt{0.625^2 + 0.429^2}}$$

$$= +2.40 sin\psi_y = \frac{Z_{\beta p}}{\sqrt{Z_{\beta p}^2 + Y_{\beta p}^2}} = 0.905$$

$$cos\psi_y = \frac{Y_{\beta p}}{\sqrt{Z_{\beta p}^2 + Y_{\beta p}^2}} = 0.424$$

Thus,

$$\psi_y = \tan^{-1}\left\{\frac{0.905}{0.424}\right\} = 64.897° + 180° = 244.897°$$

Example 5.2:

Determine the ψ angle for the members of space frame shown in Figure 5.16.

SOLUTION

Mark the global axis and local axes for all the members.

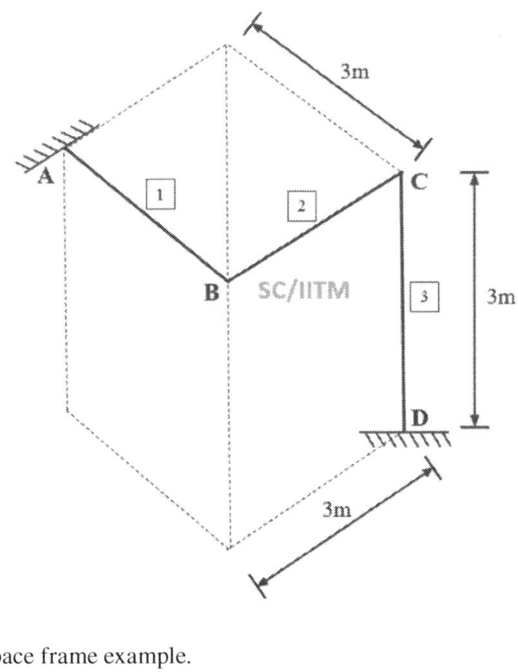

FIGURE 5.16 Space frame example.

i. *Coordinates of joints:*

Joint	X	Y	Z
A	0	0	0
B	3	0	0
C	3	0	–3
D	3	–3	–3

ii. *ψ angle:*

Member	Length (m)	Joint		Direction Cosines			Type of Transformation	ψ angle (Degrees)
		j	k	C_x	C_y	C_z		
AB	3	A	B	1	0	0	Y-Z-X	$\psi_y = 0$
BC	3	B	C	0	0	–1	Y-Z-X	$\psi_y = 0$
CD	3	D	C	0	1	0	Z-Y-X	$\psi_z = 90$

Example 5.3:

Analyze the three-dimensional space frame as shown in Figure 5.17 using the stiffness method of analysis.

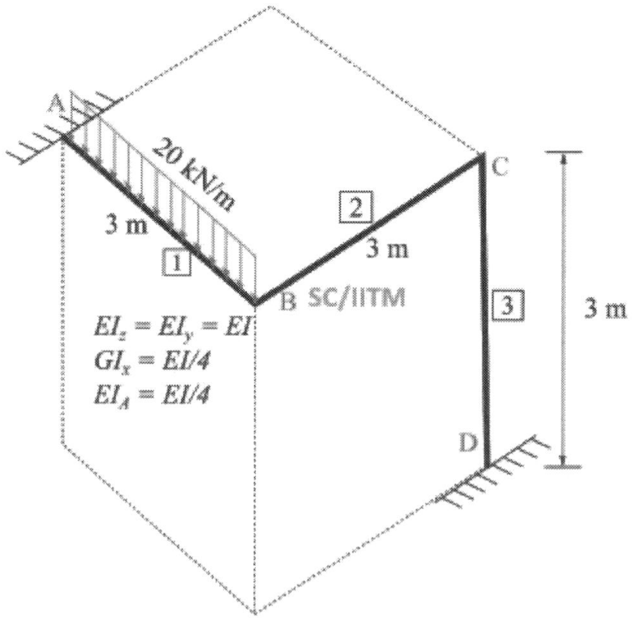

FIGURE 5.17 Space frame example.

SOLUTION

Mark the local and reference axes system.

i. *Coordinates of joints:*

Joint	X	Y	Z
A	0	0	0
B	3	0	0
C	3	0	–3
D	3	–3	–3

ii. ψ *angle:*

Member	Length (m)	Joint j	Joint K	Direction Cosines C_x	C_y	C_z	Type of Transformation	ψ angle (Degrees)
AB	3	A	B	1	0	0	Y-Z-X	$\psi_y = 0$
BC	3	B	C	0	0	–1	Y-Z-X	$\psi_y = 0$
CD	3	D	C	0	1	0	Z-Y-X	$\psi_z = 90$

iii. *Marking unrestrained and restrained degrees-of-freedom:*
The unrestrained and restrained degrees-of-freedom are shown in Figure 5.18.

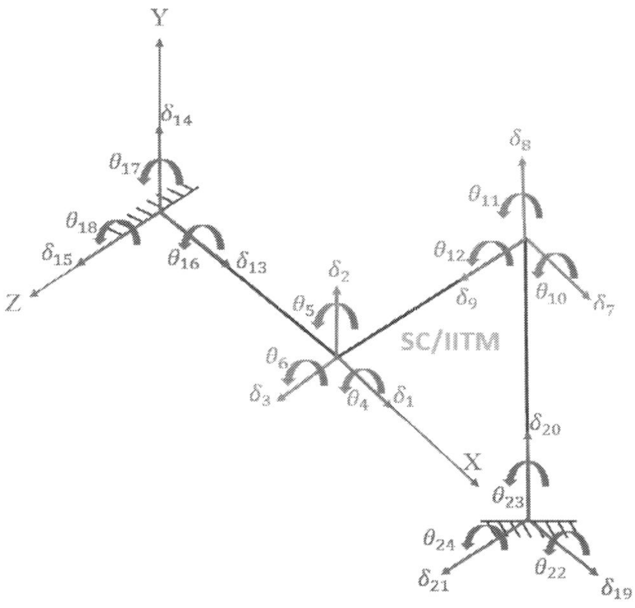

FIGURE 5.18 Unrestrained and restrained degrees-of-freedom.

Unrestrained degrees-of-freedom: 12
$(\delta_1, \delta_2, \delta_3, \theta_4, \theta_5, \theta_6, \delta_7, \delta_8, \delta_9, \theta_{10}, \theta_{11}, \theta_{12})$
Restrained degrees-of-freedom: 12
$(\delta_{13}, \delta_{14}, \delta_{15}, \theta_{16}, \theta_{17}, \theta_{18}, \delta_{19}, \delta_{20}, \delta_{21}, \theta_{22}, \theta_{23}, \theta_{24})$

iv. *Estimation of transformation matrix:*
We already know that,

$$C = \begin{bmatrix} C_{11} & C_{12} & C_{13} \\ C_{21} & C_{22} & C_{23} \\ C_{31} & C_{32} & C_{33} \end{bmatrix}$$

The direction cosine matrices for all the three members are,

$$C_1 = \begin{bmatrix} 1 & 0 & 0 \\ 0 & 1 & 0 \\ 0 & 0 & 1 \end{bmatrix}, C_2 = \begin{bmatrix} 0 & 0 & -1 \\ 0 & 1 & 0 \\ 1 & 0 & 0 \end{bmatrix}, C_3 = \begin{bmatrix} 0 & 1 & 0 \\ 0 & 0 & 1 \\ 1 & 0 & 0 \end{bmatrix}$$

Now, the transformation matrix is given by,

$$[T_i] = \begin{bmatrix} [C_i] & [0] & [0] & [0] \\ [0] & [C_i] & [0] & [0] \\ [0] & [0] & [C_i] & [0] \\ [0] & [0] & [0] & [C_i] \end{bmatrix}_{12\times12}$$

Global labels:

$$\text{AB} = [13, 14, 15, 16, 17, 18, 1, 2, 3, 4, 5, 6]$$

$$\text{BC} = [1, 2, 3, 4, 5, 6, 7, 8, 9, 10, 11, 12]$$

$$\text{CD} = [19, 20, 21, 22, 23, 24, 7, 8, 9, 10, 11, 12]$$

v. *Fixed end moments and joint load vector:*
Load is acting only on the member AB. Hence, the fixed end moments along other two members remain zero.

$$(FEM)_{AB} = \begin{Bmatrix} 0 \\ 30 \\ 0 \\ 0 \\ 0 \\ 15 \\ 0 \\ 30 \\ 0 \\ 0 \\ 0 \\ -15 \end{Bmatrix}$$

The joint load vector is the reversal of the fixed end moments. The transpose of the joint load vector is given below:

$$[J_L]^T = \{0 \;\; -30 \;\; 0 \;\; 0 \;\; 0 \;\; 15 \;\; 0 \;\; 0 \;\; 0 \;\; 0 \;\; 0 \;\; 0 \;\; -30 \;\; 0 \;\; 0 \;\; 0 \;\; -15 \;\; 0 \;\; 0 \;\; 0 \;\; 0 \;\; 0 \;\; 0\}$$

vi. *Stiffness matrix:*

$$[K_{AB}] = EI\begin{bmatrix} 0.083 & 0 & 0 & 0 & 0 & 0 & -0.083 & 0 & 0 & 0 & 0 & 0 \\ 0 & 0.444 & 0 & 0 & 0 & 0.667 & 0 & -0.444 & 0 & 0 & 0 & 0.667 \\ 0 & 0 & 0.444 & 0 & -0.667 & 0 & 0 & 0 & 0.444 & 0 & 0.667 & 0 \\ 0 & 0 & 0 & 0.083 & 0 & 0 & 0 & 0 & 0 & -0.083 & 0 & 0 \\ 0 & 0 & -0.667 & 0 & 1.333 & 0 & 0 & 0 & 0.667 & 0 & 0.667 & 0 \\ 0 & 0.667 & 0 & 0 & 0 & 1.333 & 0 & -0.667 & 0 & 0 & 0 & 0.667 \\ -0.083 & 0 & 0 & 0 & 0 & 0 & -0.083 & 0 & 0 & 0 & 0 & 0 \\ 0 & -0.444 & 0 & 0 & 0 & -0.667 & 0 & 0.444 & 0 & 0 & 0 & -0.667 \\ 0 & 0 & -0.444 & 0 & 0.667 & 0 & 0 & 0 & 0.444 & 0 & 0.667 & 0 \\ 0 & 0 & 0 & -0.083 & 0 & 0 & 0 & 0 & 0 & 0.083 & 0 & 0 \\ 0 & 0 & -0.667 & 0 & 0.667 & 0 & 0 & 0 & 0.667 & 0 & 1.333 & 0 \\ 0 & 0.667 & 0 & 0 & 0 & 0.667 & 0 & -0.667 & 0 & 0 & 0 & 1.333 \end{bmatrix}$$

$$[\bar{K}_{AB}] = EI\begin{bmatrix} 0.083 & 0 & 0 & 0 & 0 & 0 & -0.083 & 0 & 0 & 0 & 0 & 0 \\ 0 & 0.444 & 0 & 0 & 0 & 0.667 & 0 & -0.444 & 0 & 0 & 0 & 0.667 \\ 0 & 0 & 0.444 & 0 & -0.667 & 0 & 0 & 0 & 0.444 & 0 & 0.667 & 0 \\ 0 & 0 & 0 & 0.083 & 0 & 0 & 0 & 0 & 0 & -0.083 & 0 & 0 \\ 0 & 0 & -0.667 & 0 & 1.333 & 0 & 0 & 0 & 0.667 & 0 & 0.667 & 0 \\ 0 & 0.667 & 0 & 0 & 0 & 1.333 & 0 & -0.667 & 0 & 0 & 0 & 0.667 \\ -0.083 & 0 & 0 & 0 & 0 & 0 & -0.083 & 0 & 0 & 0 & 0 & 0 \\ 0 & -0.444 & 0 & 0 & 0 & -0.667 & 0 & 0.444 & 0 & 0 & 0 & -0.667 \\ 0 & 0 & -0.444 & 0 & 0.667 & 0 & 0 & 0 & 0.444 & 0 & 0.667 & 0 \\ 0 & 0 & 0 & -0.083 & 0 & 0 & 0 & 0 & 0 & 0.083 & 0 & 0 \\ 0 & 0 & -0.667 & 0 & 0.667 & 0 & 0 & 0 & 0.667 & 0 & 1.333 & 0 \\ 0 & 0.667 & 0 & 0 & 0 & 0.667 & 0 & -0.667 & 0 & 0 & 0 & 1.333 \end{bmatrix}$$

$$[K_{BC}] = [K_{AB}]$$

$$[\bar{K}_{BC}] = EI\begin{bmatrix} 0.444 & 0 & 0 & 0 & -0.667 & 0 & -0.444 & 0 & 0 & 0 & -0.667 & 0 \\ 0 & 0.444 & 0 & 0.667 & 0 & 0 & 0 & -0.444 & 0 & 0.667 & 0 & 0 \\ 0 & 0 & 0.083 & 0 & 0 & 0 & 0 & 0 & -0.083 & 0 & 0 & 0 \\ 0 & 0.667 & 0 & 1.333 & 0 & 0 & 0 & -0.667 & 0 & 0.667 & 0 & 0 \\ -0.667 & 0 & 0 & 0 & 1.333 & 0 & 0.667 & 0 & 0 & 0 & 0.667 & 0 \\ 0 & 0 & 0 & 0 & 0 & 0.083 & 0 & 0 & 0 & 0 & 0 & -0.083 \\ -0.444 & 0 & 0 & 0 & 0.667 & 0 & 0.444 & 0 & 0 & 0 & 0.667 & 0 \\ 0 & -0.444 & 0 & -0.667 & 0 & 0 & 0 & 0.444 & 0 & -0.667 & 0 & 0 \\ 0 & 0 & -0.083 & 0 & 0 & 0 & 0 & 0 & 0.083 & 0 & 0 & 0 \\ 0 & 0.667 & 0 & 0.667 & 0 & 0 & 0 & -0.667 & 0 & 1.333 & 0 & 0 \\ -0.667 & 0 & 0 & 0 & 0.667 & 0 & 0.667 & 0 & 0 & 0 & 1.333 & 0 \\ 0 & 0 & 0 & 0 & 0 & -0.083 & 0 & 0 & 0 & 0 & 0 & 0.083 \end{bmatrix}$$

$$\left[K_{CD}\right]=\left[K_{AB}\right]$$

$$\left[\bar{K}_{CD}\right]=EI$$

$$\begin{bmatrix}
0.444 & 0 & 0 & 0 & 0 & -0.667 & -0.444 & 0 & 0 & 0 & 0 & -0.667 \\
0 & 0.083 & 0 & 0 & 0 & 0 & 0 & -0.083 & 0 & 0 & 0 & 0 \\
0 & 0 & 0.444 & 0.667 & 0 & 0 & 0 & 0 & -0.444 & 0.667 & 0 & 0 \\
0 & 0 & 0.667 & 1.333 & 0 & 0 & 0 & 0 & -0.667 & 0.667 & 0 & 0 \\
0 & 0 & 0 & 0 & 0.083 & 0 & 0 & 0 & 0 & 0 & -0.083 & 0 \\
-0.667 & 0 & 0 & 0 & 0 & 1.333 & 0.667 & 0 & 0 & 0 & 0 & 0.667 \\
-0.444 & 0 & 0 & 0 & 0 & 0.667 & 0.444 & 0 & 0 & 0 & 0 & 0.667 \\
0 & -0.083 & 0 & 0 & 0 & 0 & 0 & 0.083 & 0 & 0 & 0 & 0 \\
0 & 0 & -0.444 & -0.667 & 0 & 0 & 0 & 0 & 0.444 & -0.667 & 0 & 0 \\
0 & 0 & 0.667 & 0.667 & 0 & 0 & 0 & 0 & -0.667 & 1.333 & 0 & 0 \\
0 & 0 & 0 & 0 & -0.083 & 0 & 0 & 0 & 0 & 0 & 0.083 & 0 \\
-0.667 & 0 & 0 & 0 & 0 & 0.667 & 0.667 & 0 & 0 & 0 & 0 & 1.333
\end{bmatrix}$$

The total stiffness matrix can be formed by assembling the global stiffness matrices of all the members, from which the unrestrained stiffness matrix can be partitioned.

$$\left[\boldsymbol{K_{uu}}\right]=\boldsymbol{EI}$$

$$\begin{bmatrix}
0.528 & 0 & 0 & 0 & -0.667 & 0 & -0.444 & 0 & 0 & 0 & -0.667 & 0 \\
0 & 0.889 & 0 & 0.667 & 0 & -0.667 & 0 & -0.444 & 0 & 0.667 & 0 & 0 \\
0 & 0 & 0.528 & 0 & 0.667 & 0 & 0 & 0 & -0.083 & 0 & 0 & 0 \\
0 & 0.667 & 0 & 1.417 & 0 & 0 & 0 & -0.667 & 0 & 0.667 & 0 & 0 \\
-0.667 & 0 & 0.667 & 0 & 2.667 & 0 & 0.667 & 0 & 0 & 0 & 0.667 & 0 \\
0 & -0.667 & 0 & 0 & 0 & 1.417 & 0 & 0 & 0 & 0 & 0 & -0.833 \\
-0.444 & 0 & 0 & 0 & 0.667 & 0 & 0.889 & 0 & 0 & 0 & 0.667 & 0.667 \\
0 & -0.444 & 0 & -0.667 & 0 & 0 & 0 & 0.528 & 0 & -0.667 & 0 & 0 \\
0 & 0 & -0.083 & 0 & 0 & 0 & 0 & 0 & 0.528 & -0.667 & 0 & 0 \\
0 & 0.667 & 0 & 0.667 & 0 & 0 & 0 & -0.667 & -0.667 & 2.667 & 0 & 0 \\
-0.667 & 0 & 0 & 0 & 0.667 & 0 & 0.667 & 0 & 0 & 0 & 1.417 & 0 \\
0 & 0 & 0 & 0 & 0 & -0.083 & 0.667 & 0 & 0 & 0 & 0 & 1.417
\end{bmatrix}$$

vii. *Calculation of end moments and reactions:*

$$\{M_1\}=\begin{Bmatrix} -0.045 \\ 55.725 \\ -1.543 \\ -2.820 \\ 4.346 \\ 73.232 \\ 0.045 \\ 4.275 \\ 1.543 \\ 2.820 \\ 0.284 \\ 3.943 \end{Bmatrix},\{M_2\}=\begin{Bmatrix} -0.045 \\ -4.275 \\ -1.543 \\ -2.820 \\ -0.284 \\ -3.943 \\ 0.045 \\ 4.275 \\ 1.543 \\ -10.004 \\ 0.420 \\ 3.943 \end{Bmatrix},\{M_3\}=\begin{Bmatrix} 0.045 \\ 4.275 \\ 1.543 \\ -5.374 \\ 0.420 \\ 3.807 \\ -0.045 \\ -4.275 \\ -1.543 \\ 10.004 \\ -0.420 \\ -3.943 \end{Bmatrix}$$

The member end reactions and moments are shown below:

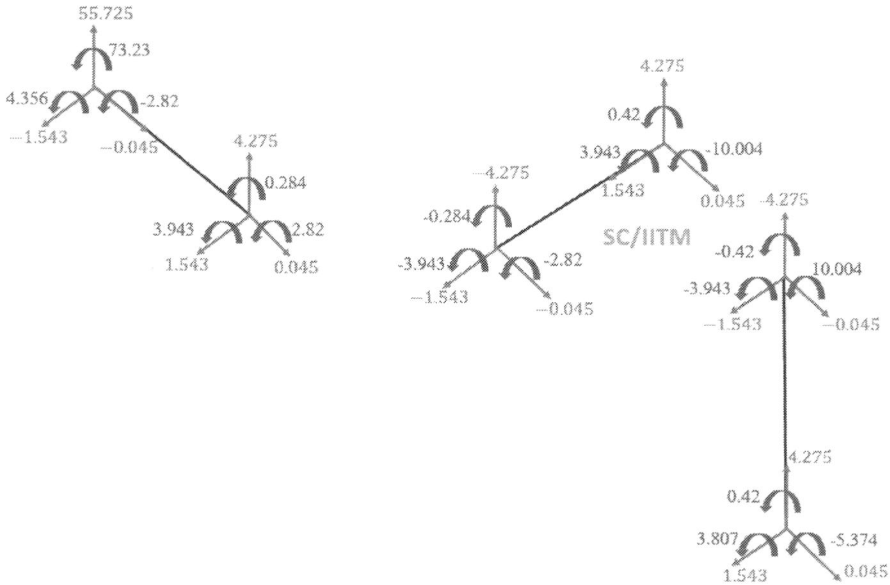

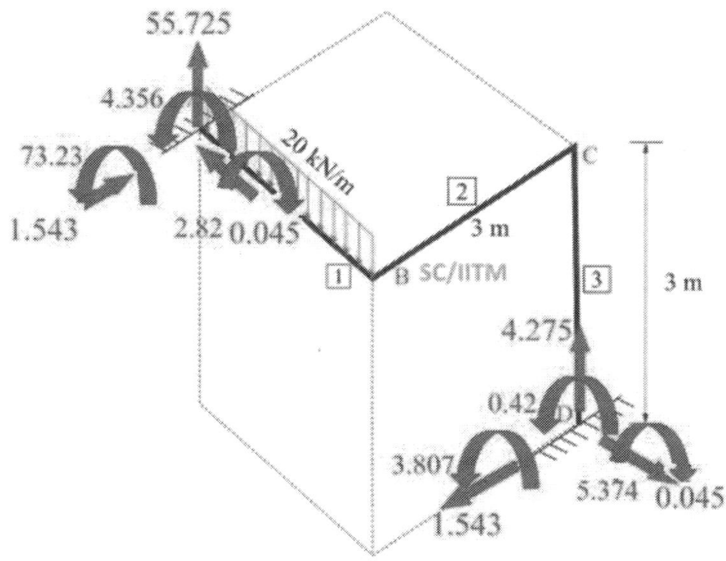

MATLAB program:

```
%% 3D analysis of space frame
clc;
clear;
%% Input
n = 3; % number of members
EI = [1 1 1]; %Flexural rigidity
EIy = EI;
EIz = EI;
GI = [0.25 0.25 0.25].*EI; %Torsional constant
EA = [0.25 0.25 0.25].*EI; %Axial rigidity
L = [3 3 3]; % length in m
nj = n+1; % Number of Joints
codm = [0 0 0; 3 0 0; 3 0 -3; 3 -3 -3]; %Coordinate wrt X,Y.Z: size=nj,3
dc = [1 0 0; 0 0 -1; 0 1 0]; % Direction cosines for each member
tytr = [1 1 2]; % Type of transformation fo each member
psi = [0 0 90]; % Psi angle in degrees for each member
% C matrix
c1 = [1 0 0; 0 1 0; 0 0 1]; % C matrix for member 1
c2 = [0 0 -1; 0 1 0; 1 0 0]; % C matrix for member 2
c3 = [0 1 0; 0 0 1; 1 0 0]; % C matrix for member 3
uu = 12; % Number of unrestrained Degrees-of-freedom
ur = 12; % Number of restrained Degrees-of-freedom
uul = [1 2 3 4 5 6 7 8 9 10 11 12]; % global labels of unrestrained dof
url = [13 14 15 16 17 18 19 20 21 22 23 24]; % global labels of
restrained dof
l1 = [13 14 15 16 17 18 1 2 3 4 5 6]; % Global labels for member 1
l2 = [1 2 3 4 5 6 7 8 9 10 11 12]; % Global labels for member 2
l3 = [19 20 21 22 23 24 7 8 9 10 11 12]; % Global labels for member 3
l= [l1; l2; l3];
dof = uu + ur; % Degrees-of-freedom
Ktotal = zeros (dof);
fem1= [0; 30; 0; 0; 0; 15; 0; 30; 0; 0; 0; -15]; % Local Fixed
end moments of member 1
fem2= [0; 0; 0; 0; 0; 0; 0; 0; 0; 0; 0; 0]; % Local Fixed end
moments of member 2
fem3= [0; 0; 0; 0; 0; 0; 0; 0; 0; 0; 0; 0]; % Local Fixed end
moments of member 3
%% Transformation matrix
T1 = zeros(12);
T2 = zeros(12);
T3 = zeros(12);
for i = 1:3
    for j = 1:3
        T1(i,j)=c1(i,j);
        T1(i+3,j+3)=c1(i,j);
        T1(i+6,j+6)=c1(i,j);
        T1(i+9,j+9)=c1(i,j);
        T2(i,j)=c2(i,j);
```

```
        T2(i+3,j+3)=c2(i,j);
        T2(i+6,j+6)=c2(i,j);
        T2(i+9,j+9)=c2(i,j);
        T3(i,j)=c3(i,j);
        T3(i+3,j+3)=c3(i,j);
        T3(i+6,j+6)=c3(i,j);
        T3(i+9,j+9)=c3(i,j);
    end
end
%% Getting Type of transformation and Psi angle
for i = 1:n
    if tytr(i) ==1
        fprintf ('Member Number =');
        disp (i);
        fprintf ('Type of transformation is Y-Z-X \n');
    else
        fprintf ('Member Number =');
        disp (i);
        fprintf ('Type of transformation is Z-Y-X \n');
    end
    fprintf ('Psi angle=');
    disp (psi(i));
end
%% Stiffness coefficients for each member
sc1 = EA./L;
sc2 = 6*EIz./(L.^2);
sc3 = 6*EIy./(L.^2);
sc4 = GI./L;
sc5 = 2*EIy./L;
sc6 = 12*EIz./(L.^3);
sc7 = 12*EIy./(L.^3);
sc8 = 2*EIz./L;
%% stiffness matrix 6 by 6
for i = 1:n
        Knew = zeros (dof);
        k1 = [sc1(i); 0; 0; 0; 0; 0; -sc1(i); 0; 0; 0; 0; 0];
        k2 = [0; sc6(i); 0; 0; 0; sc2(i); 0; -sc6(i); 0; 0; 0; sc2(i)];
        k3 = [0; 0; sc7(i); 0; -sc3(i); 0; 0; 0; -sc7(i); 0; -sc3(i); 0];
        k4 = [0; 0; 0; sc4(i); 0; 0; 0; 0; 0; -sc4(i); 0; 0];
        k5 = [0; 0; -sc3(i); 0; (2*sc5(i)); 0; 0; 0; sc3(i); 0;
sc5(i); 0];
        k6 = [0; sc2(i); 0; 0; 0; (2*sc8(i)); 0; -sc2(i); 0;
0; 0; sc8(i)];
        k7 = -k1;
        k8 = -k2;
        k9 = -k3;
        k10 = -k4;
        k11 = [0; 0; -sc3(i); 0; sc5(i); 0; 0; 0; sc3(i); 0;
(2*sc5(i)); 0];
```

```
        k12 = [0; sc2(i); 0; 0; 0; sc8(i); 0; -sc2(i); 0; 0; 0;
(2*sc8(i))];
        K = [k1 k2 k3 k4 k5 k6 k7 k8 k9 k10 k11 k12];
        fprintf ('Member Number =');
        disp (i);
        fprintf ('Local Stiffness matrix of member, [K] = \n');
        disp (K);
        if i == 1
            T = T1;
        elseif i == 2
            T = T2;
        else
            T = T3;
        end
        Ttr = T';
        Kg = Ttr*K*T;
        fprintf ('Transformation matrix, [T] = \n');
        disp (T);
        fprintf ('Global Matrix, [K global] = \n');
        disp (Kg);

        for p = 1:12
            for q = 1:12
                Knew((l(i,p)),(l(i,q))) =Kg(p,q);
            end
        end
        Ktotal = Ktotal + Knew;
        if i == 1
            Tt1= T;
            Kg1=Kg;
            fembar1= Tt1'*fem1;
        elseif i == 2
            Tt2 = T;
            Kg2 = Kg;
            fembar2= Tt2'*fem2;
        else
            Tt3 = T;
            Kg3 = Kg;
            fembar3= Tt3'*fem3;
        end
end
fprintf ('Stiffness Matrix of complete structure, [Ktotal] = \n');
disp (Ktotal);
Kunr = zeros(12);
for x=1:uu
    for y=1:uu
        Kunr(x,y)= Ktotal(x,y);
    end
end
fprintf ('Unrestrained Stiffness sub-matix, [Kuu] = \n');
```

```
disp (Kunr);
KuuInv= inv(Kunr);
fprintf ('Inverse of Unrestrained Stiffness sub-matix, [KuuInverse] = \n');
disp (KuuInv);
%% Creation of joint load vector
jl= [0; -30; 0; 0; 0; 15; 0; 0; 0; 0; 0; 0; 0; -30; 0; 0; 0; -15;
0; 0; 0; 0; 0; 0]; % values given in kN or kNm
jlu = jl(1:12,1); % load vector in unrestrained dof
delu = KuuInv*jlu;
fprintf ('Joint Load vector, [Jl] = \n');
disp (jl);
fprintf ('Unrestrained displacements, [DelU] = \n');
disp (delu);
delr = [0; 0; 0; 0; 0; 0; 0; 0; 0; 0; 0; 0];
del = zeros (dof,1);
del = [delu; delr];
deli= zeros (12,1);
for i = 1:n
    for p = 1:12
        deli(p,1) = del((l(i,p)),1) ;
    end
    if i == 1
            delbar1 = deli;
            mbar1= (Kg1 * delbar1)+fembar1;
            fprintf ('Member Number =');
            disp (i);
            fprintf ('Global displacement matrix [DeltaBar] = \n');
            disp (delbar1);
            fprintf ('Global End moment matrix [MBar] = \n');
            disp (mbar1);
     elseif i == 2
            delbar2 = deli;
            mbar2= (Kg2 * delbar2)+fembar2;
            fprintf ('Member Number =');
            disp (i);
            fprintf ('Global displacement matrix [DeltaBar] = \n');
            disp (delbar2);
            fprintf ('Global End moment matrix [MBar] = \n');
            disp (mbar2);
    else
            delbar3 = deli;
            mbar3= (Kg3 * delbar3)+fembar3;
            fprintf ('Member Number =');
            disp (i);
            fprintf ('Global displacement matrix [DeltaBar] = \n');
            disp (delbar3);
            fprintf ('Global End moment matrix [MBar] = \n');
            disp (mbar3);
     end
end
```

```
%% check
mbar = [mbar1'; mbar2'; mbar3'];
jf = zeros(dof,1);
for a=1:n
    for b=1:12 % size of k matrix
        d = l(a,b);
        jfnew = zeros(dof,1);
        jfnew(d,1)=mbar(a,b);
        jf=jf+jfnew;
    end
end
fprintf ('Joint forces = \n');
disp (jf);
```

Example 5.4:

Analyze the three-dimensional space frame as shown below:

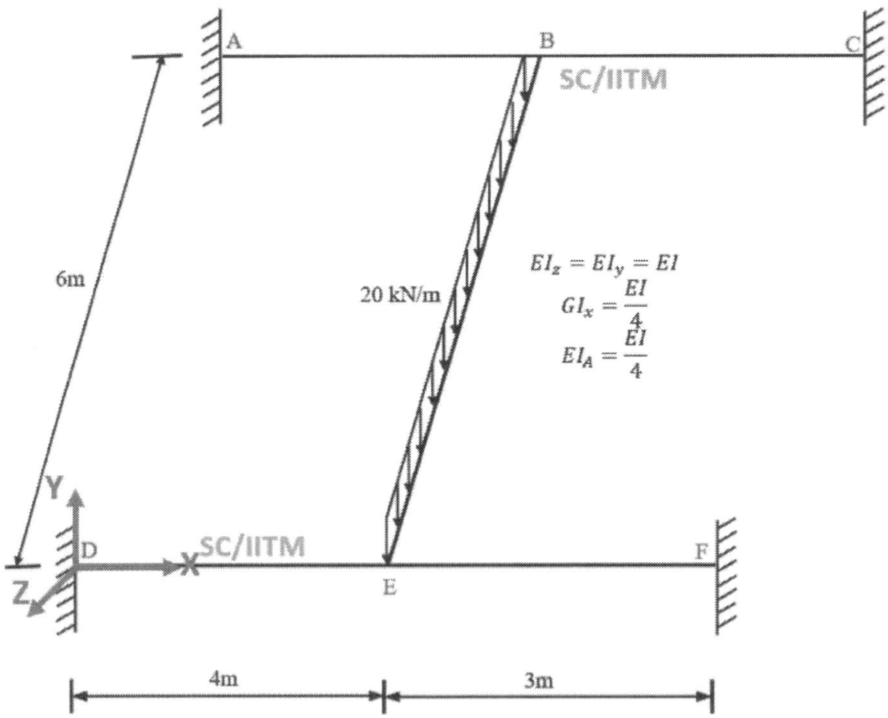

SOLUTION

Mark the local and reference axes system as shown

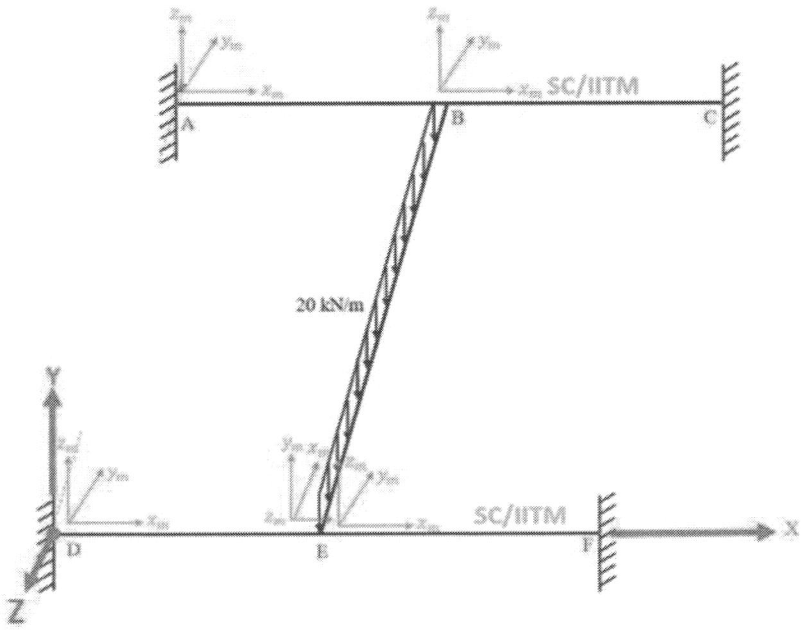

i. *Coordinates of joints:*

Joint	X	Y	Z
A	0	0	–6
B	4	0	–6
C	7	0	–6
D	0	0	0
E	4	0	0
F	7	0	0

ii. *ψ angle:*

		Joint		Direction Cosines			Type of	ψ angle
Member	Length (m)	j	k	C_x	C_y	C_z	Transformation	(Degrees)
AB	4	A	B	1	0	0	Y-Z-X	$\psi_y = 90$
BC	3	B	C	1	0	0	Y-Z-X	$\psi_y = 90$
DE	4	D	E	1	0	0	Y-Z-X	$\psi_y = 90$
EF	3	E	F	1	0	0	Y-Z-X	$\psi_y = 90$
EB	6	E	B	0	0	–1	Y-Z-X	$\psi_z = 0$

iii. *Marking unrestrained and restrained degrees-of-freedom:*
The unrestrained and restrained degrees of freedom are marked as shown in

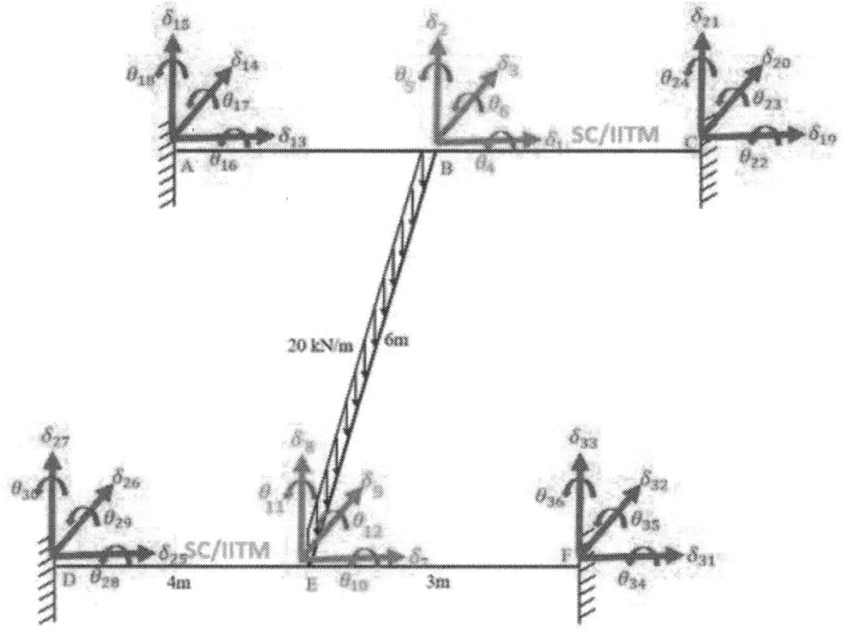

Unrestrained degrees-of-freedom: 12
$(\bar{\delta}_1, \bar{\delta}_2, \bar{\delta}_3, \bar{\theta}_4, \bar{\theta}_5, \bar{\theta}_6, \bar{\delta}_7, \bar{\delta}_8, \bar{\delta}_9, \bar{\theta}_{10}, \bar{\theta}_{11}, \bar{\theta}_{12})$
Restrained degrees-of-freedom: 24
$(\delta_{13}, \delta_{14}, \delta_{15}, \theta_{16}, \theta_{17}, \theta_{18}, \delta_{19}, \delta_{20}, \delta_{21}, \theta_{22}, \theta_{23}, \theta_{24})$
$(\delta_{25}, \delta_{26}, \delta_{27}, \theta_{28}, \theta_{29}, \theta_{30}, \delta_{31}, \delta_{32}, \delta_{33}, \theta_{34}, \theta_{35}, \theta_{36})$

iv. *Estimation of transformation matrix:*
We already know that,

$$C = \begin{bmatrix} C_{11} & C_{12} & C_{13} \\ C_{21} & C_{22} & C_{23} \\ C_{31} & C_{32} & C_{33} \end{bmatrix}$$

The direction cosine matrices for all three members are,

$$C_{AB} = \begin{bmatrix} 1 & 0 & 0 \\ 0 & 0 & -1 \\ 0 & 1 & 0 \end{bmatrix}, C_{EB} = \begin{bmatrix} 0 & 0 & -1 \\ 0 & 1 & 0 \\ 1 & 0 & 0 \end{bmatrix}$$

$$C_{AB} = C_{BC} = C_{DE} = C_{EF}$$

Now, the transformation matrix is given by,

$$[T_i]=\begin{bmatrix}[C_i] & [0] & [0] & [0]\\ [0] & [C_i] & [0] & [0]\\ [0] & [0] & [C_i] & [0]\\ [0] & [0] & [0] & [C_i]\end{bmatrix}_{12\times 12}$$

Global labels:

$$AB=[13,14,15,16,17,18,1,2,3,4,5,6]$$

$$BC=[1,2,3,4,5,6,19,20,21,22,23,24]$$

$$DE=[25,26,27,28,29,30,7,8,9,10,11,12]$$

$$EF=[7,8,9,10,11,12,31,32,33,34,35,36]$$

$$EB=[7,8,9,10,11,12,1,2,3,4,5,6]$$

v. *Fixed end moments and joint load vector:*
Load is acting only on the member EB. Hence, the fixed end moments along other two members remain zero.

Member EB:

$$V_A=\frac{20\times 6}{2}=60\,kN$$

$$V_B=60\,kN$$

$$M_{EB}=\frac{20\times 6^2}{12}=60\,kNm$$

$$M_{BE}=60\,kNm$$

$$(FEM)_{EB}=\begin{Bmatrix}0\\0\\60\\-60\\0\\0\\0\\0\\60\\60\\0\\0\end{Bmatrix}$$

The joint load vector is the reversal of the fixed end moments. The transpose of the joint load vector is given below:

$$[J_L]^T = \{0 \quad 0 \quad -60 \quad 60 \quad 0 \quad 0 \quad 0 \quad 0 \quad -60 \quad -60 \quad [0]\}_{36\times 1}$$

vi. *Stiffness matrix:*

$K_{AB} = EI$

13	14	15	16	17	18	1	2	3	4	5	6	
0.063	0	0	0	0	0	–0.063	0	0	0	0	0	13
0	0.188	0	0	0	0.375	0	–0.188	0	0	0	0.375	14
0	0	0.188	0	–0.375	0	0	0	–0.188	0	–0.375	0	15
0	0	0	0.063	0	0	0	0	0	–0.063	0	0	16
0	0	–0.375	0	1	0	0	0	0.375	0	0.5	0	17
0	0.375	0	0	0	1	0	–0.375	0	0	0	0.5	18
–0.063	0	0	0	0	0	0.063	0	0	0	0	0	1
0	–0.188	0	0	0	–0.375	0	0.188	0	0	0	–0.375	2
0	0	–0.188	0	0.375	0	0	0	0.188	0	0.375	0	3
0	0	0	–0.063	0	0	0	0	0	0.063	0	0	4
0	0	–0.375	0	0.5	0	0	0	0.375	0	1	0	5
0	0.375	0	0	0	0.5	0	–0.375	0	0	0	1	6

$$\left[\bar{K}_{AB}\right] = \left[K_{AB}\right]$$

$K_{BC} = EI$

1	2	3	4	5	6	19	20	21	22	23	24	
0.083	0	0	0	0	0	–0.083	0	0	0	0	0	1
0	0.444	0	0	0	0.667	0	–0.444	0	0	0	0.667	2
0	0	0.444	0	–0.667	0	0	0	–0.444	0	–0.667	0	3
0	0	0	0.083	0	0	0	0	0	–0.083	0	0	4
0	0	–0.667	0	1.333	0	0	0	0.667	0	0.667	0	5
0	0.667	0	0	0	1.333	0	–0.667	0	0	0	0.667	6
–0.083	0	0	0	0	0	0.083	0	0	0	0	0	19
0	–0.444	0	0	0	–0.667	0	0.444	0	0	0	–0.667	20
0	0	–0.444	0	0.667	0	0	0	0.444	0	0.667	0	21
0	0	0	–0.083	0	0	0	0	0	0.083	0	0	22
0	0	–0.667	0	0.667	0	0	0	0.667	0	1.333	0	23
0	0.667	0	0	0	0.667	0	–0.667	0	0	0	1.333	24

$$\left[\bar{K}_{BC}\right] = \left[K_{BC}\right]$$

$K_{DE} = EI$

25	26	27	28	29	30	7	8	9	10	11	12	
0.083	0	0	0	0	0	–0.083	0	0	0	0	0	25
0	0.444	0	0	0	0.667	0	–0.444	0	0	0	0.667	26
0	0	0.444	0	–0.667	0	0	0	–0.444	0	–0.667	0	27
0	0	0	0.083	0	0	0	0	0	–0.083	0	0	28
0	0	–0.667	0	1.333	0	0	0	0.667	0	0.667	0	29
0	0.667	0	0	0	1.333	0	–0.667	0	0	0	0.667	30
–0.083	0	0	0	0	0	0.083	0	0	0	0	0	7
0	–0.444	0	0	0	–0.667	0	0.444	0	0	0	–0.667	8
0	0	–0.444	0	0.667	0	0	0	0.444	0	0.667	0	9
0	0	0	–0.083	0	0	0	0	0	0.083	0	0	10
0	0	–0.667	0	0.667	0	0	0	0.667	0	1.333	0	11
0	0.667	0	0	0	0.667	0	–0.667	0	0	0	1.333	12

$$\left[\overline{K}_{DE}\right]=\left[K_{DE}\right]$$

$K_{EF} = EI$

7	8	9	10	11	12	31	32	33	34	35	36	
0.083	0	0	0	0	0	−0.083	0	0	0	0	0	7
0	0.444	0	0	0	0.667	0	−0.444	0	0	0	0.667	8
0	0	0.444	0	−0.667	0	0	0	−0.444	0	−0.667	0	9
0	0	0	0.083	0	0	0	0	0	−0.083	0	0	10
0	0	−0.667	0	1.333	0	0	0	0.667	0	0.667	0	11
0	0.667	0	0	0	1.333	0	−0.667	0	0	0	0.667	12
−0.083	0	0	0	0	0	0.083	0	0	0	0	0	31
0	−0.444	0	0	0	−0.667	0	0.444	0	0	0	−0.667	32
0	0	−0.444	0	0.667	0	0	0	0.444	0	0.667	0	33
0	0	0	−0.083	0	0	0	0	0	0.083	0	0	34
0	0	−0.667	0	0.667	0	0	0	0.667	0	1.333	0	35
0	0.667	0	0	0	0.667	0	−0.667	0	0	0	1.333	36

$$\left[\overline{K}_{EF}\right]=\left[K_{EF}\right]$$

$K_{EB} = EI$

7	8	9	10	11	12	1	2	3	4	5	6	
0.083	0	0	0	0	0	−0.083	0	0	0	0	0	7
0	0.444	0	0	0	0.667	0	−0.444	0	0	0	0.667	8
0	0	0.444	0	−0.667	0	0	0	−0.444	0	−0.667	0	9
0	0	0	0.083	0	0	0	0	0	−0.083	0	0	10
0	0	−0.667	0	1.333	0	0	0	0.667	0	0.667	0	11
0	0.667	0	0	0	1.333	0	−0.667	0	0	0	0.667	12
−0.083	0	0	0	0	0	0.083	0	0	0	0	0	1
0	−0.444	0	0	0	−0.667	0	0.444	0	0	0	−0.667	2
0	0	−0.444	0	0.667	0	0	0	0.444	0	0.667	0	3
0	0	0	−0.083	0	0	0	0	0	0.083	0	0	4
0	0	−0.667	0	0.667	0	0	0	0.667	0	1.333	0	5
0	0.667	0	0	0	0.667	0	−0.667	0	0	0	1.333	6

$$\left[\overline{K}_{EB}\right]=\left[K_{EB}\right]$$

The total stiffness matrix can be formed by assembling the global stiffness matrices of all the members, from which the unrestrained stiffness matrix can be partitioned.

$[K_{UU}] = EI$

1	2	3	4	5	6	7	8	9	10	11	12	
0.042	0	0	0	0	0	−0.042	0	0	0	0	0	1
0	0.056	0	0	0	0.167	0	−0.056	0	0	0	0.167	2
0	0	0.056	0	−0.167	0	0	0	−0.056	0	−0.167	0	3
0	0	0	0.042	0	0	0	0	0	−0.042	0	0	4
0	0	−0.167	0	0.667	0	0	0	0.167	0	0.333	0	5
0	0.167	0	0	0	0.667	0	−0.167	0	0	0	0.333	6
−0.042	0	0	0	0	0	0.042	0	0	0	0	0	7
0	−0.056	0	0	0	−0.167	0	0.056	0	0	0	−0.167	8
0	0	−0.056	0	0.167	0	0	0	0.056	0	0.167	0	9
0	0	0	−0.042	0	0	0	0	0	0.042	0	0	10
0	0	−0.167	0	0.333	0	0	0	0.167	0	0.667	0	11
0	0.167	0	0	0	0.333	0	−0.167	0	0	0	0.667	12

vii. *Calculation of end moments and reactions:*

$$\{M_1\}=\begin{Bmatrix}0\\0\\21.028\\-16.364\\-39.898\\0\\0\\0\\-21.028\\16.364\\-44.215\\0\end{Bmatrix},\{M_2\}=\begin{Bmatrix}0\\0\\-36.412\\21.818\\51.739\\0\\0\\0\\36.412\\-21.818\\57.496\\0\end{Bmatrix},\{M_3\}=\begin{Bmatrix}0\\0\\22.978\\16.364\\-43.564\\0\\0\\0\\-22.978\\-16.363\\-48.348\\0\end{Bmatrix},\{M_4\}=\begin{Bmatrix}0\\0\\-39.582\\-21.818\\56.183\\0\\0\\0\\39.582\\21.818\\62.562\\0\end{Bmatrix},\{M_5\}=\begin{Bmatrix}0\\60\\2.5599\\38.1818\\-7.8354\\0\\0\\60\\-2.5599\\-38.1818\\-7.524\\0\end{Bmatrix}$$

The member end forces and moments are shown in Figure 5.19. The final end moments and reactions are shown in Figure 5.20.

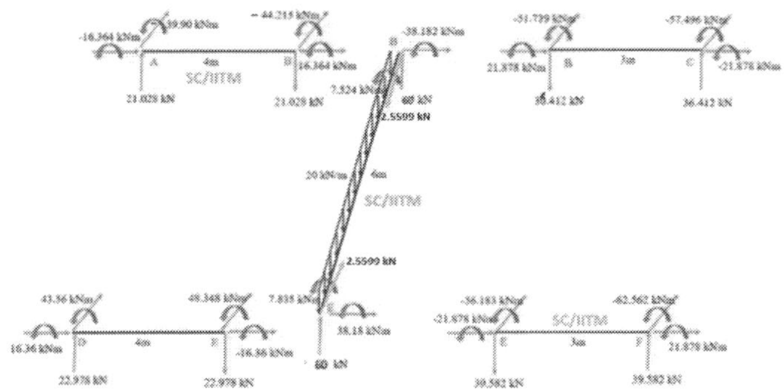

FIGURE 5.19 Member end forces.

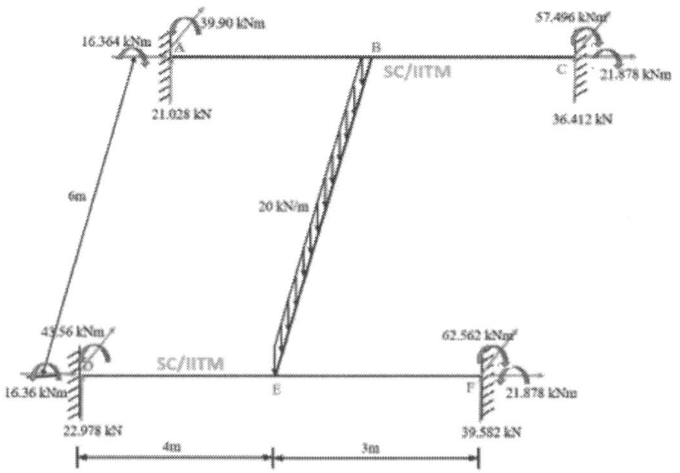

FIGURE 5.20 Find end reactions.

MATLAB program:

```
%% 3D analysis of space frame
clc;
clear;
%% Input
n = 5; % number of members
EI = [1 1 1 1 1]; %Flexural rigidity
EIy = EI;
EIz = EI;
GI = [0.25 0.25 0.25 0.25 0.25].*EI; %Torsional constant
EA = [0.25 0.25 0.25 0.25 0.25].*EI; %Axial rigidity
L = [4 3 4 3 6]; % length in m
nj = n+1; % Number of Joints
codm = [0 0 -6; 4 0 -6; 7 0 -6; 0 0 0; 4 0 0; 7 0 0];
%Coordinate wrt X,Y.Z: size=nj,3
dc = [1 0 0; 1 0 0; 1 0 0; 1 0 0; 0 0 -1]; % Direction cosines for each
member
tytr = [1 1 1 1 1]; % Type of transformation for each member
psi = [90 90 90 90 0]; % Psi angle in degrees for each member


% C matrix
c1 = [1 0 0; 0 0 -1; 0 1 0]; % C matrix for member 1
c2 = [1 0 0; 0 0 -1; 0 1 0]; % C matrix for member 2
c3 = [1 0 0; 0 0 -1; 0 1 0]; % C matrix for member 3
c4 = [1 0 0; 0 0 -1; 0 1 0]; % C matrix for member 4
c5 = [0 0 -1; 0 1 0; 1 0 0]; % C matrix for member 5


uu = 12; % Number of unrestrained Degrees-of-freedom
ur = 24; % Number of restrained Degrees-of-freedom
uul = [1 2 3 4 5 6 7 8 9 10 11 12]; % global labels of unrestrained dof
url = [13 14 15 16 17 18 19 20 21 22 23 24 25 26 27 28 29 30 31
32 33 34 35 36]; % global labels of restrained dof


l1 = [13 14 15 16 17 18 1 2 3 4 5 6]; % Global labels for member 1
l2 = [1 2 3 4 5 6 19 20 21 22 23 24]; % Global labels for member 2
l3 = [25 26 27 28 29 30 7 8 9 10 11 12]; % Global labels for member 3
l4 = [7 8 9 10 11 12 31 32 33 34 35 36]; % Global labels for member 4
l5 = [7 8 9 10 11 12 1 2 3 4 5 6]; % Global labels for member 5
l= [l1; l2; l3; l4; l5];
dof = uu + ur; % Degrees-of-freedom
Ktotal = zeros (dof);


fem1= [0; 0; 0; 0; 0; 0; 0; 0; 0; 0; 0; 0]; % Local Fixed end moments
of member 1
fem2= [0; 0; 0; 0; 0; 0; 0; 0; 0; 0; 0; 0]; % Local Fixed end moments
of member 2
fem3= [0; 0; 0; 0; 0; 0; 0; 0; 0; 0; 0; 0]; % Local Fixed end moments
of member 3
```

```
fem4= [0; 0; 0; 0; 0; 0; 0; 0; 0; 0; 0; 0]; % Local Fixed end moments
of member 4
fem5= [0; 0; 60; 60; 0; 0; 0; 0; 60; -60; 0; 0]; % Local Fixed
end moments of member 5

%% Transformation matrix
T1 = zeros(12);
T2 = zeros(12);
T3 = zeros(12);
T4 = zeros(12);
T5 = zeros(12);
for i = 1:3
    for j = 1:3
        T1(i,j)=c1(i,j);
        T1(i+3,j+3)=c1(i,j);
        T1(i+6,j+6)=c1(i,j);
        T1(i+9,j+9)=c1(i,j);
        T2(i,j)=c2(i,j);
        T2(i+3,j+3)=c2(i,j);
        T2(i+6,j+6)=c2(i,j);
        T2(i+9,j+9)=c2(i,j);
        T3(i,j)=c3(i,j);
        T3(i+3,j+3)=c3(i,j);
        T3(i+6,j+6)=c3(i,j);
        T3(i+9,j+9)=c3(i,j);
        T4(i,j)=c4(i,j);
        T4(i+3,j+3)=c4(i,j);
        T4(i+6,j+6)=c4(i,j);
        T4(i+9,j+9)=c4(i,j);
        T5(i,j)=c5(i,j);
        T5(i+3,j+3)=c5(i,j);
        T5(i+6,j+6)=c5(i,j);
        T5(i+9,j+9)=c5(i,j);
    end
end

%% Getting Type of transformation and Psi angle
for i = 1:n
    if tytr(i) ==1
        fprintf ('Member Number =');
        disp (i);
        fprintf ('Type of transformation is Y-Z-X \n');
    else
        fprintf ('Member Number =');
        disp (i);
        fprintf ('Type of transformation is Z-Y-X \n');
    end
    fprintf ('Psi angle=');
    disp (psi(i));
end
```

```
%% Stiffness coefficients for each member
sc1 = EA./L;
sc2 = 6*EIz./(L.^2);
sc3 = 6*EIy./(L.^2);
sc4 = GI./L;
sc5 = 2*EIy./L;
sc6 = 12*EIz./(L.^3);
sc7 = 12*EIy./(L.^3);
sc8 = 2*EIz./L;


%% stiffness matrix 6 by 6
for i = 1:n
        Knew = zeros (dof);
        k1 = [sc1(i); 0; 0; 0; 0; 0; -sc1(i); 0; 0; 0; 0; 0];
        k2 = [0; sc6(i); 0; 0; 0; sc2(i); 0; -sc6(i); 0; 0; 0; sc2(i)];
        k3 = [0; 0; sc7(i); 0; -sc3(i); 0; 0; 0; -sc7(i); 0; -sc3(i); 0];
        k4 = [0; 0; 0; sc4(i); 0; 0; 0; 0; 0; -sc4(i); 0; 0];
        k5 = [0; 0; -sc3(i); 0; (2*sc5(i)); 0; 0; 0; sc3(i);
0; sc5(i); 0];
        k6 = [0; sc2(i); 0; 0; 0; (2*sc8(i)); 0; -sc2(i); 0;
0; 0; sc8(i)];
        k7 = -k1;
        k8 = -k2;
        k9 = -k3;
        k10 = -k4;
        k11 = [0; 0; -sc3(i); 0; sc5(i); 0; 0; 0; sc3(i); 0;
(2*sc5(i)); 0];
        k12 = [0; sc2(i); 0; 0; 0; sc8(i); 0; -sc2(i); 0; 0; 0;
(2*sc8(i))];
        K = [k1 k2 k3 k4 k5 k6 k7 k8 k9 k10 k11 k12];
        fprintf ('Member Number =');
        disp (i);
        fprintf ('Local Stiffness matrix of member, [K] = \n');
        disp (K);
        if i == 1
            T = T1;
        elseif i == 2
            T = T2;
        else
            T = T3;
        end
        Ttr = T';
        Kg = Ttr*K*T;
        fprintf ('Global Matrix, [K global] = \n');

        for p = 1:12
            for q = 1:12
                Knew((l(i,p)),(l(i,q))) =Kg(p,q);
            end
        end
```

```
        Ktotal = Ktotal + Knew;
        if i == 1
            Tt1= T;
            Kg1=Kg;
            fembar1= Tt1'*fem1;
        elseif i == 2
            Tt2 = T;
            Kg2 = Kg;
            fembar2= Tt2'*fem2;
        elseif i ==3
            Tt3 = T;
            Kg3 = Kg;
            fembar3= Tt3'*fem3;
        elseif i ==4
            Tt4 = T;
            Kg4 = Kg;
            fembar4= Tt4'*fem4;
         else
            Tt5 = T;
            Kg5 = Kg;
            fembar5= Tt5'*fem5;
        end
end
fprintf ('Stiffness Matrix of complete structure, [Ktotal] = \n');
disp (Ktotal);
Kunr = zeros(12);
for x=1:uu
    for y=1:uu
        Kunr(x,y)= Ktotal(x,y);
    end
end
fprintf ('Unrestrained Stiffness sub-matix, [Kuu] = \n');
disp (Kunr);
KuuInv= inv(Kunr);
fprintf ('Inverse of Unrestrained Stiffness sub-matix,
[KuuInverse] = \n');
disp (KuuInv);


%% Creation of joint load vector
jl= [0; 0; -60; 60; 0; 0; 0; 0; -60; -60; 0; 0; 0; 0; 0; 0; 0; 0; 0; 0; 0;
0; 0; 0; 0; 0; 0; 0; 0; 0; 0; 0; 0; 0; 0]; % values given in kN or kNm
jlu = jl(1:12,1); % load vector in unrestrained dof
delu = KuuInv*jlu;
fprintf ('Joint Load vector, [Jl] = \n');
disp (jl');
fprintf ('Unrestrained displacements, [DelU] = \n');
disp (delu');
delr = [0; 0; 0; 0; 0; 0; 0; 0; 0; 0; 0; 0; 0; 0; 0; 0; 0; 0;
0; 0; 0; 0; 0; 0];
del = zeros (dof,1);
```

```
del = [delu; delr];
deli= zeros (12,1);
for i = 1:n
    for p = 1:12
        deli(p,1) = del((l(i,p)),1) ;
    end
    if i == 1
            delbar1 = deli;
            mbar1= (Kg1 * delbar1)+fembar1;
            fprintf ('Member Number =');
            disp (i);
            fprintf ('Global displacement matrix [DeltaBar] = \n');
            disp (delbar1');
            fprintf ('Global End moment matrix [MBar] = \n');
            disp (mbar1');
        elseif i == 2
            delbar2 = deli;
            mbar2= (Kg2 * delbar2)+fembar2;
            fprintf ('Member Number =');
            disp (i);
            fprintf ('Global displacement matrix [DeltaBar] = \n');
            disp (delbar2');
            fprintf ('Global End moment matrix [MBar] = \n');
            disp (mbar2');
        elseif i ==3
            delbar3 = deli;
            mbar3= (Kg3 * delbar3)+fembar3;
            fprintf ('Member Number =');
            disp (i);
            fprintf ('Global displacement matrix [DeltaBar] = \n');
            disp (delbar3');
            fprintf ('Global End moment matrix [MBar] = \n');
            disp (mbar3');
        elseif i ==4
            delbar4 = deli;
            mbar4= (Kg4 * delbar4)+fembar4;
            fprintf ('Member Number =');
            disp (i);
            fprintf ('Global displacement matrix [DeltaBar] = \n');
            disp (delbar4');
            fprintf ('Global End moment matrix [MBar] = \n');
            disp (mbar4');
        else
            delbar5 = deli;
            mbar5= (Kg5 * delbar5)+fembar5;
            fprintf ('Member Number =');
            disp (i);
            fprintf ('Global displacement matrix [DeltaBar] = \n');
            disp (delbar5');
            fprintf ('Global End moment matrix [MBar] = \n');
            disp (mbar5');
```

```
        end
end

%% check
mbar = [mbar1'; mbar2'; mbar3'; mbar4'; mbar5'];
jf = zeros(dof,1);
for a=1:n
    for b=1:12 % size of k matrix
        d = l(a,b);
        jfnew = zeros(dof,1);
        jfnew(d,1)=mbar(a,b);
        jf=jf+jfnew;
    end
end
fprintf ('Joint forces = \n');
disp (jf');
```

6 Unsymmetric Bending

6.1 INTRODUCTION

In members generally idealized as one-dimensional, it is comfortable to compute the stresses in the cross-section for the loads applied at the prefixed points, as the bending takes place parallel to the plane of applied moment. While using the simple bending equation, it is assumed that the neutral axis of the cross-section is perpendicular to the loading plane.

In Figure 6.1, the YY axis is the trace of the plane of the applied moment. The bending moment in the YY axis is said to be zero, and mathematically,

$$\sum M_y = \int_A \sigma_x dA = 0 \tag{6.1}$$

and $\int zy\, dA = 0$ only when the ZZ and YY axes are inertia's principal axes.

For symmetrical bending,

- It is essential that the plane containing none of the principal axes of inertia, the plane of the applied moment, and the plane of deflection should coincide.
- It is also obvious that the neutral axis will coincide with other principal axes of inertia.

When the trace of the plane of the applied moment does not coincide with any of the principal axes of inertia, this type of bending is called *Unsymmetrical bending* or *non-uniform bending*. Under such conditions in unsymmetrical bending, the neutral axis is no longer perpendicular to the trace of the applied moment plane, and the deflection

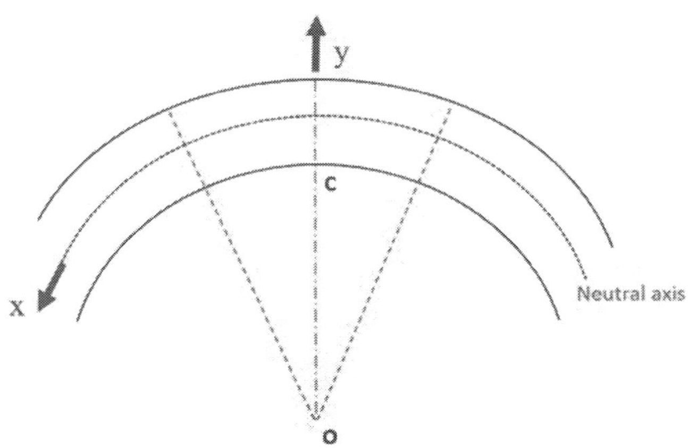

FIGURE 6.1 Uniform bending.

DOI: 10.1201/9781003639152-6

curve is not plane. The significant consequence of unsymmetrical bending is that the symmetric members about a vertical axis with thin-walled sections will twist under transverse loads. The reasons for the unsymmetrical bending are as follows:

- The section is symmetrical, but the line of action of the load is inclined to both the principal axes.
- The section is unsymmetrical, and the line of action of the load is along any centroidal axis.

In the case of unsymmetrical bending, the applied moment will cause bending about both the principal axes of inertia, which should be located to calculate the stresses at any point in the cross-section. If the moment acts on the plane of symmetry, then the conventional simple bending equation can be used to calculate the stresses.

$$\sigma_b = \frac{M}{I} y \tag{6.2}$$

However, if the load acts on another plane, it becomes unsymmetric, and one cannot use the conventional flexure equation to obtain the stresses. In such a case, the following procedure has to be followed to estimate the stresses:

Step 1: To transform the problem of unsymmetric bending to uniplanar bending
Consider a cross-section of the beam under the action of a bending moment M.

Let,

ZZ and YY = coordinate axes passing through the centroid.
UU and VV = principal axes inclined at an angle α to the ZZ and YY axes, respectively.

The axes are shown in the figure below:

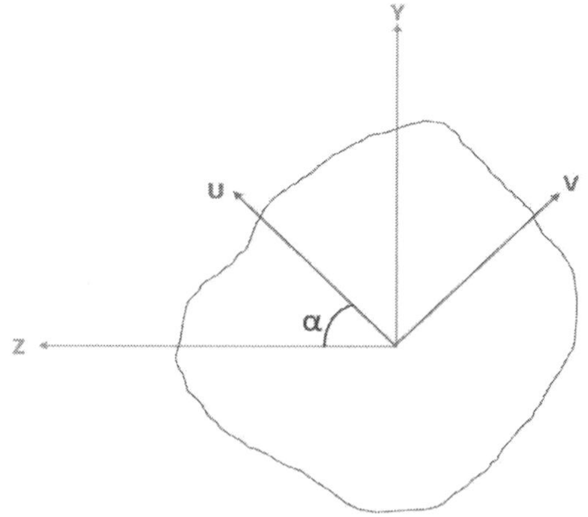

To locate the principal axes of inertia, the following relationship is valid:

$$u = z cos\alpha + y sin\alpha \tag{6.3a}$$

$$v = -z sin\alpha + y cos\alpha \tag{6.3b}$$

Where the angle is measured in the positive coordinate, the moment of inertia about the UU and VV axes should be calculated to estimate the bending stresses, as discussed below:

$$\begin{aligned} I_u &= \int_A v^2 dA \\ &= \int_A (-z sin\alpha + y cos\alpha)^2 dA \\ &= \int_A (z^2 sin^2\alpha + y^2 cos^2\alpha - 2z sin\alpha cos\alpha) dA \\ &= sin^2\alpha \int_A z^2 dA + cos^2\alpha \int_A y^2 dA - sin2\alpha \int_A yz dA \\ &= I_y sin^2\alpha + I_z cos^2\alpha - I_{yz} sin^2\alpha \end{aligned}$$

$$\begin{aligned} cos2\alpha &= 1 - 2sin^2\alpha \\ &= 2cos^2\alpha - 1 \end{aligned}$$

Substituting in the above equation,

$$\begin{aligned} I_u &= \frac{I_y}{2}(1 - cos2\alpha) + \frac{I_Z}{2}(1 - cos2\alpha) - I_{yZ} sin2\alpha \\ &= \frac{I_Y + I_Z}{2} + \frac{I_Z + I_Y}{2} cos2\alpha - I_{YZ} sin2\alpha \\ I_u &= \frac{I_Y + I_Z}{2} + \frac{I_Z + I_Y}{2} cos2\alpha - I_{YZ} sin2\alpha \end{aligned}$$

$$\begin{aligned} I_v &= \int_A u^2 dA \\ &= \int_A (z cos\alpha + y sin\alpha)^2 dA \\ &= \int_A (z^2 cos^2\alpha + y^2 sin^2\alpha + 2zy sin\alpha cos\alpha) dA \\ I_v &= cos^2\alpha \int_A z^2 dA + sin^2\alpha \int_A y^2 dA + sin2\alpha \int_A yz dA \\ &= cos^2\alpha I_y + sin^2\alpha I_z + I_{zy} sin^2\alpha \\ &= (1 + cos2\alpha)\frac{I_y}{2} + \frac{1 - cos2\alpha I_z}{2} + I_{zy} sin2\alpha \end{aligned}$$

$$I_v = \frac{I_z - I_y}{2} - \frac{I_z - I_y}{2} cos2\alpha + I_{zy} sin2\alpha$$

$$I_u + I_v = I_z + I_y$$

$$I_{uv} = \int_A (uv)$$

$$= \int_A (zcos\alpha + ysin\alpha)(-zsin\alpha + cos\alpha) dA$$

$$= \int_A (-z^2 sin\alpha cos\alpha - yzsin^2\alpha - yzcos^2\alpha - y^2 sin\alpha cos\alpha) dA$$

$$\int_A z^2 dA = I_y$$

$$\int_A yz^2 dA = I_z$$

$$\int_A yz^2 dA = I_{yz}$$

Using the above relationship,

$$I_{uv} = -I_y sin\alpha cos\alpha + I_z sin\alpha cos\alpha + I_{yz} (cos\alpha^2 - sin\alpha^2)$$

$$I_{uv} = \frac{I_z + I_y}{2} sin2\alpha + I_{yz} cos2\alpha$$

for u,v being the principal axes of inertia,

$$I_{uv} = 0$$

$$I_{uv} = \frac{I_z + I_y sin2\alpha}{2} + I_{yz} cos2\alpha = 0$$

$$tan(2\alpha) = -\frac{2I_{yz}}{I_z - I_y}$$

Step 2: To determine the bending stress at any point in the cross-section

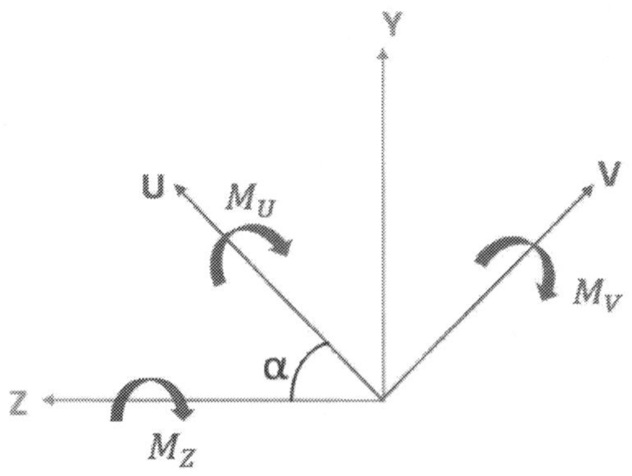

The moment about the principal axes is given by,

$$M_u = M_z cos\alpha$$

$$M_v = M_z sin\alpha$$

Stress at any point p(u,v),

$$\frac{M_u}{I_v}(v) - compressive\,stress$$

$$\frac{M_u}{I_v}(u) - tensile\,stress$$

Consider tensile stress as follows:

$$\sigma_p = \frac{M_u}{I_u}v + \frac{M_v}{I_u}u$$

$$= -\left[\frac{M_u}{I_u}v + \frac{M_v}{I_u}u\right]$$

(negative sign indicates compressive stress).

The nature of the resultant bending stress will always depend upon the quadrant in which it lies. So, the signs of u and v must be considered while determining the resultant stress. It is also noted that the maximum stress will occur at a point at the most significant distance from the neutral axis. On one side of the neutral axis, all the points will carry the stresses of the exact nature and opposite nature on the other.

6.2 UNIPLANAR BENDING

Bending is one of the most common forms of loading in structural elements such as beams, columns, and frames. When a structural member is subjected to bending, internal stresses develop, leading to deformations. Uniplanar bending refers to bending a structural member within a single plane. This is the most fundamental type of bending and serves as the basis for understanding more complex bending behavior, such as unsymmetric bending. In steel design, uniplanar bending occurs when external loads act in a way that produces bending moments about only one of the principal axes of the cross-section. This simplifies analysis and design, as the bending behavior can be studied independently in one plane without considering interactions between axes. In a structural member, bending occurs when a force or moment is applied perpendicular to the longitudinal axis of the element. The primary characteristics of uniplanar bending are:

1. Loads and moments act in a single plane (either about the major or minor principal axis).
2. Deflection and stresses develop only within that plane, without interaction from the perpendicular direction.

3. Stress distribution follows the standard bending equation:

$$\sigma = \frac{M}{I} y$$

where:

1. σ = bending stress at a point in the cross-section,
2. M = applied bending moment,
3. I = moment of inertia about the neutral axis,
4. y = distance from the neutral axis.

The neutral axis remains perpendicular to the bending plane, ensuring a predictable and symmetric stress distribution.

Uniplanar bending assumes that loads align with one principal axis. However, in cases where loading is not aligned with principal axes or occurs obliquely, bending moments develop about both axes simultaneously. This leads to unsymmetric bending, which requires a more complex analysis involving moment interaction equations and shear center considerations. Thus, understanding uniplanar bending provides a strong foundation for addressing more advanced steel structural analysis and design topics.

6.3 EXERCISE PROBLEMS

Example 6.1:

Find the stresses on the simply supported beam shown in the figure.

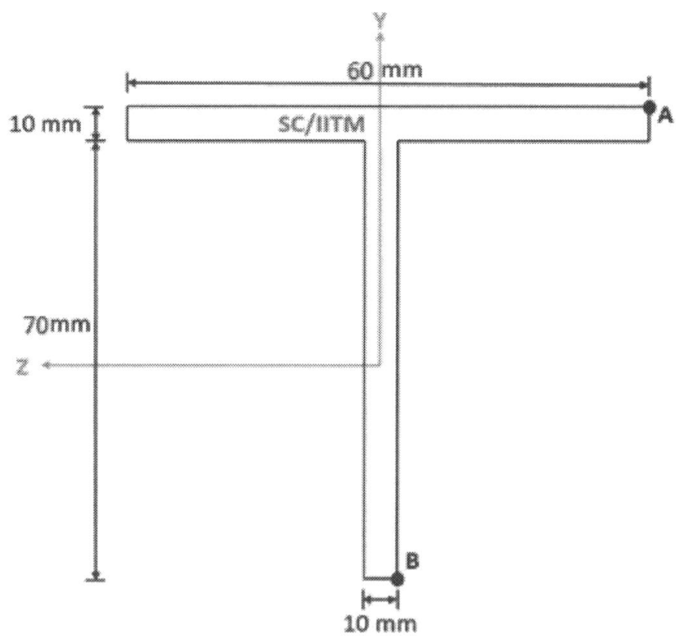

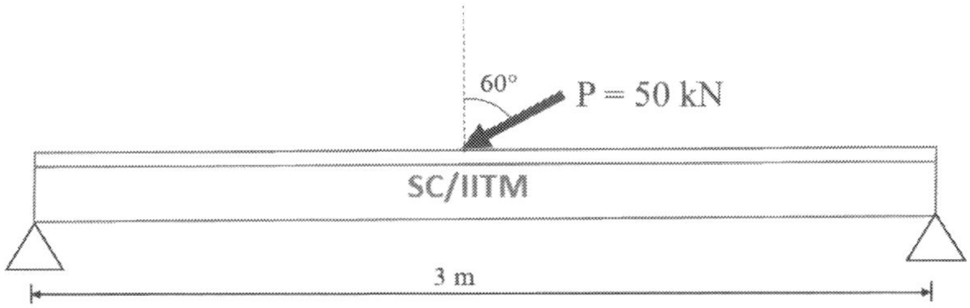

$$\bar{y} = \frac{\sum ay}{\sum a}$$

$$\bar{y} = \frac{(60\times10\times5)+(70\times10\times45)}{(60\times10)+(70\times10)} = 26.54\,mm \text{ From the top of the flange}$$

$$I_Y = \frac{10\times60^3}{12} + \frac{70\times10^3}{12}$$

$$= 1.858\times10^5\,mm^4$$

$$I_Z = \frac{60\times10^3}{12} + 60\times10\times(21.54)^2 + \frac{10\times70^3}{12} + 70\times10\times(18.46)^2$$

$$= 8.078\times10^5\,mm^4$$

$$\text{Moment about Z} - \text{axis} = \text{load}\times\text{perpendicular distance} = 50\times3/4 = 37.5\text{kNm}.$$

MATLAB code:

```
%% INPUT
% calculate the values of the moment of inertia of the section
in mm^4
Iy = 1.858e5;
Iz = 8.078e5;
Mz = 37.5; % Moment about Z axis in kNm
ybar = 26.54; % location of centroidal axis in mm
zbar = 0;
te = 60; %angle of inclination of load
%% Principal axis location
Iu = Iz;
Iv = Iy;
%% Stress calculation
Mu = Mz*cosd(te);Mv = -Mz*sind(te);
% calculation for point A
ua = 30;
```

```
va = 26.54;
sa = -(Mu*(10^6)*va/Iu)+(Mv*(10^6)*ua/Iv);

% calculation for point B
ub = 5;
vb = -53.46;
sb = -(Mu*(10^6)*vb/Iu)+(Mv*(10^6)*ub/Iv);

fprintf('Stress at point A = %6.2f N/mm^2 \n',sa);
fprintf('Stress at point B = %6.2f N/mm^2 \n \n',sb);
```

MATLAB output:

Stress at point A = $-5859.72\ \frac{N}{mm^2}$

Stress at point B = $366.92\ \frac{N}{mm^2}$

Example 6.2:

A roof truss ISLC 125 purlin is placed at an angle of 45°. Purlins are placed at 500 mm c/c along the principal rafter within 3.0 m spacing. Assume the purlin will act as a simply supported beam with a uniformly distributed load. Find the maximum stresses at A and B due to the member's self-weight. The dead load in the member is 2 kN/m^2.

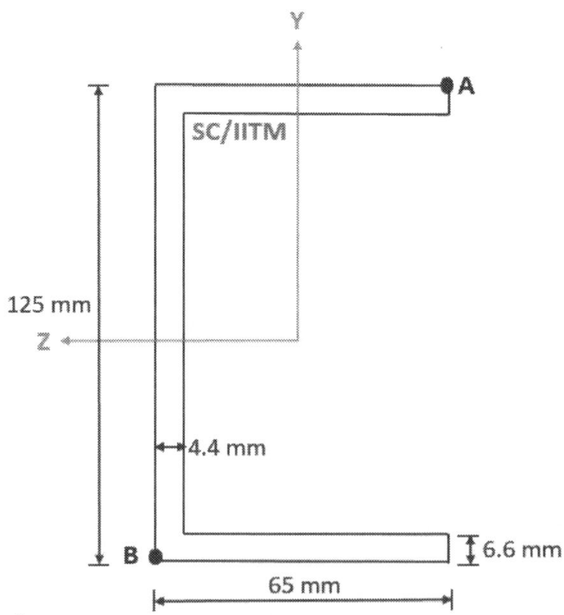

Dead Load on the Member

1. Given dead load per unit area = 2 kN/m^2
2. The spacing between purlins along the principal rafter = 0.5 m
3. So, the uniformly distributed load from dead load: (0.5 × 2) = 1 kN/m

Self-weight of the Purlin

1. **ISLC 125** section has a unit weight of **10.7 kg/m**
2. Converting to kN/m: 10.7 × 10/1000 = 0.107 kN/m

Total Load per Unit Length

The total distributed load on the purlin is: 1.0 + 0.107 = 1.107 kN/m

$$\text{Moment} = \frac{wl^2}{8} = \frac{1.107 + (3)^2}{8} = 1.245\,kNm$$

$$\alpha = 45^\circ$$

$$M_U = M_Z \cos\alpha = 0.88\,kNm$$

$$M_V = -M_Z \sin\alpha = -0.88\,kNm$$

For ISLC 125,

$$\mathrm{I_U = I_Z = 352.25 \times 10^4\,mm^4}$$

$$\mathrm{I_V = I_Y = 58.57 \times 10^4\,mm^4}$$

Point A: (43.54 mm, 62.5 mm)

Point B: (–21.46 mm, –62.5 mm)

MATLAB code:

```
%% Unsymmetrical bending - Channel section
clc;
clear;

%% INPUT
% calculate the values of the moment of inertia of the section
in mm^4
Iy = 57.2e4;
Iz = 356.8e4;
Mz = 1.245; % Moment about Z axis in kNm
ybar = 62.5; % location of centroidal axis in mm
zbar = 20.4;
te = 45; %angle of inclination of load

%% Principal axis location
Iu = Iz;
Iv = Iy;

%% Stress calculation
Mu = Mz*cosd(te);
Mv = -Mz*sind(te);
```

```
% calculation for point A
ua = 44.6;
va = 62.5;
sa = -(Mu*(10^6)*va/Iu)+(Mv*(10^6)*ua/Iv);

% calculation for point B
ub = -20.4;
vb = -62.5;
sb = -(Mu*(10^6)*vb/Iu)+(Mv*(10^6)*ub/Iv);

fprintf('Stress at point A = %6.2f N/mm^2 \n',sa);
fprintf('Stress at point B = %6.2f N/mm^2 \n \n',sb);
```

MATLAB output:

Stress at point A $= -81.06 \frac{N}{mm^2}$

Stress at point B $= 36.03 \frac{N}{mm^2}$

Example 6.3:

Consider a simply supported beam of length 3.0 m with a uniformly distributed load of 10 kN/m. The beam has a rectangular cross-section of 100 mm × 150 mm. The applied load makes an angle of 30° with the vertical plane. Calculate the stresses at points A (top right corner of the section) and B (bottom left corner of the section). The neutral axis of the section is also located.

$$\text{Moment} = \frac{wl^2}{8} = \frac{10*(3)^2}{8} = 11.25\,kNm$$

$$\alpha = 30°$$

$$M_U = M_Z \cos\alpha = 27.735\,kNm$$

$$M_V = -M_Z \sin\alpha = -41.6\,kNm$$

Point A: (50 mm, 75mm)

Point B: (–50 mm, –75 mm)

MATLAB code:

```
%% Unsymmetrical bending - rectangular section
clc;
clear;

%% INPUT
% Calculate the values of the moment of inertia of the section
in mm^4
Iy = 2.813e7;
Iz = 1.25e7;
Mz = 11.25; % Moment about Z axis in kNm
te = 30; %angle of inclination of load
```

```
%% Principal axis location
Iu = Iz;
Iv = Iy;

%% Stress calculation
Mu = Mz*cosd(te);
Mv = -Mz*sind(te);

% calculation for point A - on flange top
ua = 50;
va = 75;
sa = -(Mu*(10^6)*va/Iu)+(Mv*(10^6)*ua/Iv);

% calculation for point B - on web bottom
ub = -50;
vb = -75;
sb = -(Mu*(10^6)*vb/Iu)+(Mv*(10^6)*ub/Iv);

fprintf('Stress at point A = %6.2f N/mm^2 \n',sa);
fprintf('Stress at point B = %6.2f N/mm^2 \n \n',sb);

% location of neutral axis
vbyu=Iu*Mv/(Iv*Mu);
be=atand(vbyu);
fprintf('Beta = %6.2f degrees \n',be);
```

MATLAB output:

Stress at point A = −68.45 $\frac{\text{N}}{\text{mm}^2}$

Stress at point B = 68.45 $\frac{\text{N}}{\text{mm}^2}$

Beta = −14.39 degrees

Location of Neutral Axis

$$\sigma_X = -\left(\frac{M_U}{I_U} v - \frac{M_V}{I_V} u\right)$$

To locate the neutral axis, substituting σ_X = zero,

$$\frac{v}{u} = \tan \beta$$

where β is the inclination between the neutral axis and U.

$$\beta = -14.39°$$

Example 6.4:

A simply supported beam of section 190 mm × 50 mm spans 3.0 m. It rests on the support such that 190 mm face makes an angle of 45° with the horizontal. It carries a load of 100 kN at the mid-span. Find the stresses at points A and B. Also, locate the neutral axis of the section.

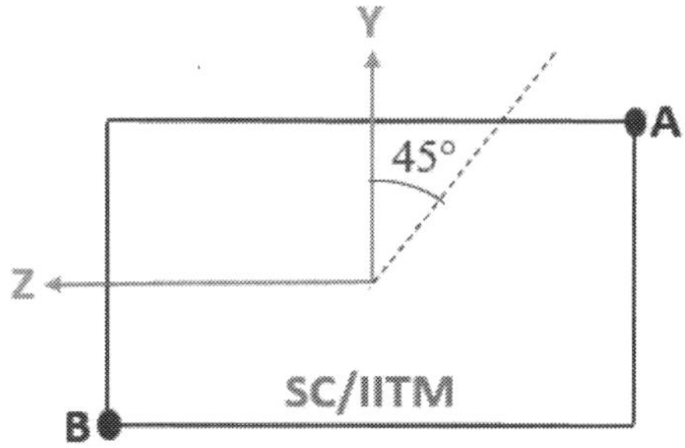

$$I_Y = \frac{50 \times 190^3}{12} = 2.858 \times 10^7 \ mm^4$$

$$I_Z = \frac{190 \times 50^3}{12} = 0.198 \times 10^7 \ mm^4$$

$$M_z = \frac{wl}{4} = 75kNm$$

$$\theta = 45°$$

Point A: (95 mm, 25 mm)

Point B: (–95 mm, –25 mm)

MATLAB code:

```
%% Unsymmetr0069cal bending - Rectangular section
clc;
clear;

%% INPUT
% calculate the values of moment of inertia of the section in
mm^4
Iy = 2.858e7;
Iz = 0.198e7;
Mz = 75; % Moment about Z axis in kNm
te = 45; %angle of inclination of load

%% Principal axis location
Iu = Iz;
Iv = Iy;

%% Stress calculation
Mu = Mz*cosd(te);
Mv = -Mz*sind(te);
```

```
% calculation for point A - on flange top
ua = 95;
va = 25;
sa = -(Mu*(10^6)*va/Iu)+(Mv*(10^6)*ua/Iv);

% calculation for point B - on web bottom
ub = -95;
vb = -25;
sb = -(Mu*(10^6)*vb/Iu)+(Mv*(10^6)*ub/Iv);

fprintf('Stress at point A = %6.2f N/mm^2 \n',sa);
fprintf('Stress at point B = %6.2f N/mm^2 \n \n',sb);

% location of neutral axis
vbyu=Iu*Mv/(Iv*Mu);
be=atand(vbyu);
fprintf('Beta = %6.2f degrees \n',be);
```

MATLAB output:

Stress at point A = $-845.89\ \frac{N}{mm^2}$

Stress at point B = $845.89\ \frac{N}{mm^2}$

Beta = -3.96 degrees

Example 6.5:

Consider a cantilever beam with two concentrated loads of 10 kN and 7 kN at 2.0 m and 3.0 m from the left end, acting at 30° and 45°, respectively, at the centroid of the T section. Find the stresses at points A, B, C, and D.

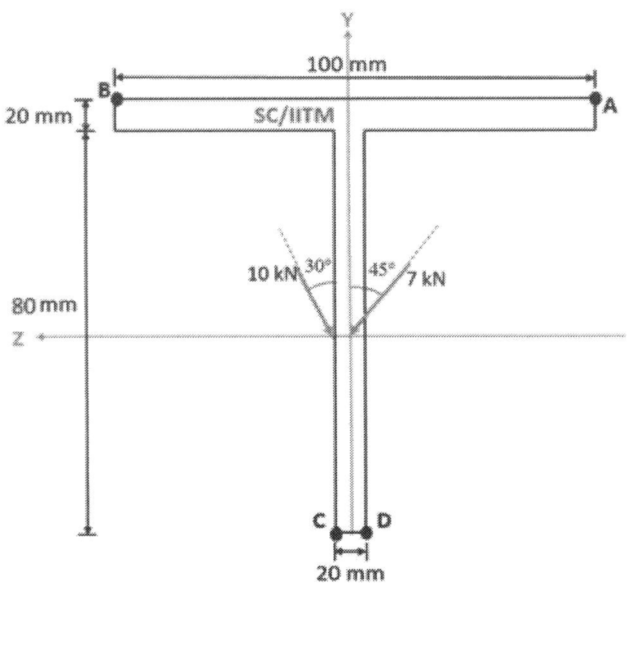

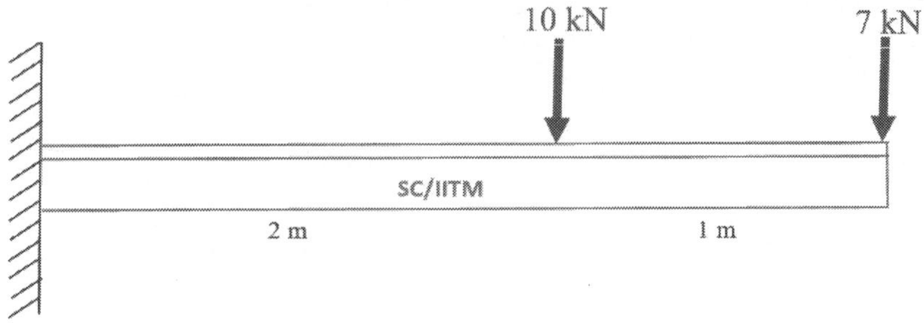

$$\bar{y} = \frac{\Sigma ay}{\Sigma a}$$

$$\bar{y} = \frac{(100\times 20\times 10)+(80\times 20\times 60)}{(100\times 20)+(80\times 20)} = 32.22\,mm$$

$$I_Y = \frac{20\times 100^3}{12} + \frac{80\times 20^3}{12}$$

$$= 1.72\times 10^6\,mm^4$$

$$I_Z = \frac{100\times 20^3}{12} + 100\times 20\times (22.22)^2 + \frac{20\times 80^3}{12} + 80\times 20\times (40-12.22)^2$$

$$= 3.412\times 10^6\,mm^4$$

$$M_U = M_Z \cos\alpha$$

$$M_U = \left[(10\times 2\times \sin 30)+(7\times 3\times \cos 45)\right] = 32.17\,kNm$$

$$M_V = -M_Z \sin\alpha$$

$$M_V = -\left[(10\times 2\times \sin 30)+(7\times 3\times \cos 45)\right] = -24.85\,kNm$$

Point A: (50 mm, 32.22 mm)

Point B: (–50 mm, 32.22 mm)

Point C: (–10 mm, –67.78 mm)

Point D: (10 mm, –67.78 mm)

MATLAB code:

```
  %% Unsymmetrical bending - T section
clc;
clear;

%% INPUT
% calculate the values of moment of inertia of the section in
mm^4
Iy = 1.72e6;
Iz = 3.412e6;
Mz = 37.5; % Moment about Z axis in kNm
```

```
ybar = 32.22; % location of centroidal axis in mm
zbar = 0;
te = 60; %angle of inclination of load

%% Principal axis location
Iu = Iz;
Iv = Iy;

%% Stress calculation
Mu = 32.17; % in kNm
Mv = 24.85;

% calculation for point A - on flange top right
ua = 50;
va = 32.22;
sa = -(Mu*(10^6)*va/Iu)+(Mv*(10^6)*ua/Iv);

% calculation for point B - on flange top left
ub = -50;
vb = 32.22;
sb = -(Mu*(10^6)*vb/Iu)+(Mv*(10^6)*ub/Iv);

% calculation for point C - on web bottom left
uc = -10;
vc = 67.78;
sc = -(Mu*(10^6)*vc/Iu)+(Mv*(10^6)*uc/Iv);

% calculation for point D - on web bottom right
ud = 10;
vd = -67.78;
sd = -(Mu*(10^6)*vd/Iu)+(Mv*(10^6)*ud/Iv);

fprintf('Stress at point A = %6.2f N/mm^2 \n',sa);
fprintf('Stress at point B = %6.2f N/mm^2 \n',sb);
fprintf('Stress at point C = %6.2f N/mm^2 \n',sc);
fprintf('Stress at point D = %6.2f N/mm^2 \n',sd);
```

MATLAB output:

Stress at point A = 418.60 $\frac{\text{N}}{\text{mm}^2}$

Stress at point B = −1026.17 $\frac{\text{N}}{\text{mm}^2}$

Stress at point C = −783.54 $\frac{\text{N}}{\text{mm}^2}$

Stress at point D = 783.54 $\frac{\text{N}}{\text{mm}^2}$

Example 6.6:

Find the stresses at points A and B for the following.

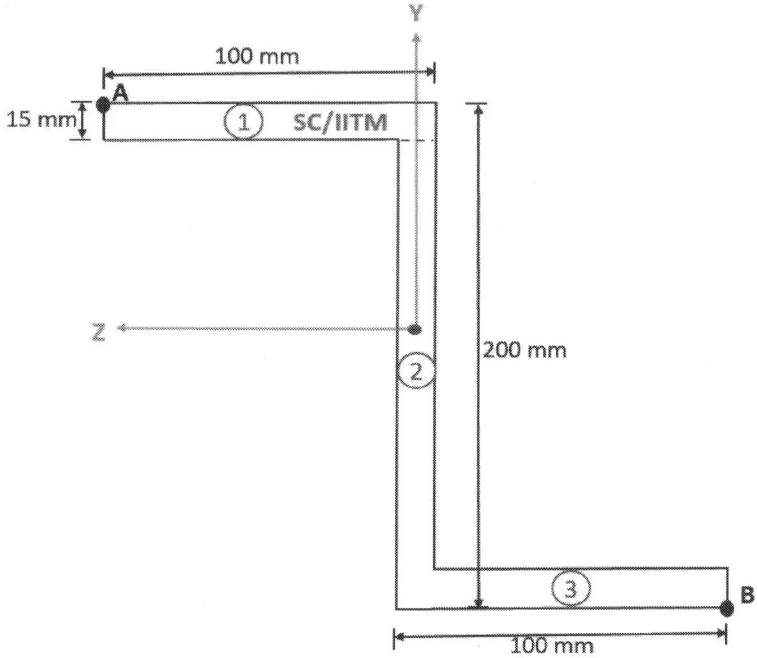

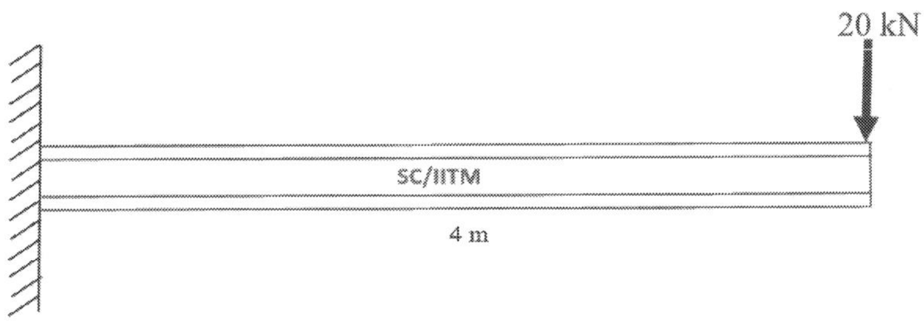

$$I_Y = \left\{ \left[\frac{50 \times 100^3}{12} + \left(100 \times 15 \times 42.5^2\right) \right] \times 2 \right\} + \frac{170 \times 15^3}{12}$$

$$= 7.97 \times 10^6 \ mm^4$$

$$I_Z = \left\{ \left[\frac{100 \times 15^3}{12} + 100 \times 15 \times (92.5)^2 \right] \times 2 \right\} + \frac{15 \times 170^3}{12}$$

$$= 3.187 \times 10^7 \ mm^4$$

$$I_{YZ} = \int zy\,dA$$
$$= \left[(100 \times 15 \times 42.5 \times 92.5)\right] + \left[100 \times 15 \times (-42.5) \times (-92.5)\right]$$
$$= 1.179 \times 10^7\ mm^4$$

$$\text{Moment about Z} - \text{axis} = \text{load} \times \text{perpendicular distance} = 20 \times 4 = 80\,\text{kNm}.$$

MATLAB code:

```
%% Unsymmetrical bending - Z section
clc;
clear;

%% INPUT
% calculate the values of the moment of inertia of the section
in mm^4
Iy = 7.97e6;
Iz = 3.187e7;
Iyz = 1.179e7;
Iuv = 0;
Mz = -80; % Moment about Z axis in kNm

%% Principal axis location
tt = (-2*Iyz)/(Iz-Iy);
al = (atand(tt)+180)/2;
Iu = ((Iy+Iz)/2)+(((Iz-Iy)*cosd(2*al))/2)-(Iy*sind(2*al));
Iv = (Iz+Iy)-Iu;

%% Stress calculation
Mu = Mz*cos(al);
Mv = -Mz*sin(al);
ra = Mv*Iu/(Mu*Iv);
be = atand(ra);

% calculation for point A - on flange top
ya = 92.5;
za = 100;
ua = (za*cos(al))+(ya*sin(al));
va = -(za*sin(al))+(ya*cos(al));
sa = -(Mu*(10^6)*va/Iu)+(Mv*(10^6)*ua/Iv);

% calculation for point B - on web bottom
yb = -92.5;
zb = -100;
ub = (zb*cos(al))+(yb*sin(al));
vb = -(zb*sin(al))+(yb*cos(al));
sb = -(Mu*(10^6)*vb/Iu)+(Mv*(10^6)*ub/Iv);

fprintf('Stress at point A = %6.2f N/mm^2 \n',sa);
fprintf('Stress at point B = %6.2f N/mm^2 \n \n',sb);
```

MATLAB output:

```
Stress at point  A = 407.93 N/mm^2

Stress at point  B = −407.93 N/mm^2
```

Example 6.7:

For a beam cross-section shown in figure, locate the point of the neutral axis and maximum stresses at points A and B under the action of the external moment of 1 MNm.

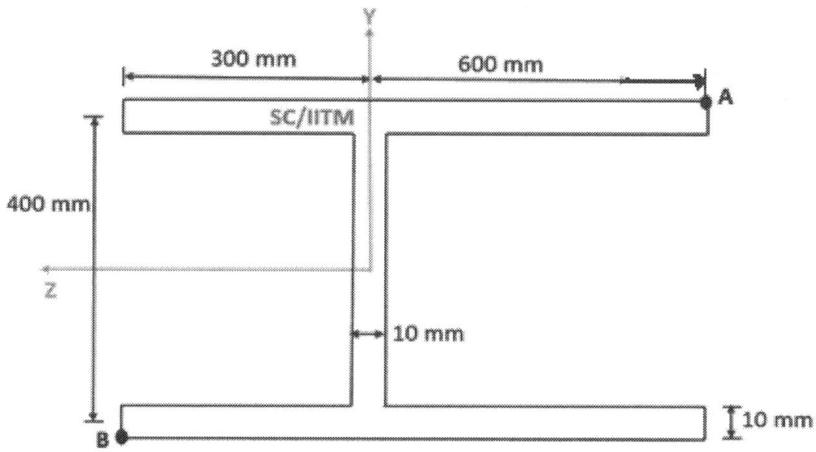

$$\bar{x} = \frac{\sum ax}{\sum a}$$

$$\bar{x} = \frac{(90\times100\times450)\times2+(390\times10\times300)}{(90\times10)\times2+(390\times10)} = 123.28\,mm$$

$$I_Y = \left\{\left[\frac{10\times900^3}{12}+\left(900\times10\times\langle450-423.28\rangle^2\right)\right]\times2\right\}+\frac{390\times10^3}{12}+\left(390\times10\times123.28^2\right)$$

$$= 1.287\times10^9\,mm^4$$

$$I_Z = \left\{\left[\frac{900\times10^3}{12}+900\times10\times(200)^2\right]\times2\right\}+\frac{10\times390^3}{12}$$

$$= 7.696\times10^8\,mm^4$$

Point A: (–476.72 mm, 205 mm)
Point B: (423.28 mm, –205 mm)

MATLAB code:

```
%% Unsymmetrical bending - Rectangular section
clc;
clear;
```

```
%% INPUT
% calculate the values of the moment of inertia of the section
in mm^4
Iy = 1.287e9;
Iz = 7.696e8;
Mz = 1000; % Moment about Z axis in kNm
te = 30; %angle of inclination of load

%% Principal axis location
Iu = Iz;
Iv = Iy;

%% Stress calculation
Mu = Mz*cosd(te);
Mv = -Mz*sind(te);

% calculation for point A - on flange top
ua = -476.72;
va = 205;
sa = -(Mu*(10^6)*va/Iu)+(Mv*(10^6)*ua/Iv);

% calculation for point B - on web bottom
ub = 423.28;
vb = -205;
sb = -(Mu*(10^6)*vb/Iu)+(Mv*(10^6)*ub/Iv);

fprintf('Stress at point A = %6.2f N/mm^2 \n',sa);
fprintf('Stress at point B = %6.2f N/mm^2 \n \n',sb);

% location of neutral axis
vbyu=Iu*Mv/(Iv*Mu);
be=atand(vbyu);
fprintf('Beta = %6.2f degrees \n',be);
```

MATLAB output:

Stress at point A $= -45.48\ \frac{\text{N}}{\text{mm}^2}$

Stress at point B $= 66.24\ \frac{\text{N}}{\text{mm}^2}$

Beta $= -19.05$ degrees

6.4 CURVED BEAMS

The curved beams are classified based on the initial curvature into:

1. Beams with small initial curvature: The ratio of the initial curvature radius and the section's depth is greater than 10.
2. Beams with large initial curvature: The ratio of the initial curvature radius and the section's depth is less than or equal to 10.

6.4.1 Small Initial Curvature

Bending of curved beams with small initial curvature is shown in Figure 6.2.

Let,

$d\varphi'$ be the angle subtended after deformation.
R be the initial radius of curvature.
$d\varphi$ be the angle subtended at the center of curvature by the portion "abcd."
R′ be the radius of curvature after moment M is applied.

If R′<R, then the applied moment will tend to close the curvature. Let us consider a fiber PQ at a distance y from the neutral axis.

Then,

The original length of the fiber = (R+y) $d\varphi$
Length of the fiber after application of the moment = (R′+y) $d\varphi'$
Change in length of the fiber = (R′+y) $d\varphi'$ – (R+y) $d\varphi$

$$\text{Strain, } \epsilon = \frac{(R'+y)d\varphi' - (R+y)d\varphi}{(R+y)d\varphi}$$

As the length of the fiber at the neutral axis remains unchanged,

$$ds = R\,d\varphi = R'd\varphi'$$

Substituting the above equation in the strain equation,

$$\varepsilon = \frac{y(d\varphi' - d\varphi)}{(R+y)d\varphi}$$

In the above equation, *y* may be neglected, since y≪R.

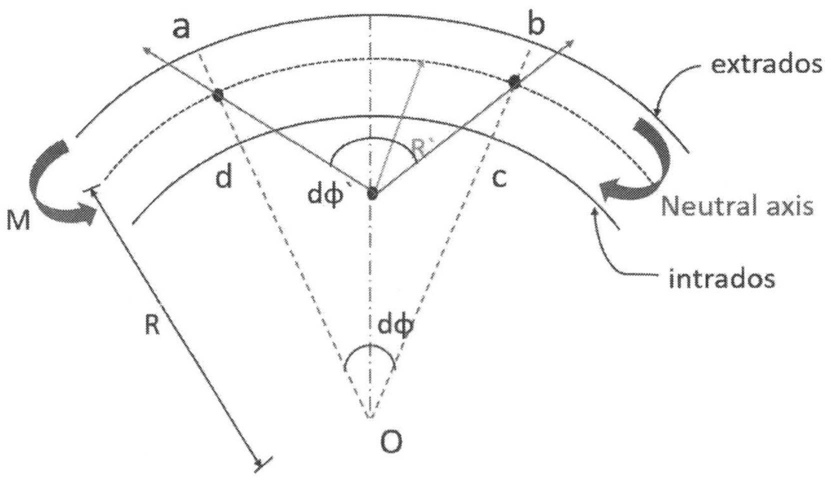

FIGURE 6.2 Curved beam with small initial curvature.

Hence,

$$\varepsilon = \frac{y(\mathrm{d}\varphi' - \mathrm{d}\varphi)}{\mathrm{R}\,\mathrm{d}\varphi}$$

On simplification,

$$\varepsilon = \frac{y\,\mathrm{d}\varphi'}{R\,\mathrm{d}\varphi} - \frac{y\,\mathrm{d}\varphi}{R\,\mathrm{d}\varphi} = y\left(\frac{\mathrm{d}\varphi'}{\mathrm{ds}} - \frac{\mathrm{d}\varphi}{\mathrm{ds}}\right)$$

$$\varepsilon = y\left(\frac{1}{R'} - \frac{1}{R}\right)$$

Substituting $\varepsilon = \frac{\sigma}{E}$,

$$y\left(\frac{1}{R'} - \frac{1}{R}\right) = \frac{\sigma}{E}$$

$$\frac{\sigma}{y} = E\left(\frac{1}{R'} - \frac{1}{R}\right)$$

The assumptions made in deriving the equation are as follows:

1. Every cross-section of the curved beam remains plane and perpendicular to the centroidal axis, before and after the application of the external moment.
2. To satisfy the above condition, it is to be agreed that the net force acting on any cross-section of the curved beam should be zero. If the net force is not equal to zero, then it may result in warping.

Mathematically,

$$\int_A \sigma\, dA = 0$$

$$\int Ey\left(\frac{1}{R'} - \frac{1}{R}\right) dA = 0$$

$$E\left(\frac{1}{R'} - \frac{1}{R}\right)\int y dA = 0$$

Since $E\left(\frac{1}{R'} - \frac{1}{R}\right) \neq 0$

$\int y dA = 0$ which implies that the geometric axis of the curved beam should coincide with the neutral axis of the curved beam. As the curved beam is in equilibrium under the applied moment, it can be stated that,

$$\int \sigma y dA = M$$

Substituting for σ in the above equation,

$$E\left(\frac{1}{R'}-\frac{1}{R}\right)\int y^2 dA = M$$

Since $\int y^2 dA = I$,

$$E\left(\frac{1}{R'}-\frac{1}{R}\right)I = M$$

Thus,

$$\frac{M}{I}=\frac{\sigma}{y}=E\left(\frac{1}{R'}-\frac{1}{R}\right)$$

Deflection of the curved beam with small initial curvature

For deriving the deflection of the curved beam, the Castigliano's theorem is used. The strain energy of the curved beam is given by,

$$U=\int\frac{1}{2}M\Delta d\varphi'$$

Where dφ` is the change in the angle produced by the moment *M* at the center of curvature.

From the following equation,

$$\frac{M}{I}=E\left(\frac{1}{R'}-\frac{1}{R}\right)$$

Multiplying *ds* on both sides of the equation,

$$\frac{M}{I}ds=E\left(\frac{1}{R'}-\frac{1}{R}\right)ds$$

Since the length of the fiber remains unchanged,

$$\mathrm{ds}=\mathrm{R\,d}\varphi=\mathrm{R'\,d}\varphi'$$

$$\frac{M}{I}ds=E\left(\mathrm{d}\varphi-\mathrm{d}\varphi'\right)ds$$

$$\frac{M}{EI}ds=\Delta\mathrm{d}\varphi$$

Thus, the strain of the curved beam is given by,

$$U=\int\frac{1}{2}\frac{M^2 ds}{EI}$$

By partially differentiating U with respect to p,

$$\Delta = \frac{\partial U}{\partial p} = \int \frac{M}{EI} \frac{\partial M}{\partial p} ds$$

Similarly, the angular rotation may be obtained by partially differentiating U with respect to the angular momentum M_o,

$$\theta = \frac{\partial U}{\partial M_o} = \int \frac{M}{EI} \frac{\partial M}{\partial M_o} ds$$

6.4.2 Large Initial Curvature

Consider a strip in the cross-section of the curved beam; the concave side is called *intrados,* and the convex side is called *extrados*. The important points to be noted are:

1. Stress variation is nonlinear across the cross-section; it is hyperbolic. Experimental studies showed that stress on the concave side is larger than that of the convex side.
2. Neutral axis will shift toward the center of curvature.
3. Neutral axis will not pass through the centroid of the cross-section since, $\sigma_{concave\ side} \neq \sigma_{convex\ side}$

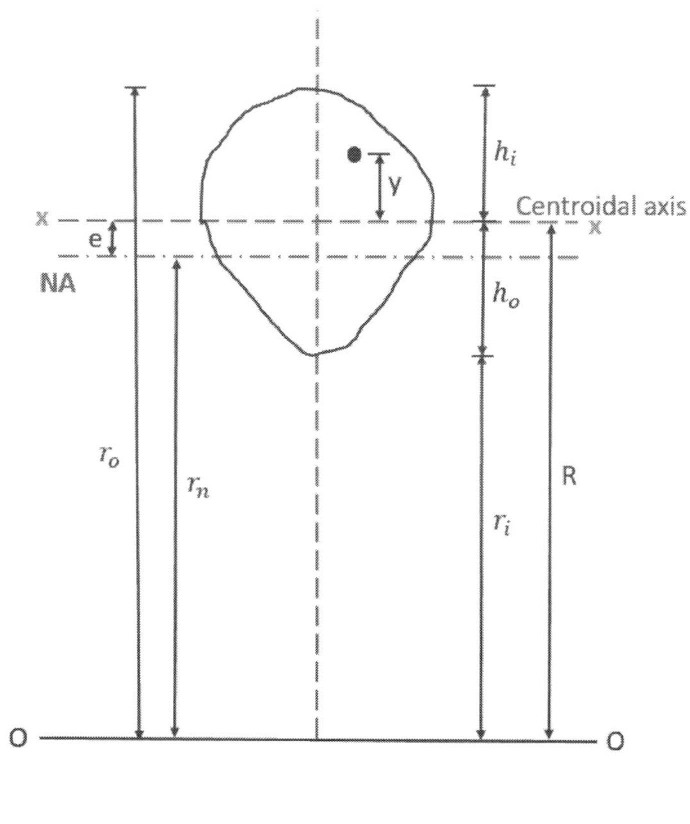

Let us consider a curved beam with radius R. It is subjected to a moment, M, which tends to decrease the radius of curvature. The cross-section details are shown in the above figure. It causes tension in the extrados and compression in the intrados.

Consider an element ABDC, subtending an angle $(d\varphi)$ at the center. Under the applied moment, M, this element deforms to AB $D'C'$ for curvature. It is assumed that AB is fixed and only plane CD rotates, and thus, face $\overline{CD}$ rotates to $C'D'$ subtending an angle $\Delta d\varphi$.

Initial lengths of the fiber $\overline{PQ}$ at distance y from the centroidal axis are given by

$$(R+y)d\varphi$$

Change in length of the fiber on the application of the moment, M is given by

$$(y+e)\Delta d\varphi$$

Where,

e is the distance of the neutral axis measured from the centroidal axis
M is applied in such a manner that the neutral axis is shifted toward the center of curvature

$$\text{Strain}, \epsilon = \frac{(y+e)\Delta d\varphi}{(R+y)d\varphi}$$

Assuming that the longitudinal fibers do not undergo any deformation, stress is given by:

$$\sigma = E\left(\frac{y+e}{R+y}\right)$$

It shows that the stress distribution is nonlinear and hyperbolic.

The basic assumption made is that every section normal to the centroid axis remains plane and perpendicular before and after application of moment M.

$$\text{Total compressive force} = \text{total tensile force}$$

Since the average stress on the concave side is more than the convex side, the neutral axis will shift toward the center of curvature.

Equate the sum of internal forces to zero at the cross-section.

Mathematically,

$$\int_A \sigma dA = 0$$

Substituting the expression for stress, we get

$$\int_A E\frac{Ad\varphi}{d\varphi}\left(\frac{y+e}{y+R}\right)dA = 0$$

$$E\frac{\Delta d\varphi}{d\varphi}\int_A\left(\frac{y+e}{y+R}\right)dA=0$$

Since $E\frac{\Delta d\varphi}{d\varphi}\neq 0$ and it is a constant,

$$\int_A\left(\frac{y+e}{y+R}\right)dA=0$$

$$\int_A\left(\frac{y}{R+y}\right)dA+e\int_A\left(\frac{1}{y+R}\right)dA=0$$

We know that the first integral term, $\int_A\frac{y}{R+y}dA=mA$, where m is a constant depending on the geometry of the X section. The quantity mA is termed as the modified area of the cross-section, which is modified due to the application of moment, M.

The second integral term is $e\int_A\frac{1}{(R+y)}dA$

$$e\int_A\frac{1}{(R+y)}dA=e\int\frac{R+y-y}{R}\cdot\frac{1}{(R+y)}dA$$

$$=\frac{e}{R}\int_A\frac{R+y}{R+y}dA-\int_A\frac{ydA}{R+y}=\frac{eA}{R}-\frac{e}{R}(mA)$$

Thus,

$$mA+\frac{eA}{R}-\frac{emA}{R}=0$$

$$m+\frac{e}{R}-\frac{em}{R}=0$$

$$m=e\left(\frac{m}{R}-\frac{1}{R}\right)$$

$$m=\frac{e}{R}(m-1)$$

$$e=\left(\frac{m}{m-1}\right)R$$

where m is the geometry property of the section.

It is to be noted that the applied moment in the cross-section should be equal to the resisting moment for any equation.

Hence, by applying the following condition,

$$\int_A(\sigma dA)y=M$$

Substitution for stress, $\int_A\frac{y^2+ye}{R+y}dA$ is to be evaluated.

$$\int_A\frac{y^2dA}{R+y}+e\int_A\frac{y}{y+R}dA=\int_A\left(y-\frac{Ry}{R+y}\right)dA+e\int_A\frac{y}{y+R}dA$$

We know that,

$$\int_A y dA = 0$$

Hence,

$$\int_A \left(\frac{y^2 + ye}{R+y} \right) dA = -R\int_A \left(\frac{y}{R+y} \right) dA + e\int_A \frac{ydA}{R+y}$$

$$= -R(mA) + emA = -mA(R-e)$$

$$-E\left(\frac{\Delta d\varphi}{d\varphi} \right) = \frac{M}{mA(R-e)}$$

Since, $m = \dfrac{e}{R-e}$

$$E\frac{\Delta d\varphi}{d\varphi} = \frac{M}{Ae}$$

On simplification,

$$e = \frac{M}{Ae}\left(\frac{y+e}{R+y} \right)$$

It is also known that,

$$\sigma = \frac{M}{(m-1)}R$$

$$\sigma = \frac{M}{AR}\frac{(m-1)}{m}\left[\frac{y + \left(\frac{m}{m-1} \right)R}{R+y} \right]$$

$$= \frac{m}{AR}\frac{(m-1)}{m}\left[\frac{(m-1)y + mR}{(m-1)(R+y)} \right]$$

$$\sigma = \frac{M}{AR}\frac{1}{m}\left[\frac{m(y+R) - y}{y+R} \right]$$

$$\sigma = \frac{M}{AR}\left[1 - \frac{1}{m}\left(\frac{y}{R+y} \right) \right]$$

The above equation is named as the *Winkler Bach equation*, in which

σ is the tensile/compressive stress at distance y from the centroidal axis (not from the NA)
M is the applied moment (causing a decrease in curvature)
A area of cross-section

m is the section properties (geometry/shape of the cross-section)
R is the radius of curvature of the unstressed curved beam

Sign convention:
y is negative when measured toward the concave side and positive when measured toward the convex side. With the sign convention, negative stress indicates compressive stress, and positive stress indicates tensile stress.

In the plastic design, to enable maximum load capacity, designers use sections with large shape factors

$$\sigma = \frac{M}{Ae}\left[\left(\frac{y+e}{y+R}\right)\right]$$

Where e is the offset of the neutral axis from the center of gravity measured toward the center of curvature. Specific stress equations for intrados and extrados to be used to find the maximum stress are the extreme fibers as follows:

$$\sigma_{intrados} = \frac{M}{Ae}\left[\frac{-h_i+e}{R-h_i}\right] = -\frac{M}{Ae}\left[\frac{h_i+e}{r_i}\right]$$

$$\sigma_{extrados} = \frac{M}{Ae}\left[\frac{h_o+e}{R+h_o}\right] = \frac{M}{Ae}\left[\frac{(e+h_o)}{r_o}\right]$$

From the above equations, it is clear that stress is a function of the parameter 'm'. We already know that

$$mA = \int_A \frac{y}{R+y}\,dA$$

6.4.3 Simplified Equations to Estimate Stresses

$$\sigma = K\frac{Mh}{I}$$

Where K is a factor to be used for intrados and extrados as below:

$$K_{\text{intrados}} = \frac{\frac{M}{Mh}\frac{(h_i-e)}{r_i}}{\frac{Mh_i}{2I}}$$

$$K_{\text{extrados}} = \frac{\frac{h_o+e}{r_0}}{\frac{Mh_0}{2I}}$$

k_i *and* k_0 are called stress correction factors, as shown in Table 6.1.

TABLE 6.1
Stress Correction Factors for Various Cross-Sections

Cross-Section	R/h	Factors K_i	K_o	e
Circular	1.2	3.41	0.54	0.224R
	1.4	2.40	0.60	0.151R
	1.6	1.96	0.65	0.108R
	1.8	1.75	0.68	0.084R
	2.0	1.62	0.71	0.009R
	3.0	1.33	0.79	0.030R
	4.0	1.23	0.84	0.016R
	6.0	1.14	0.89	0.007R
Rectangular	1.2	2.89	0.57	0.305R
	1.4	2.13	0.63	0.204R
	1.6	1.79	0.67	0.149R
	1.8	1.63	0.70	0.112R
	2.0	1.52	0.73	0.090R
	3.0	1.30	0.81	0.041R
	4.0	1.20	0.85	0.021R
	6.0	1.12	0.90	0.0093R

6.5 CRANE HOOKS

Consider an open section comprising a 100 MN load as shown in figure. Compute the stresses at points A and B in the curved beam.

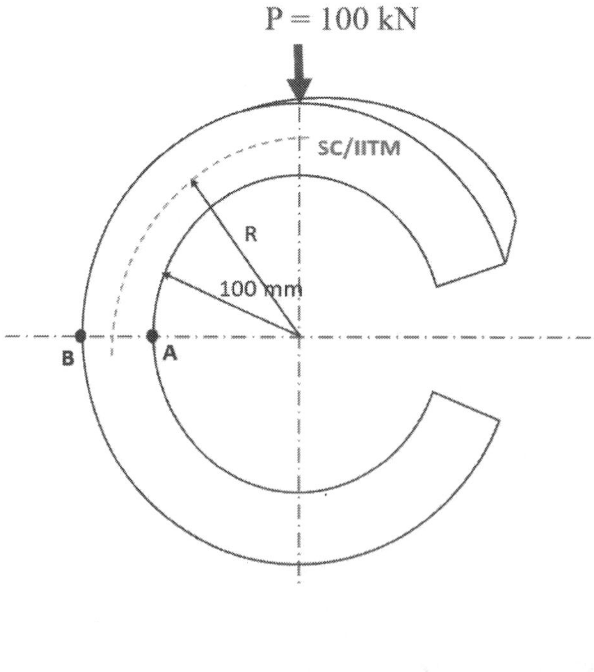

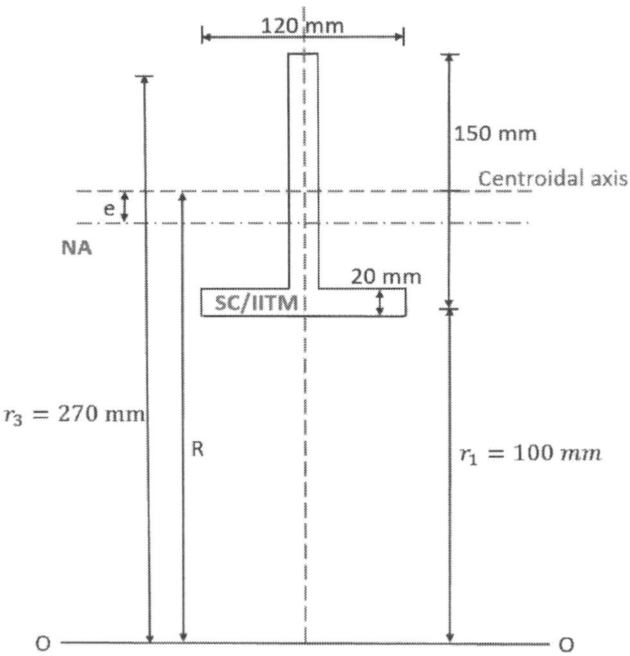

MATLAB code:

```
% Curved beam - T section
clc;
clear;

%% INPUT
p=100; % load acting on the beam in kN
%Geometric properties
r1=100;
r2=120;
r3=270;
r=157.22;
h1=112.78;
h2=57.22;
h3=37.22;
b1=120;
b2=20;
a= 5400; % cross sectional area in mm^2
m=1-(r/a)*((b1*log(r2/r1))+(b2*log(r3/r2))); % sectional
property (no unit)
e=m*r/(m-1); % eccentricity in mm
```

```
%% Section AB on the centroidal axis
sd=-p*1000/a; % direct stress in N/mm^2
M=p*r/1000; % moment at CG in kNm
si=-(M*1000000*(h2-e)/(a*e*r1)); % stress at intrados in N/
mm^2
so=(M*1000000*(h1+e)/(a*e*r3)); % stress at extrados in N/mm^2
sa=sd+si; % total stress at intrados
sb=sd+so; % total stress at extrados
fprintf('SOLUTION: \n');
fprintf('Stress at point A = %6.2f N/mm^2 \n',sa);
fprintf('Stress at point B = %6.2f N/mm^2 \n',sb);
```

MATLAB output:

Stress at point A $= -97.04 \frac{\text{N}}{\text{mm}^2}$

Stress at point B $= 70.84 \frac{\text{N}}{\text{mm}^2}$

Example 6.8:

A circular ring of the trapezoidal cross-section has a horizontal slit, as shown in the figure. Find the stresses at A and B.

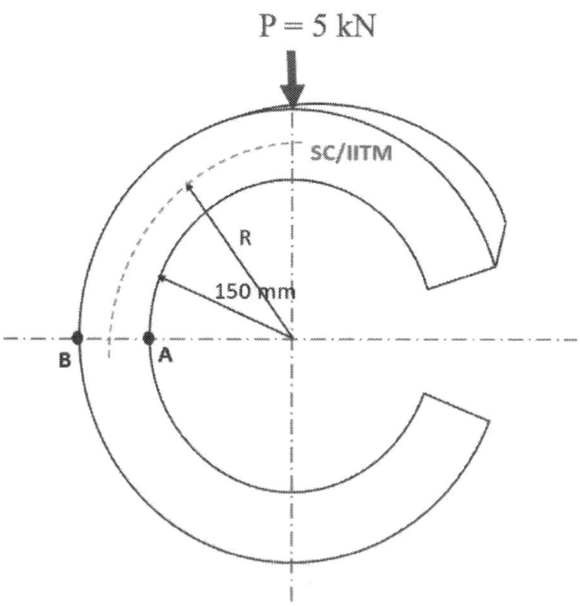

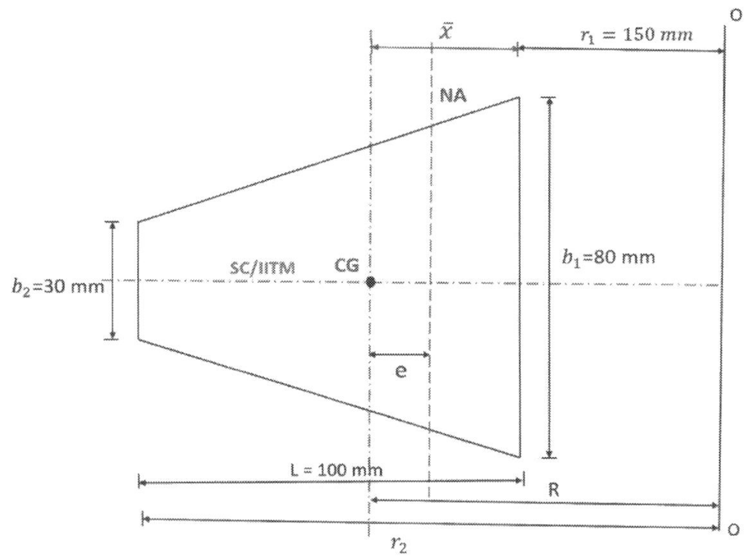

MATLAB code:

```
% Curved beam with a trapezoidal cross section
clc;
clear;

%% INPUT
h=100; % height of the section in mm
b1=80; % breadth of the section in mm
b2=30;
r1=150; % inner radius of the beam in mm
p=5; % load action on the beam in kN

%% Geometric properties
b3=(b1-b2)/2;
x=((2*h*b3/3)+(b2*h))/(b1+b2); % Location of neutral axis
hi=x;
ho=h-x;
R=r1+x; % radius of the curved beam in mm
r2=r1+h; % outer radius of the curved beam in mm
a=(b1+b2)*h/2; % cross sectional area in mm^2
m=1-((R/a)*((b2+((r2*(b1-b2))/(r2-r1)))*log(r2/r1)-(b1-b2)));
% sectional property
e=m*R/(m-1); % eccentricity in mm
I=(2*(((b3*(h)^3)/36)+(0.5*b3*h*((h/3)-x)*((h/3)-x))))+(((b2*
(h)^3)/12)+(b2*h*((h/2)-x)*((h/2)-x))); % moment of inertia in
mm^2
```

```
%% Section AB on the centroidal axis
sd=-p*1000/a; % direct stress in N/mm^2
M=p*R/1000; % moment at CG in kNm
si=-(M*1000000*(hi-e)/(a*e*r1)); % stress at intrados in
N/mm^2
so=(M*1000000*(ho+e)/(a*e*r2)); % stress at extrados in N/mm^2
sa=sd+si; % total stress at intrados
sb=sd+so; % total stress at extrados
fprintf('SOLUTION: \n');
fprintf('Stress at point A = %6.2f N/mm^2 \n',sa);
fprintf('Stress at point B = %6.2f N/mm^2 \n',sb);
```

MATLAB output:

SOLUTION:

Stress at point A $= -12.35 \frac{N}{mm^2}$

Stress at point B $= 10.05 \frac{N}{mm^2}$

Example 6.9:

Find the stresses developed at points A and B of the curved beam with a circular cross-section shown in the figure.

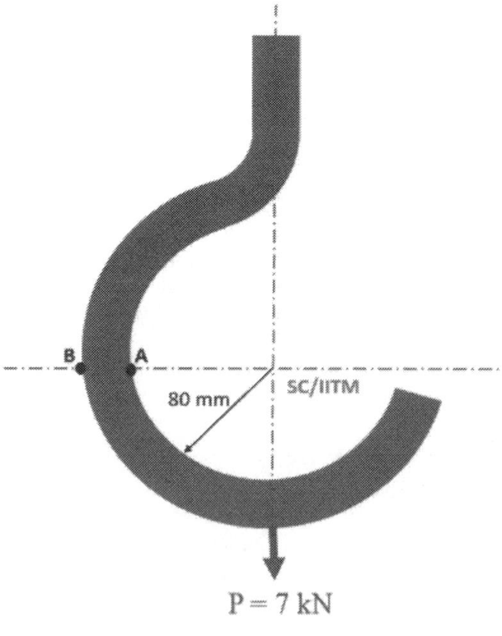

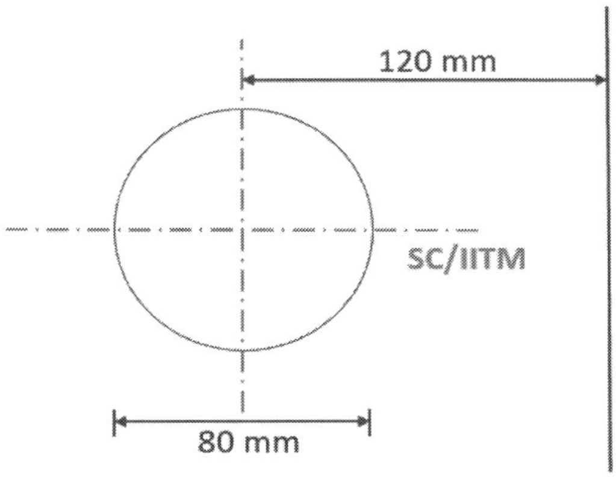

MATLAB code:

```
% Crane hook problem with Circular cross-section
clc;
clear;
%% INPUT

r1=80; % inner radius of the beam in mm
d=80; % diameter of the section in mm
p=7; % load action on the beam in kN

%% Geometric properties
x=d/2; % Location of neutral axis
hi=x;
ho=d-x;
R=r1+x; % radius of the curved beam in mm
r2=r1+d; % outer radius of the curved beam in mm
a=3.14*d*d/4; % cross sectional area in mm^2
m=1-(2*((R/d)^2))+(2*(R/d)*sqrt(((R/d)^2)-1)); % sectional
property
e=m*R/(m-1); % eccentricity in mm

%% Section AB - stress calculation - Winkler Bach equation
sd=p*1000/a; % direct stress in N/mm^2
M=-p*R/1000; % moment at CG in kNm
ri=r1;
ro=r2;
si=-M*1000000*(hi-e)/(a*e*ri); % Stress at intrados
so=M*1000000*(ho+e)/(a*e*ro); % stress at extrados
sa=sd+si; % total stress in intrados
sb=sd+so; % total stress at extrados
fprintf('SOLUTION FROM WINKLER BACH EQUATION: \n');
fprintf('Stress at point A = %6.2f N/mm^2 \n',sa);
fprintf('Stress at point B = %6.2f N/mm^2 \n \n',sb);
```

MATLAB output:

SOLUTION FROM WINKLER BACH EQUATION:

Stress at point A = 4.77 $\frac{\text{N}}{\text{mm}^2}$

Stress at point B = −2.39 $\frac{\text{N}}{\text{mm}^2}$

Example 6.10:

Find the stresses at points A and B of the crane hook of the trapezoidal cross-section under the action of 70 kN load as shown in figure.

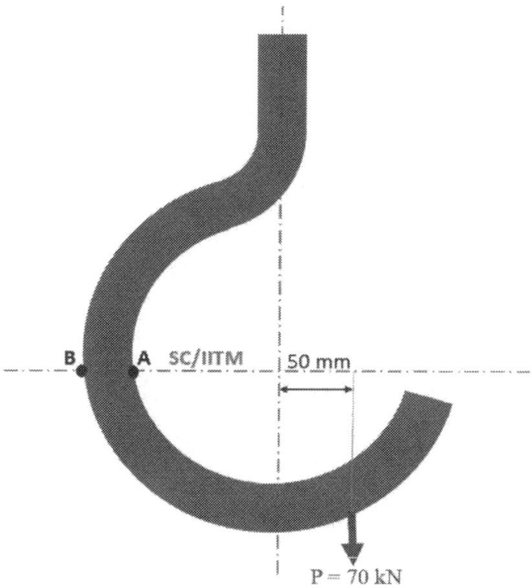

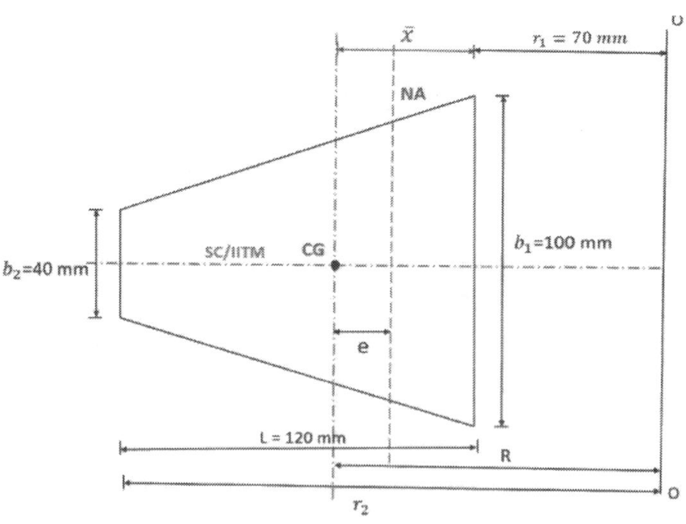

MATLAB code:

```
% Crane hook problem with a trapezoidal cross section
clc;
clear;

%% INPUT
h=120; % height of the section in mm
b1=100; % breadth of the section in mm
b2=40;
r1=70; % inner radius of the beam in mm
p=70; % load action on the beam in kN

%% Geometric properties
b3=(b1-b2)/2;
x=((2*h*b3/3)+(b2*h))/(b1+b2); % Location of neutral axis
hi=x;
ho=h-x;
R=r1+x; % radius of the curved beam in mm
r2=r1+h; % outer radius of the curved beam in mm
a=(b1+b2)*h/2; % cross-sectional area in mm^2
m=1-((R/a)*((b2+((r2*(b1-b2))/(r2-r1)))*log(r2/r1)-(b1-b2)));
% sectional property
e=m*R/(m-1); % eccentricity in mm
I=(2*(((b3*(h)^3)/36)+(0.5*b3*h*((h/3)-x)*((h/3)-x))))+(((b2*(
h)^3)/12)+(b2*h*((h/2)-x)*((h/2)-x))); % moment of inertia in
mm^2

%% Section AB - stress calculation - Winkler Bach equation
sd=p*1000/a; % direct stress in N/mm^2
M=-p*(R+50)/1000; % moment at CG in kNm
ri=r1;
ro=r2;
si=-M*1000000*(hi-e)/(a*e*ri); % Stress at intrados
so=M*1000000*(ho+e)/(a*e*ro); % stress at extrados
sa=sd+si; % total stress in intrados
sb=sd+so; % total stress at extrados
fprintf('SOLUTION FROM WINKLER BACH EQUATION: \n');
fprintf('Stress at point A = %6.2f N/mm^2 \n',sa);
fprintf('Stress at point B = %6.2f N/mm^2 \n \n',sb);
```

MATLAB output:

SOLUTION FROM WINKLER BACH EQUATION:

Stress at point A = 102.87 $\frac{N}{mm^2}$

Stress at point B = −55.65 $\frac{N}{mm^2}$

7 Stability of Structural Systems

7.1 CONDITIONS FOR STABILITY

As discussed earlier, stability is affected significantly under compressive forces. Further, geometric stability is more critical than material stability, as the former may challenge the functional requirements of the structural member. Therefore, stability refers to a stable state of equilibrium and is defined as the ability of any structural system to remain (or continue to remain) in its geometric form, which is capable of performing the intended function even if the geometric position is disturbed by external forces. By this definition, compliant offshore structures are said to be stable as they can perform their intended function at the disturbed geometric position. For example, a TLP is said to be in a disturbing position under the combined effect of offset and set-down. As long as this change in geometric position does not affect its load-disbursing capacity, the TLP is stable. Therefore, structures should not be necessary to remain (or continue to remain) in their original geometric form to classify them as stable. They may remain stable even under the deformed geometric position if they continue successfully performing their intended function.

The three criteria for checking stability are (i) Euler's static criterion, (ii) Liapunov's dynamic criterion, and (iii) the Potential energy stability criterion. Euler's static criterion is applicable under the nontrivial equilibrium state. It evaluates the stability of a structural system by examining the optimum geometric configuration of the system other than the original (initially straight) configuration at which the structural system can still disburse the applied load (P_{cr}). Structures are examined under the given boundary conditions and initially perfect straight geometry. Euler's criterion evaluates whether the structure can carry the load (maybe in a lesser magnitude, Pcr where P>Pcr) instead of remaining in a state where it cannot have any load. It is interesting to note that the load-carrying capacity of the structural member is reduced from P to P_{cr}. Still, the critical fact is that the structural member can carry at least P_{cr} even at a changed geometric form, which is relatively weak compared to the initial form. This load value with a lesser magnitude than the initially intended load (P) is termed a critical or buckling load. It can be easily seen from the standard literature that P_{cr} is easily computed from the boundary conditions, cross-sectional properties of the member, and slenderness ratio. However, instability occurs when two or more adjacent equilibrium positions correspond to different mode shapes.

For assessing the stability of offshore compliant structures, it is reasonably simple to disagree with the application of Euler's criterion, as the boundary conditions of the member do not permit the examination of Euler's criterion of stability. They are either entirely floating under hydrodynamic stability or compliant, and hence,

DOI: 10.1201/9781003639152-7

stability depends on the tethers' high-magnitude pretension. For example, in the case of TLPs, high pretension imposed on tethers ensures stability and recentering of the platform under lateral loads. The functional working of the platform is not lost even under the deformed position of the platform under wave loads. Please note that this condition is actually present even though the offset values are large (about 10% of the initial pretension of tethers), causing large deformation. Liapunov's condition examines the stability of the structural system under dynamic excitations. If a member is subjected to a continuously varying disturbing force, it is necessary to determine whether the equilibrium condition under the dynamic forces is satisfied. As explained in the literature (Chandrasekaran, 2015, 2016, 2017), stability can be influenced by varying amplitudes of the exciting force and its excitation period. While the former can affect the member's load-carrying capacity, which is the strength-dependent criterion, the latter can result in an unconditional response at the near-resonance state of vibration. Hence, Lyapunov's stability assessment focuses on the structural system's dynamic response behavior instead of purely assessing its load-carrying capacity, as in the case of Euler's criterion.

Liapunov's condition is more significant for structures that perform their intended function under varying external forces. The most challenging part in the context of stability is that the structure shall also assume a different geometric position with time. Structural systems like ships, offshore compliant platforms, and floating production platforms fall under this category of stability check. Stability calculations of ships focus on estimating the center of gravity, the center of buoyancy, the metacenter of vessels, and their interaction. The saving part of such systems is that they are designed to remain hydrostatically stable at any instant. In the case of structures that are permitted to undergo large displacements, as in the case of TLPs, then the geometric design ensures proper recentering, which means that the structural system continuously tries to regain its original geometric position with the help of dedicated members present within the system; in case of TLP, it is tethered. A potential energy stability criterion applies to structural systems for which the system's potential energy ceases to be the minimum, which is more relevant to conservative systems.

7.2 BUCKLING AND INSTABILITY

Structural stability loss is termed instability, which can be assessed based on (i) material properties, (ii) geometric configuration, and (iii) the nature and magnitude of loads acting on the structure. Assessment based on material strength highlights where the applied load exceeds the accepted strength level of the material, classifying the structure as unstable. Assessment based on the geometric configuration explores the functional success of the member even under the deformed (displaced) position, while that based on the nature and magnitude of the loads assesses dynamic instability. In general terms, loss of stability is expressed by a load-deformation relationship. Figure 7.1 shows an unstable situation; even with a slight disturbance, the ball can roll over, causing instability.

A significant change in the geometric configuration of a structure or any structural component under compression, resulting in loss of ability to resist the encountered loads, is termed instability. It is essential, as this can lead to a catastrophic failure, as

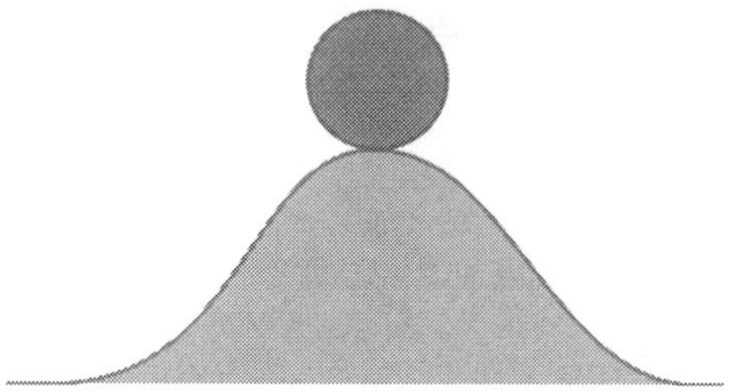

FIGURE 7.1 Unstable condition.

the load-carrying capacity of an unstable structural system is close to zero; hence, it should be appropriately accounted for in the design. Instability is a strength-related limit state. Structural systems can become unstable upon the application of excessive loads. Beyond the threshold limits, stresses developed in the member can magnify the deflections, causing failure. Any structural system whose deformation exceeds the permissible limits is termed unstable. Therefore, the stability of any structural system depends upon its ability to withstand the encountered load without undergoing excessive deformation. It is important to note that a structure found to be stable for a load applied at one section may become unstable under the same magnitude and nature of the load but applied at a different section. Stability can be illustrated with the help of the figures shown aside. The ball remains in equilibrium in all three cases, as the specific locations are shown. However, a slight disturbance imposed on the ball can change its behavior differently in the three cases. As seen in case (a), the ball can return to its original position upon removal of any disturbing external force

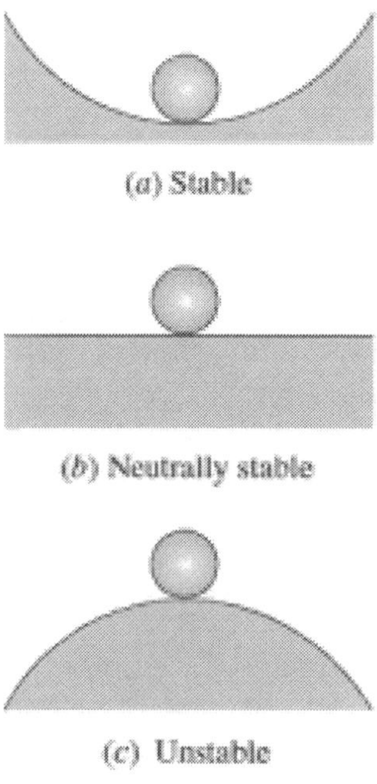

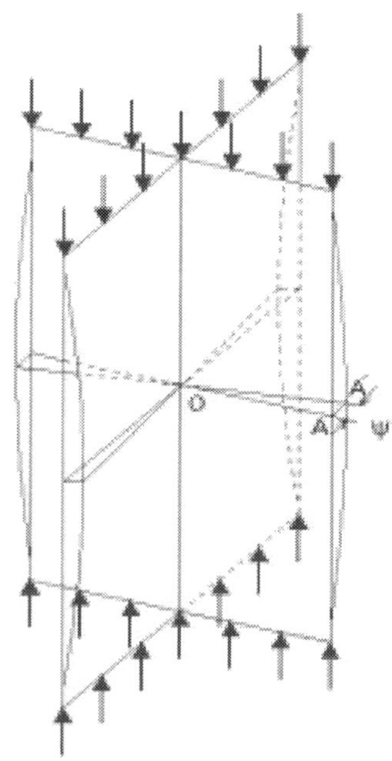

and, therefore, assumes a stable position. It corresponds to a position in the material stress-strain curve within the elastic limit. In case (b), the ball is considered neutrally stable, as the ball can continue to move under the influence of disturbing external forces. This state corresponds to a point on the ultimate strength region of the material's stress-strain curve. However, in case (c), the ball cannot return to its original position even under a minor disturbance caused by an external force. It is a classical state of instability. It is important to note that loss of stability, arising either from a change in geometry or the structural configuration, adds a new set of additional loads to the structural system. It may lead to the further classification of instability as flexural buckling (FB), torsional buckling (TB), flexural-torsional buckling (FTB), and lateral-torsional buckling (LTB).

FB can occur in any compression member undergoing deflection that arises from bending. Two factors that influence FB are a large slenderness ratio and a smaller radius of gyration. TB occurs in slender compression members that are double-symmetric in shape and arrangement, as shown in the figure. As seen in the figure, the longitudinal axis of the members is turned (bent laterally), causing the TB. It is commonly seen in built-up sections due to design inadequacy in terms of sectional properties along the length of the member. If the loads in any structural system are set to act through the shear center of the section, then no twisting or torsion of the member occurs. As the shear center is located on the axis of symmetry, cross-sections with two axes of symmetry will have a shear center located at the intersection of both axes. For sections with one axis of symmetry, for example, a channel section, the shear center can be located using the standard procedure (Chandrasekaran, 2015). While the center of gravity (or the mass center) is the point about which moments generated from the mass of the elements will be zero, the eccentricity of the center of gravity concerning the shear center will result in a couple and cause twisting of the section, referred to as TB. Distortional buckling is a unique mode of buckling where the flange of the section rotates at the intersection of the flange and the section's web. It is also referred to as local-torsional buckling. It occurs mainly due to adequate stiffening of the web compared to the flange.

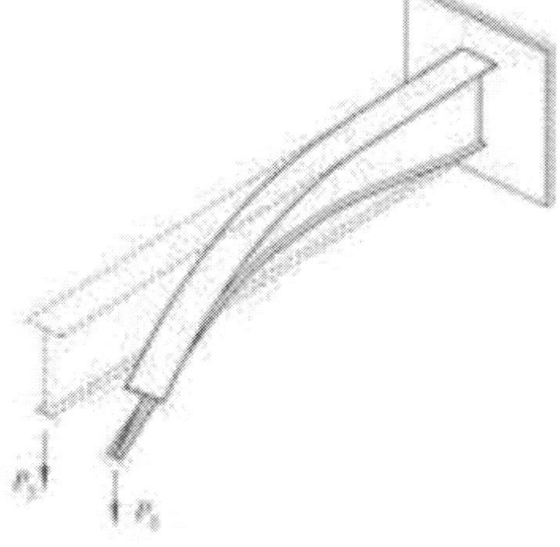

On the other hand, compression members with one axis of symmetry undergo FTB. FTB is a combination of bending and twisting of a member under compressive loads. This buckling is commonly seen in members of open cross-sections subjected to compressive loads. Examples of open sections are channel, tee, and angle sections. I am mainly familiar with these open sections because of the lower torsional stiffness. It is important to note that circular cross-sections do not experience such a buckling mode. The above figure shows a schematic view of a cantilever beam undergoing FTB.

LTB occurs in open sections as well. Consider a supported beam under a central concentrated load causing bending. While the top fiber experiences compression, the bottom fiber experiences tension. Considering the long-span condition of the beam, if the compression flange is not laterally supported, it experiences large lateral deflections. Hence, the beam will twist in addition to a sizeable lateral deflection, a combination termed LTB. Even the width of the compression flange (b_f) significantly influences the nature of the buckling mode. Sections with wide compression flange will undergo twisting in torsion, but narrow-flange sections experience lateral buckling due to their reduced bending stiffness. Box sections (with a square side), even those hollow, shall not experience LTB due to their high torsional rigidity. Figure 7.2 shows a schematic view of an I-section undergoing LTB. The flexural strength of the member undergoing LTB will be modified using a modification factor, as given below:

$$\text{Modification factor} = \frac{12.5\,M_{max}}{\left[2.5\,M_{max} + 3\,M_A + 4\,M_B + 3M_C\right]} \tag{7.1}$$

Where M_{max} is the maximum moment in the unbraced segment of the member, M_A, M_B, and M_C are the maximum values of the moments at the quarter, mid, and 3/4th points of the unbraced segment of the member. The modification factor allows nonuniform moment diagrams at the end of the braced beam segments.

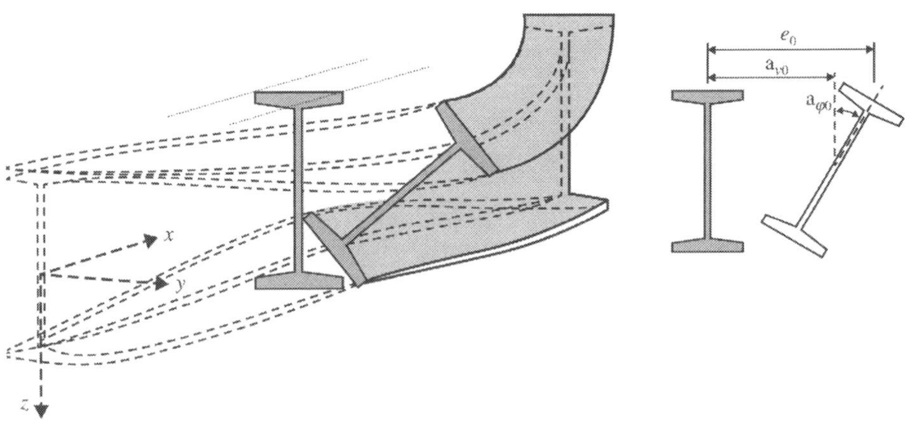

FIGURE 7.2 Lateral-torsional buckling of I-section.

If a compression member is loaded suddenly and then released, it can sustain a higher load than its static buckling load. Let us consider a long, unsupported column used as a drop hammer. The duration of the impact load (compression) at one end is the time required for a stress wave to travel along the column's length and return. It is seen that the maximum buckling occurs near the impact end as the wavelength is shorter than the length of the member; however, the stress will be much higher than that of the buckling stress of any statically loaded column. It is called dynamic buckling.

Stability analysis evaluates the failure modes regarding structural stability; each mode will yield a corresponding load under which instability occurs (or is initiated). These loads are termed critical loads. It is important to note that the structural system remains at rest before instability occurs and at rest immediately after the failure is initiated by buckling. Four classical methods, namely (i) equilibrium state approach, (ii) work approach, (iii) energy principle, and (iv) dynamic approach, help solve the buckling problems. However, in the case of dynamic instability, loss of stability is just a transition from the state of rest to the state of motion.

7.3 EULER CRITICAL LOAD

Consider an ideal column, as shown in Figure 7.3. The column is assumed to be initially straight and compressed by a concentric load, P, as shown in the figure.

The column is pinned at both the supports and is uniformly slender. Further, it is assumed to be laterally restrained in position at both supports. The column section

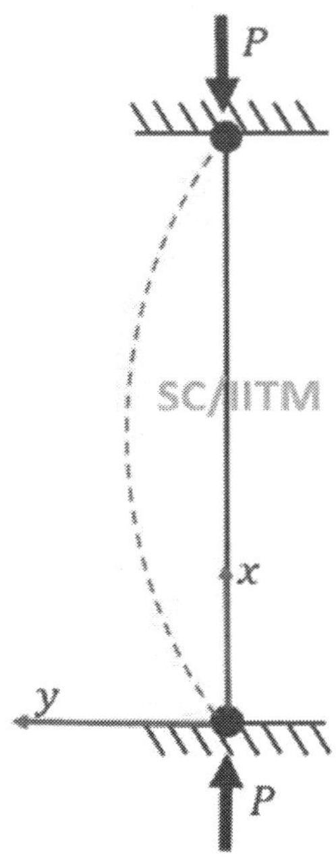

FIGURE 7.3 Euler column.

is assumed to be of negligible weight and perfectly elastic. Stresses developed by the axial forces are assumed to be within the proportional limit of the column material. If the applied force, P, is less than the critical value, then the column will continue to remain straight and undergo only axial compression. Under this state, the column is said to be in *stable equilibrium*. Under such conditions, if a lateral load is applied at any point, for example, at the mid-height of the column, it will result in lateral deflection. However, importantly, the column will return to its original position in terms of geometry, shape, and size regarding its cross-section.

However, upon a continuous (at a steady rate) increase of the axial load, P, the straight form of equilibrium tends to become gradually unstable. Under this condition, even a lateral of a minimal magnitude can cause lateral deflection, which will not disappear upon removing the lateral load; note that this was not the case when P is less than the axial capacity of the member, as discussed earlier. One can define a critical load based on the two sets of explanations. The critical load is the axial load, which must maintain (or continue to maintain) the member in its initial straight position. This critical load can be computed based on the elastic curve equation of a beam.

$$\frac{d^2y}{d^2x} = \frac{M}{EI} \tag{7.2}$$

Where M is the bending moment, I is the moment of inertia, and E is the modulus of elasticity. Figure 7.4 shows the free-body diagram of the column member under the applied load.

Concerning the figure, the equilibrium of the free-body diagram is written as follows:

$$EI\frac{d^2y}{d^2x} = M = -Py \tag{7.3}$$

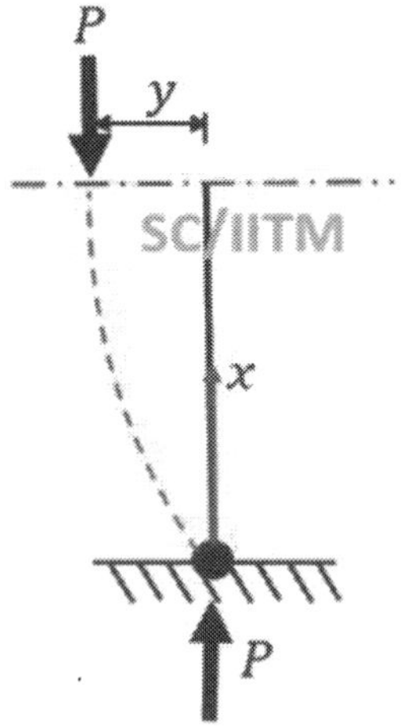

FIGURE 7.4 Free-body diagram of the column member.

$$EI\frac{d^2y}{d^2x}+Py=0 \tag{7.4}$$

$$y=A\sin\left[\frac{\alpha x}{L}\right]+B\cos\left[\frac{\alpha x}{L}\right] \tag{7.5}$$

where $\alpha = L\sqrt{\frac{P}{EI}}$

For the boundary condition: @ x = 0; y = 0, B = 0

Hence, Eq. (7.5) becomes as below:

$$y=A\sin\left[\frac{\alpha x}{L}\right] \tag{7.6}$$

Further, @ x = L, y = 0; applying this boundary condition, Eq. (7.6) becomes as below:

$$A\sin(\alpha)=0 \tag{7.7}$$

which means that either A = 0 or sin (α) = 0. If A = 0, there will be no lateral deflection. Therefore, setting sin (α) = 0

$$\alpha = \text{n}\,\pi \text{ for n} = 0,1,2,3,\ldots. \tag{7.8}$$

Substituting Eq. (7.8), we get:

$$n\pi = L\sqrt{\frac{P}{EI}} \tag{7.9}$$

Squaring,

$$n^2\pi^2 = L^2\frac{P}{EI} \tag{7.10}$$

$$P=\frac{n^2\pi^2EI}{L^2}\text{ for n}=1,2,3,4,\ldots. \tag{7.11}$$

n = 0 is meaningless, as this will cause no axial load (P = 0)

In Eq. (7.11), P is called an Euler critical load.

$$P_E=\frac{n^2\pi^2EI}{L^2}\text{ for n}=1,2,3,4,\ldots. \tag{7.12}$$

7.4 BEAM ELEMENT

A beam element, as shown in Figure 7.5, is a fundamental stability analysis element. A few sign conventions are required before deriving the stiffness matrix of the standard beam element.

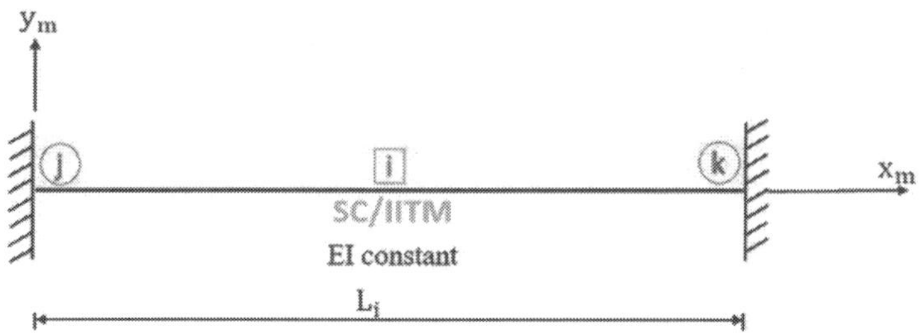

FIGURE 7.5 Standard beam element.

- The end moment, joint rotation, and joint moments, which are anticlockwise, are considered to be positive.
- Upward force (or displacement) of the joint is considered a positive value.
- Force or axial displacement toward the right of the joint is also considered as positive.
- Upward-end shear at the ends of the beam is positive.
- The right direction force at the ends of the beam is positive.

Consider a fixed beam undergoing deformation due to bending, neglecting the axial deformation. The standard fixed beam is shown in Figure 7.5. End nodes of the beam are designated as j and k ends, while the length of the member is designated as L_i; subscript "i" refers to the member index. The beam has a constant EI over its entire length. The axes (x_m, y_m) are the member's local axes. It is very important to note the axis system. The Axis system is such that it has an origin at the jth end; x_m is directed toward a kth end. y_m is counterclockwise 90° to the x_m axis. Therefore, the (x_m, y_m) plane defines the plane of bending the beam element.

Neglecting the axial deformation, one should identify both the translational and rotational displacements at each end of the beam, as shown in Figure 7.6. Suitable subscripts are used for denoting the rotational and translational displacements. The displacements at the jth end and kth end are (θ_p,δ_r) and (θ_q,δ_s), respectively. All these displacements happen in the x_m and y_m planes, and there is no out-of-plane bending.

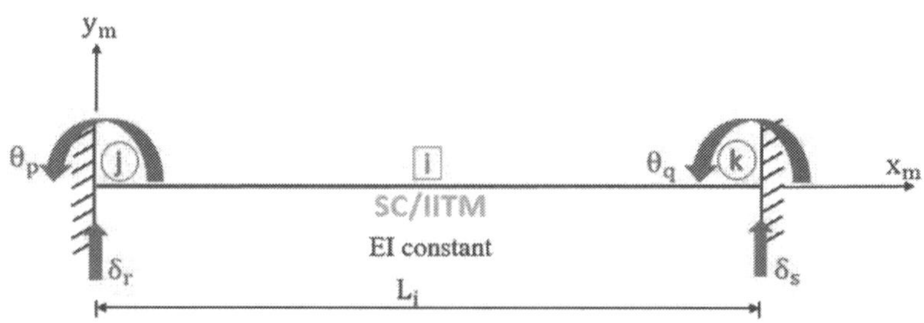

FIGURE 7.6 Rotational and translational moments in the standard beam.

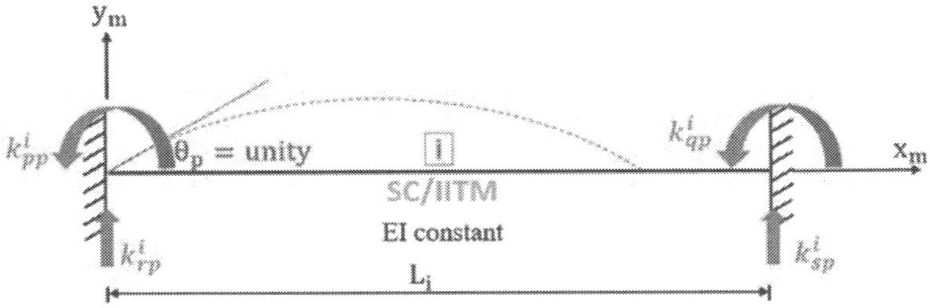

FIGURE 7.7 Unit rotation at the jth end of standard beam.

By classical definition, the stiffness coefficient, k_{ij}, is the force in the ith degree of freedom by imposing a unit displacement (either translational or rotational) in the jth degree of freedom by keeping all other degrees of freedom restrained. As seen in the above figure, there are four degrees of freedom (2 rotations and 2 translations). One should give unit displacement (or rotation) in each degree of freedom to find the forces (or moments) in the respective degrees of freedom by restraining the remaining degrees of freedom. Imposing unit displacement represents $\delta_r = 1$ or $\delta_s = 1$, and that of unit rotation implies $\theta_p = 1$ or $\theta_q = 1$.

Let us apply unit rotation at the jth end, keeping all other degrees of freedom restrained as shown in Figure 7.7. It will invoke the members with the end forces, $k^i_{pp}, k^i_{qp}, k^i_{rp}, k^i_{sp}$ as seen in the figure. k^i_{pp} is the force in pth degree-of-freedom by giving unit displacement in pth degree-of-freedom; superscript "i" refers to the ith member. Similarly, k^i_{qp} is the force in qth degree-of-freedom by giving unit displacement in pth degree-of-freedom in the ith member. The second subscript in all the notations is common, which is 'p', indicating that the unit displacement (in this case, it is unit rotation since p is a rotational degree-of-freedom) is applied at pth degree. The stiffness coefficients derived column-wise correspond to the first column of the stiffness matrix. Similarly, applying unit rotation at the kth end of the ith member, as shown in Figure 7.8, develops the stiffness coefficients $(k^i_{pq}, k^i_{qq}, k^i_{rq}, k^i_{sq})$.

The stiffness coefficients are obtained by applying unit displacements at the jth end and kth end, as shown in Figures 7.9 and 7.10, respectively, which yields the stiffness coefficients at the jth end as $(k^i_{pr}, k^i_{qr}, k^i_{rr}, k^i_{sr})$ and at kth end as $(k^i_{ps}, k^i_{qs}, k^i_{rs}, k^i_{ss})$,

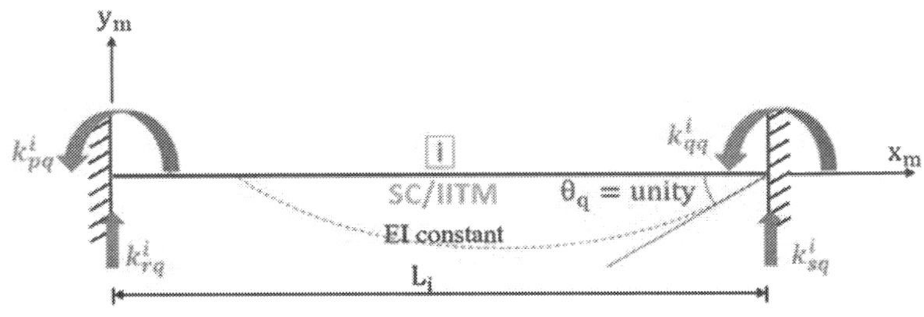

FIGURE 7.8 Unit rotation at the kth end of standard beam.

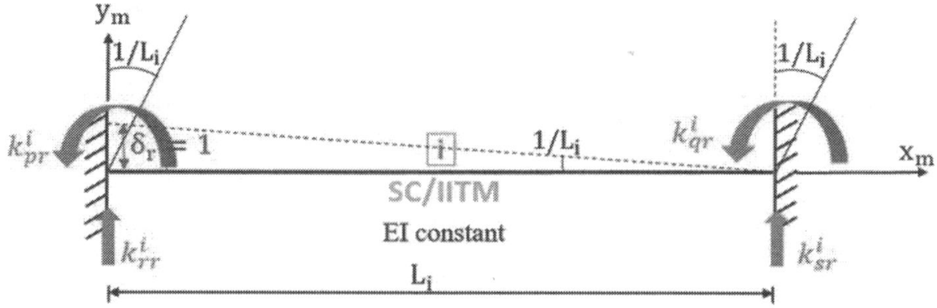

FIGURE 7.9 Unit displacement at the jth end of standard beam.

respectively. As shown in the figures, a tangent can be drawn by connecting the deflected position of the beam at which the unit rotation is applied and the initial position of the beam at the other end. It can be inferred that the beam has undergone a rotation of ($1/L_i$), where L_i is the length of the ith beam element. The rotation at the ends of the beam is equal to ($1/L_i$).

The corresponding end reactions (moment, shear) for the ith beam element are required to be estimated under the arbitrary end displacements (θ_p, δ_r) and (θ_q, δ_s) at the jth and kth ends of the beam. They are estimated by maintaining the equilibrium of the restrained member. The governing equations are as follows:

$$m_p^i = k_{pp}^i \theta_p + k_{pq}^i \theta_q + k_{pr}^i \delta_r + k_{ps}^i \delta_s \tag{7.13}$$

$$m_q^i = k_{qp}^i \theta_p + k_{qq}^i \theta_q + k_{qr}^i \delta_r + k_{qs}^i \delta_s \tag{7.14}$$

$$p_r^i = k_{rp}^i \theta_p + k_{rq}^i \theta_q + k_{rr}^i \delta_r + k_{rs}^i \delta_s \tag{7.15}$$

$$p_s^i = k_{sp}^i \theta_p + k_{sq}^i \theta_q + k_{sr}^i \delta_r + k_{ss}^i \delta_s \tag{7.16}$$

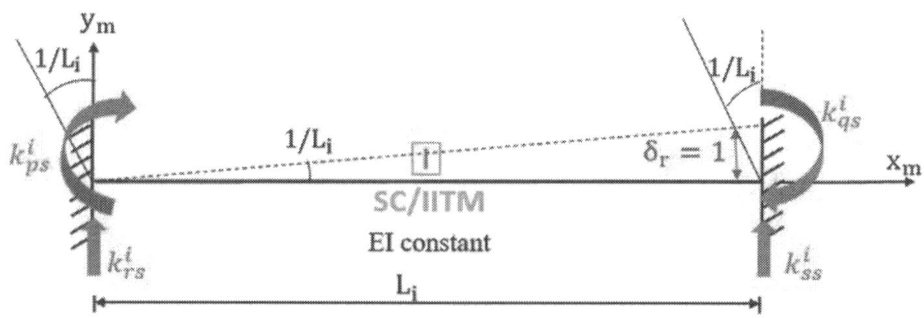

FIGURE 7.10 Unit displacement at the kth end of standard beam.

It can be seen from the above equation that the first subscript corresponds to the end at which the unit rotation (or displacement) is applied. The above set of equations gives the end moments and end shear forces for arbitrary displacements $(\theta_p, \theta_q, \delta_r, \delta_s)$, which are unity at respective degrees of freedom. These equations can be generalized as follows:

$$\{m_i\} = [k]_i \{\delta_i\} \tag{7.17}$$

$$\{m_i\} = \begin{Bmatrix} m_p \\ m_q \\ p_r \\ p_s \end{Bmatrix} \tag{7.18}$$

$$\{\delta_i\} = \begin{Bmatrix} \theta_p \\ \theta_q \\ \delta_r \\ \delta_s \end{Bmatrix} \tag{7.19}$$

$$[k]_i = \begin{bmatrix} k_{pp} & k_{pq} & k_{pr} & k_{ps} \\ k_{qp} & k_{qq} & k_{qr} & k_{qs} \\ k_{rp} & k_{rq} & k_{rr} & k_{rs} \\ k_{sp} & k_{sq} & k_{sr} & k_{ss} \end{bmatrix} \tag{7.20}$$

Figure 7.11 shows the forces at both ends of the standard beam member under unit rotation applied at the jth end of the member.

In the standard fixed beam element with unit rotation at the jth end, moments developed at the ends to control the applied unit rotation are (k^i_{pp}, k^i_{qp}), respectively.

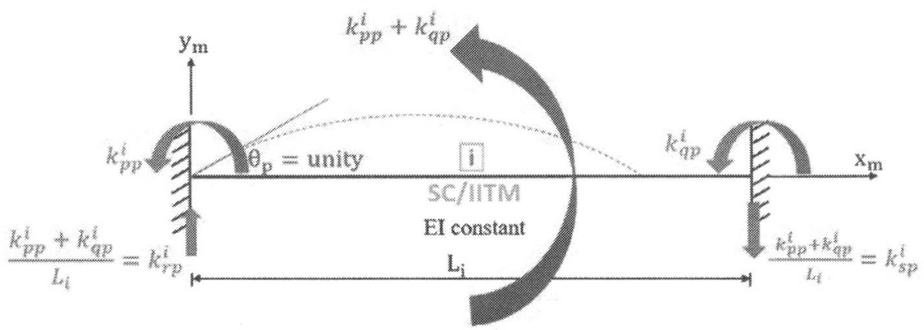

FIGURE 7.11 Rotation coefficients of standard beam.

It results in the development of an anticlockwise moment ($k^i_{pp} + k^i_{qp}$), which should be counteracted by a coupled shear. The shear forces at the ends of the beam are determined as [$\frac{k^i_{pp}+k^i_{qp}}{L_i}$]. At the jth and kth ends of the beam, the magnitude of the shear will be the same, but it will be in the opposite direction at the jth end of the member. It can be seen that the second subscript in the stiffness coefficients indicates the end at which the unit rotation is applied, and the first subscript indicates the forces in the respective degrees of freedom. Concerning the above figure, it is clear that one needs to evaluate only a set of rotational coefficients in Eq. (7.20). These rotational coefficients are ($k^i_{pp}, k^i_{pq}, k^i_{qp}, k^i_{qq}$). Knowing the rotational coefficients, end shear can be expressed as follows:

$$k^i_{rp} = \frac{k^i_{pp} + k^i_{qp}}{L_i} \tag{7.21}$$

$$k^i_{sp} = -\frac{k^i_{pp} + k^i_{qp}}{L_i} \tag{7.22}$$

The negative sign in Eq. (7.22) is because the direction of k^i_{sp} is opposite to the end shear developed by the restraining moments, as shown in Figure 7.11. For the unit rotation applied at the kth end, end shear, in terms of the rotational coefficients, is given by:

$$k^i_{rq} = \frac{k^i_{pq} + k^i_{qq}}{L_i} \tag{7.23}$$

$$k^i_{sq} = -\frac{k^i_{pq} + k^i_{qq}}{L_i} \tag{7.24}$$

By applying unit displacement at the jth end of the beam element, the stiffness coefficients can be expressed as:

$$k^i_{pr} = \frac{k^i_{pp} + k^i_{pq}}{L_i} \tag{7.25}$$

$$k^i_{qr} = -\frac{k^i_{qp} + k^i_{qq}}{L_i} \tag{7.26}$$

$$k^i_{rr} = \frac{k^i_{pr} + k^i_{qr}}{L_i} = \left[\frac{k^i_{pp} + k^i_{pq}}{(L_i)^2}\right] + \left[\frac{k^i_{qp} + k^i_{qq}}{(L_i)^2}\right] \tag{7.27a}$$

$$k^i_{rr} = \frac{k^i_{pp} + k^i_{pq} + k^i_{qp} + k^i_{qq}}{(L_i)^2} \tag{7.27b}$$

$$k_{sr}^{i} = -\frac{k_{pp}^{i} + k_{pq}^{i} + k_{qp}^{i} + k_{qq}^{i}}{\left(L_i\right)^2} \quad (7.28)$$

BY giving unit displacement at the kth end of the beam element, the stiffness coefficients are expressed as:

$$k_{ps}^{i} = -\frac{k_{pp}^{i} + k_{pq}^{i}}{L_i} \quad (7.29)$$

$$k_{qs}^{i} = -\frac{k_{qp}^{i} + k_{qq}^{i}}{L_i} \quad (7.30)$$

$$k_{rs}^{i} = -\frac{k_{ps}^{i} + k_{qs}^{i}}{L_i} = -\frac{k_{pp}^{i} + k_{pq}^{i} + k_{qp}^{i} + k_{qq}^{i}}{\left(L_i\right)^2} \quad (7.31)$$

$$k_{ss}^{i} = \frac{k_{pp}^{i} + k_{pq}^{i} + k_{qp}^{i} + k_{qq}^{i}}{\left(L_i\right)^2} \quad (7.32)$$

It can be seen from the above expression that out of the sixteen coefficients of the stiffness matrix given in Eq. (7.20), one needs to evaluate only the four rotational coefficients ($k_{pp}^{i}, k_{pq}^{i}, k_{qp}^{i}, k_{qq}^{i}$). The remaining coefficients can be expressed as a function of these rotational coefficients. The complete stiffness matrix is given below:

$$[k] = \begin{bmatrix} k_{pp} & k_{pq} & \frac{k_{pp}+k_{pq}}{L} & -\left(\frac{k_{pp}+k_{pq}}{L}\right) \\ k_{qp} & k_{qq} & \frac{k_{qp}+k_{qq}}{L} & -\left(\frac{k_{qp}+k_{qq}}{L}\right) \\ \frac{k_{pp}+k_{pq}}{L} & \frac{k_{pq}+k_{qq}}{L} & \frac{k_{pp}+k_{pq}+k_{qp}+k_{qq}}{L^2} & \frac{k_{pp}+k_{pq}+k_{qp}+k_{qq}}{L^2} \\ -\left(\frac{k_{pp}+k_{pq}}{L}\right) & -\left(\frac{k_{pq}+k_{qq}}{L}\right) & -\left(\frac{k_{pp}+k_{pq}+k_{qp}+k_{qq}}{L^2}\right) & -\left(\frac{k_{pp}+k_{pq}+k_{qp}+k_{qq}}{L^2}\right) \end{bmatrix} \quad (7.33)$$

7.4.1 Rotational Coefficients

Consider a simply supported beam as shown in Figures 7.12 and 7.13. Unit rotation is applied at the (j,k) ends of the beam to obtain the flexibility coefficients, respectively. The flexibility coefficients ($\delta_{jj}^{i}, \delta_{kj}^{i}$) define rotations at end j and k, respectively, of the ith member, caused due to unit moment applied at the jth end. Similarly, the

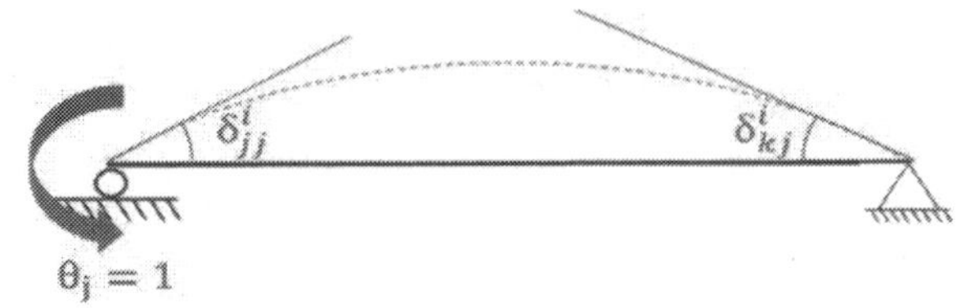

FIGURE 7.12 Unit rotation at the jth end of the simply supported beam.

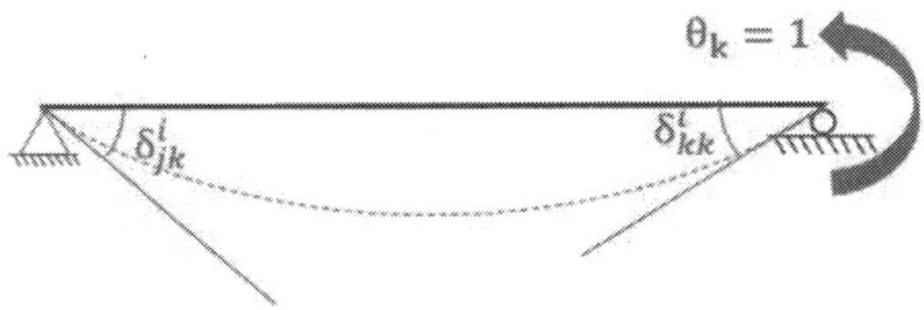

FIGURE 7.13 Unit rotation at the kth end of the simply supported beam.

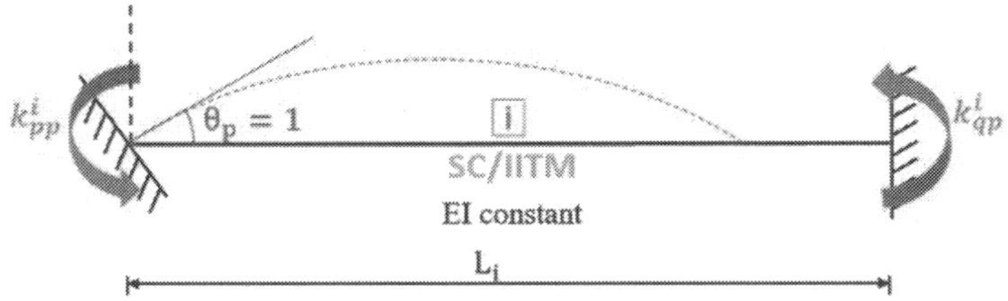

FIGURE 7.14 Unit rotation at the jth end of fixed beam.

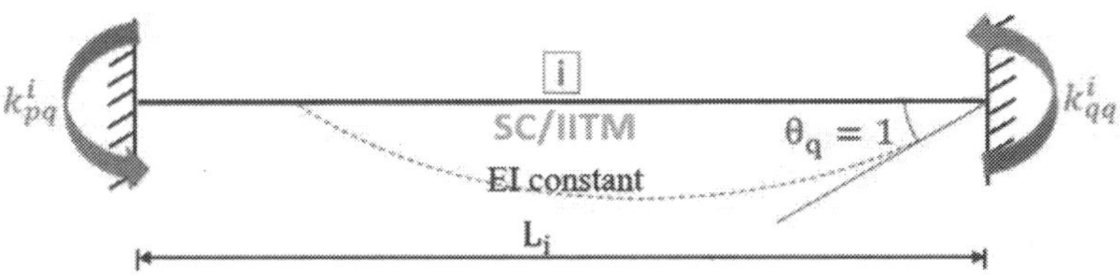

FIGURE 7.15 Unit rotation at the kth end of fixed beam.

flexibility coefficients $(\delta^i_{jk}, \delta^i_{kk})$ define rotations at the jth and kth ends of the ith member due to unit moment applied at the kth end.

Let us now consider a beam fixed at the kth end, imposed by unit rotation at the jth end (Figure 7.14) and unit rotation at the kth end as shown in Figure 7.15. The stiffness coefficients (k^i_{pp}, k^i_{qp}) define end moments required at jth and kth ends to maintain equilibrium when the jth end is subjected to unit rotation while the kth end is restrained. Similarly, the stiffness coefficients (k^i_{pq}, k^i_{qq}) define end moments required at the jth and the kth ends to maintain equilibrium when the kth end is subjected to unit rotation and the jth end is restrained.

Thus,

$$\begin{bmatrix} \delta_{jj} & \delta_{jk} \\ \delta_{kj} & \delta_{kk} \end{bmatrix} \begin{bmatrix} k_{pp} & k_{pq} \\ k_{qp} & k_{qq} \end{bmatrix} = \begin{bmatrix} 1 & 0 \\ 0 & 1 \end{bmatrix} \tag{7.34}$$

Expanding the above equation, we get:

$$k^i_{pp}\delta^i_{jj} + k^i_{qp}\delta^i_{jk} = 1 \tag{7.35a}$$

$$k^i_{pp}\delta^i_{kj} + k^i_{qp}\delta^i_{kk} = 1 \tag{7.35b}$$

$$k^i_{pq}\delta^i_{jj} + k^i_{qq}\delta^i_{jk} = 1 \tag{7.35c}$$

$$k^i_{pq}\delta^i_{kj} + k^i_{qq}\delta^i_{kk} = 1 \tag{7.35d}$$

Let us denote the flexibility matrix as $[D_r]$ and stiffness matrix as $[k_r]$. The subscript *r* refers to the rotational degrees of freedom. In order to estimate the flexibility matrix for the beam element, assume the simply supported beam with the unit moment at the jth end as shown in Figure 7.16. The anticlockwise moment is balanced by the clockwise couple created by the forces. The bending moment diagram is also shown in the figure with tension at the top and compression at the bottom.

Let us replace the loading diagram with a conjugate beam as shown in Figure 7.17. Taking moment about A,

$$V_B = \left[\left\{\frac{1}{2}L_i\left(\frac{1}{EI}\right)\right\}\frac{1}{3}L_i\right]\frac{1}{L_i} = \frac{L_i}{6EI}\left(\text{downward}\right) \tag{7.36a}$$

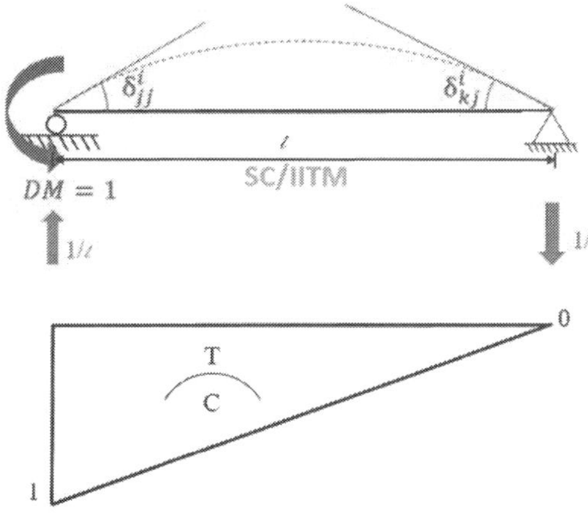

FIGURE 7.16 Simply supported beam with the unit moment at the jth end.

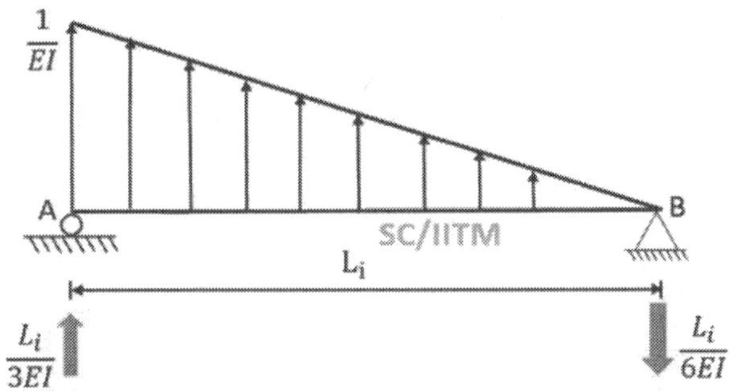

FIGURE 7.17 Conjugate beam.

$$V_A = \left\{\frac{1}{2}L_i\left(\frac{1}{EI}\right)\right\} - \frac{L_i}{6EI} = \frac{L_i}{3EI}\,(\text{upward}) \tag{7.36b}$$

The same procedure is followed for the other case to derive the following flexibility matrix.

$$D_r = \begin{bmatrix} \frac{L}{3EI} & -\frac{L}{6EI} \\ -\frac{L}{6EI} & \frac{L}{3EI} \end{bmatrix} \tag{7.37}$$

$$k_r = [D_r]^{-1} = \frac{12(EI)^2}{L^2}\begin{bmatrix} \frac{L}{3EI} & \frac{L}{6EI} \\ \frac{L}{6EI} & \frac{L}{3EI} \end{bmatrix} = \begin{bmatrix} \frac{4EI}{L} & \frac{2EI}{L} \\ \frac{2EI}{L} & \frac{4EI}{L} \end{bmatrix} \tag{7.38}$$

Thus, from the above four rotational coefficients, the whole stiffness matrix can be derived as below:

$$K_i = \begin{bmatrix} \frac{4EI}{l} & \frac{2EI}{l} & \frac{6EI}{l^2} & -\frac{6EI}{l^2} \\ \frac{2EI}{l} & \frac{4EI}{l} & \frac{6EI}{l^2} & -\frac{6EI}{l^2} \\ \frac{6EI}{l^2} & \frac{6EI}{l^2} & \frac{12EI}{l^3} & -\frac{12EI}{l^3} \\ -\frac{6EI}{l^2} & -\frac{6EI}{l^2} & -\frac{12EI}{l^3} & \frac{12EI}{l^3} \end{bmatrix} \tag{7.39}$$

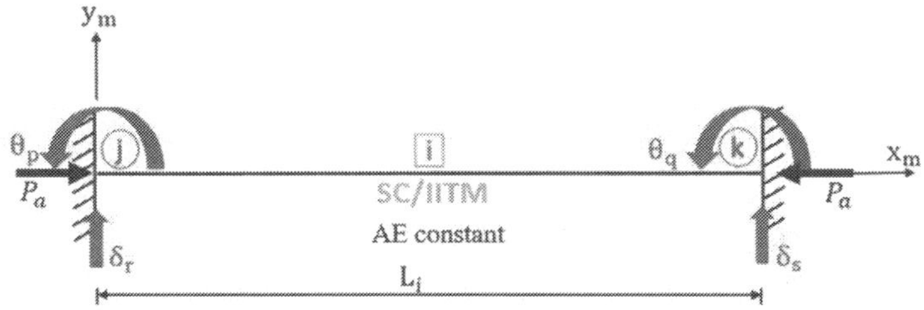

FIGURE 7.18 Fixed beam under axial compressive load.

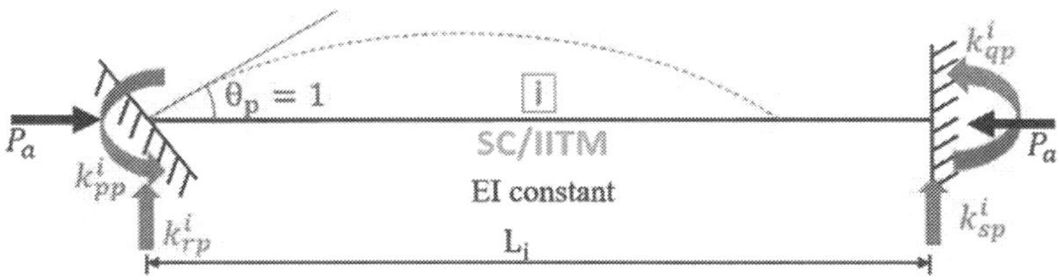

FIGURE 7.19 Unit rotation at the jth end of fixed beam under axial load.

7.5 STABILITY FUNCTIONS

Consider a beam element, both ends fixed, as shown in Figure 7.18. It is important to note that the beam element is subjected to axial compressive load, P_a, as shown in the figure.

7.5.1 Rotation Functions under Axial Compression Load

To obtain the stiffness coefficients, let us now apply unit rotation at the jth end of the member. Figure 7.19 shows the rotations and shear developed at both ends of the member due to the unit rotation applied at the jth end of the member.

From the above figure and based on the expressions derived in Section 7.6, the following expressions can be written:

$$\frac{\left(k_{pp}+k_{qp}\right)}{L_i}=k_{rp}=-k_{sp} \tag{7.40}$$

A free-body diagram, under the influence of the applied unit rotation and the axial compressive load, is shown in Figure 7.20.

With reference to figure, and applying the differential equation as given in Eq. (7.2), we get:

$$EI\frac{d^2y}{d^2x}=M \tag{7.41}$$

$$=-P_a\left(y\right)-k_{pp}+k_{rp}\left(x\right) \tag{7.42}$$

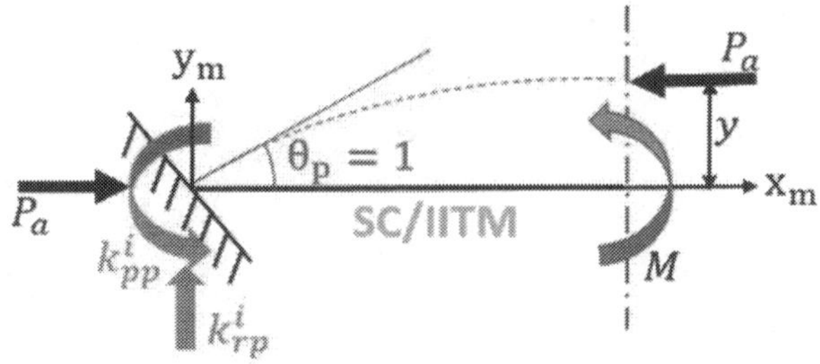

FIGURE 7.20 Free-body diagram under axial load and unit rotation at the jth end.

Substituting Eq. (7.40) in Eq. (7.42), we get:

$$EI\frac{d^2y}{d^2x} = -P_a(y) - k_{pp} + \frac{(k_{pp}+k_{qp})}{L_i}(x) \tag{7.43}$$

In the above equation, let us express the axial load (P_a) as a function of Euler load (P_E), as given below:

$$P_a = \varphi_i P_E \tag{7.44}$$

Substituting Eq. (7.12) in the above equation, we get:

$$P_a = \frac{\pi^2\varphi_i EI}{L^2} \; for\, n = 1 \tag{7.45}$$

It is important to note that buckling is happening in the plane where unit rotation is applied. Therefore, Eq. (7.43) is modified as below:

$$EI\frac{d^2y}{d^2x} = -\frac{\pi^2\varphi_i EI}{L_i^2}{}_a(y) - k_{pp} + \frac{(k_{pp}+k_{qp})}{L_i}(x) \tag{7.46}$$

Dividing by (EI) and rearranging the terms, we get:

$$\frac{d^2y}{d^2x} + \frac{\pi^2\varphi_i}{L_i^2}(y) = \frac{1}{EI}\left[(k_{pp}+k_{qp})\frac{x}{L_i} - (k_{pp})\right] \tag{7.47}$$

The general solution of Eq. (7.47) is given by:

$$y = A\sin\left[\frac{\alpha_i x}{L_i}\right] + B\cos\left[\frac{\alpha_i x}{L_i}\right] + \frac{L_i^2}{\alpha_i^2 EI}\left[(k_{pp}+k_{qp})\frac{x}{L_i} - (k_{pp})\right] \tag{7.48}$$

$$\alpha_i = \pi\sqrt{\varphi_i} \tag{7.49}$$

Applying the boundary conditions: y = 0 at x = 0, we get:

$$B = \frac{L_i^2}{\alpha_i^2 EI} k_{pp} \tag{7.50}$$

Substituting another boundary condition: y = 0 @ x = L in Eq. (7.48), we get:

$$0 = A\sin(\alpha_i) + B\cos(\alpha_i) + \frac{L_i^2}{\alpha_i^2 EI} k_{qp} \tag{7.51a}$$

$$-\left[\frac{L_i^2}{\alpha_i^2 EI}\right](k_{qp}) = A\sin(\alpha_i) + B\cos(\alpha_i) \tag{7.51b}$$

$$-\left[\frac{L_i^2}{\alpha_i^2 EI}\right](k_{qp}) - \left[\frac{L_i^2}{\alpha_i^2 EI}\right](k_{pp})\cos(\alpha_i) = A\sin(\alpha_i) \tag{7.51c}$$

$$-\left[\frac{L_i^2}{\alpha_i^2 EI}\right]\left[\frac{(k_{qp}) + (k_{pp})\cos(\alpha_i)}{\sin(\alpha_i)}\right] = A \tag{7.51d}$$

$$-\left[\frac{L_i^2}{\alpha_i^2 EI}\right]\left[k_{qp}\, cosec(\alpha_i) + k_{pp}\cot(\alpha_i)\right] = A \tag{7.51e}$$

Substituting Eqs. (7.51e and 7.50) in Eq. (7.48), we get:

$$\frac{\alpha_i^2 EI}{L_i^2} y = -\left[k_{pp}\cot(\alpha_i) + k_{qp}\, cosec(\alpha_i)\right]\sin\left[\frac{\alpha_i x}{L_i}\right] + k_{pp}\cos\left[\frac{\alpha_i x}{L_i}\right] + (k_{pp} + k_{qp})\frac{x}{L_i} - k_{pp} \tag{7.52}$$

Differentiating once, we get:

$$\begin{aligned}\frac{\alpha_i^2 EI}{L_i^2}\frac{dy}{dx} &= -\frac{\alpha_i}{L_i}\left[k_{pp}\cot(\alpha_i) + k_{qp}\, cosec(\alpha_i)\right]\cos\left[\frac{\alpha_i x}{L_i}\right] \\ &\quad -\frac{\alpha_i}{L_i} k_{pp}\sin\left[\frac{\alpha_i x}{L_i}\right] + (k_{pp} + k_{qp})\frac{1}{L_i}\end{aligned} \tag{7.53}$$

$$\begin{aligned}\frac{\alpha_i^2 EI}{L_i}\frac{dy}{dx} &= k_{pp}\left[1 - \alpha_i\sin\left[\frac{\alpha_i x}{L_i}\right] - \alpha_i\cot(\alpha_i)\cos\left[\frac{\alpha_i x}{L_i}\right]\right] \\ &\quad + k_{qp}\left[1 - \alpha_i\, cosec(\alpha_i)\cos\left[\frac{\alpha_i x}{L_i}\right]\right]\end{aligned} \tag{7.54}$$

Note: At x = 0, slope (dy/dx) is equal to θ_p, which is equal to unity in the present case. Applying the boundary condition (@ x = L, dy/dx = 0), we get:

$$0 = k_{pp}\left[1-\alpha_i \sin(\alpha_i) - \alpha_i \cot(\alpha_i)\cos(\alpha_i)\right] + k_{qp}\left[1-\alpha_i \, cosec(\alpha_i)\cos(\alpha_i)\right] \tag{7.55}$$

$$0 = k_{pp} - k_{pp}\,\alpha_i\left[\sin(\alpha_i) + \frac{\cos(\alpha_i)^2}{\sin(\alpha_i)}\right] + k_{qp} - k_{qp}\,\alpha_i\,\cot(\alpha_i) \tag{7.56}$$

The above equation is simplified to obtain the stiffness coefficient as below:

$$k_{qp} = \left[\frac{\alpha_i - \sin(\alpha_i)}{\sin(\alpha_i) - \alpha_i\cos(\alpha_i)}\right]k_{pp} \tag{7.57a}$$

Similarly, at x = 0, the slope is unity, which implies the following relationship:

$$k_{pp} = \left[\frac{\alpha_i\left(\sin(\alpha_i) - \alpha_i\cos(\alpha_i)\right)}{2\left(1-\cos(\alpha_i)\right) - \alpha_i\sin(\alpha_i)}\right]\frac{EI}{L_i} \tag{7.57b}$$

By expressing the stiffness coefficients as a function of rotation functions (r_i and c_i), we get the following set of equations:

$$r_i = \frac{\alpha_i\left(\sin(\alpha_i) - \alpha_i\cos(\alpha_i)\right)}{2\left(1-\cos(\alpha_i)\right) - \alpha_i\sin(\alpha_i)} \tag{7.58a}$$

$$c_i = \frac{\alpha_i - \sin(\alpha_i)}{\sin(\alpha_i) - \alpha_i\cos(\alpha_i)} \tag{7.58b}$$

The above equations are termed as *rotation functions* for the compressive axial load case.

7.5.2 Rotation Functions under No Axial Load

For a special case of *zero axial loads*, for which (α_i) becomes zero, one needs to apply L'Hospital rule to obtain the limit of the quotient. L'Hospital rule is briefly explained for the benefit of the readers:

Suppose f(x) and g(x) are differentiable functions and g'(x) $\neq 0$ on an open interval I, which contains (a) {except @ a}, then the following conditions apply:

$$\text{Suppose}, \lim_{x\to a} f(x) = 0; \lim_{x\to a} g(x) = 0 (\text{or}) \lim_{x\to a} f(x) = \pm\infty; \lim_{x\to a} g(x) = \pm\infty$$

then, it may reduce to a form (0/0) or (∞/∞). In such cases, the following equations hold good:

$\lim_{x\to 0} \frac{f(x)}{g(x)} = \lim_{x\to a} \frac{f'(x)}{g'(x)}$ if the limit of RHS exists. For example,

$$\lim_{x\to 0} \frac{e^x - 1}{x^2 + x} = \lim_{x\to 0} \frac{\frac{d}{dx}\left(e^x - 1\right)}{\frac{d}{dx}\left(x^2 + x\right)} = \lim_{x\to 0} \frac{e^x}{\left(2x+1\right)} = 1$$

L'Hospial Rule uses derivatives to evaluate the limits involving indeterminate forms. It states that for indeterminate functions (or forms), where the unity tends to a form (0/0) or (∞ / ∞), the limit of that form is equal to the limit of the derivatives. L'Hospital Rule may be applied as many times until the function does not reduce to the form (0/0) or (∞ / ∞).

Now, let us consider Eqs. (7.58a and 7.58b). As (α_i) approaches zero, both [$\frac{f(\alpha_i)}{g(\alpha_i)}$] approaches zero. Then, one can apply the L'Hospital rule as explained earlier, which will yield the following results:

$$r_{i@\varphi_i=0} = 4 \tag{7.59a}$$

$$c_{i@\varphi_i=0} = 0.5 \tag{7.59b}$$

Hence, at zero axial loads, the stiffness coefficients reduce to the conventional carry-over factors of the beam.

Substituting Eqs. (7.58a and 7.58b) in Eqs. (7.57a and 7.57b), we get:

$$k_{pp} = r_i \frac{EI}{L_i} \tag{7.60a}$$

$$k_{qp} = c_i\, k_{pp} \tag{7.60b}$$

$$k_{rp} = \frac{k_{pp} + k_{qp}}{L_i} = r_i \frac{EI}{L_i^2}\left(1 + c_i\right) \tag{7.60c}$$

$$k_{sp} = -k_{rp} \tag{7.60d}$$

By applying unit rotation at the kth end (please refer to Figure 7.21), another set of stiffness coefficients can be derived as follows:

$$k_{pq} = c_i\, r_i \frac{EI}{L_i} \tag{7.61a}$$

$$k_{qq} = r_i \frac{EI}{L_i} \tag{7.61b}$$

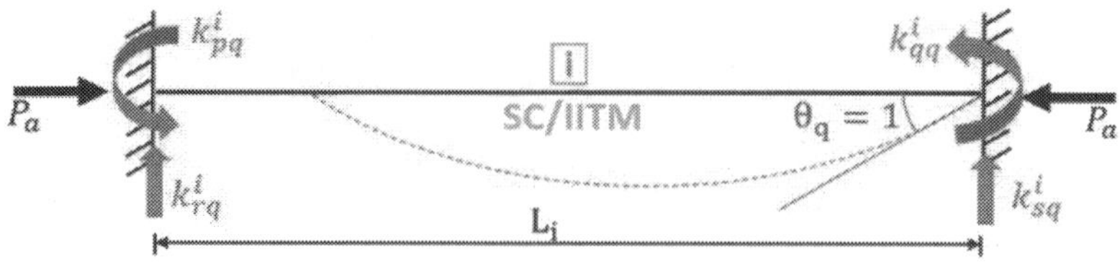

FIGURE 7.21 Unit rotation at the kth end under axial load.

7.5.3 Rotation Functions under Axial Tensile Load

If the beam member is subjected to axial tensile load, then (φ_i) becomes negative. In that case, the following condition holds good:

$$\beta_i = \pi\sqrt{-\varphi_i} = i\pi\sqrt{\varphi_i} = i\alpha_i \tag{7.62}$$

Further,

$$\sin(\beta_i) = \frac{e^{i\beta_i} - e^{-i\beta_i}}{2i} \tag{7.63a}$$

$$\cos(\beta_i) = \frac{e^{i\beta_i} + e^{-i\beta_i}}{2} \tag{7.63b}$$

Substituting Eq. (7.62) in Eq. (7.63), we get:

$$\sin(\beta_i) = \frac{e^{i^2\alpha_i} - e^{-i^2\alpha_i}}{2i} = \frac{e^{-\alpha_i} - e^{\alpha_i}}{2i} \tag{7.64a}$$

$$\cos(\beta_i) = \frac{e^{i^2\alpha_i} + e^{-i^2\alpha_i}}{2} = \frac{e^{-\alpha_i} + e^{\alpha_i}}{2} \tag{7.64b}$$

Rotation constants at the jth end can be obtained by substituting Eqs. (7.64a and 7.64b) in Eq. (7.58), along with Eq. (7.62), as given below:

$$r_i = \frac{i\alpha_i\left\{\left[\frac{e^{-\alpha_i} - e^{\alpha_i}}{2i}\right] - i\alpha_i\left[\frac{e^{-\alpha_i} + e^{\alpha_i}}{2}\right]\right\}}{2\left[1 - \frac{e^{-\alpha_i} + e^{\alpha_i}}{2}\right] - i\alpha_i\left[\frac{e^{-\alpha_i} - e^{\alpha_i}}{2i}\right]} \tag{7.65a}$$

$$c_i = \frac{i\alpha_i - \left[\frac{e^{-\alpha_i} - e^{\alpha_i}}{2i}\right]}{\frac{e^{-\alpha_i} - e^{\alpha_i}}{2i} - i\alpha_i\left[\frac{e^{-\alpha_i} + e^{\alpha_i}}{2}\right]} \tag{7.65b}$$

The above equations can be further simplified as follows:

$$r_i = \frac{\alpha_i\left\{\left[\frac{e^{-\alpha_i}-e^{\alpha_i}}{2}\right]+\alpha_i\left[\frac{e^{-\alpha_i}+e^{\alpha_i}}{2}\right]\right\}}{2\left[1-\frac{e^{-\alpha_i}+e^{\alpha_i}}{2}\right]-\alpha_i\left[\frac{e^{-\alpha_i}-e^{\alpha_i}}{2}\right]} \tag{7.66a}$$

$$c_i = \frac{\alpha_{i+}\left[\frac{e^{-\alpha_i}-e^{\alpha_i}}{2}\right]}{-\left[\frac{e^{-\alpha_i}-e^{\alpha_i}}{2}\right]-\alpha_i\left[\frac{e^{-\alpha_i}+e^{\alpha_i}}{2}\right]} \tag{7.66b}$$

Using the following hyperbolic functions as given below:

$$\sinh\alpha_i = \frac{e^{\alpha_i}-e^{\alpha_i}}{2} \tag{7.67a}$$

$$\cosh\alpha_i = \frac{e^{\alpha_i}+e^{\alpha_i}}{2} \tag{7.67b}$$

By considering the absolute values of φ_i, Eq. (7.66) will reduce to the following form:

$$r_i = \frac{\alpha_i\left(\alpha_i\cosh\alpha_i-\sinh\alpha_i\right)}{2\left(1-\cosh\alpha_i\right)+\alpha_i\sinh\alpha_i} \tag{7.68a}$$

$$c_i = \frac{\alpha_i-\sinh\alpha_i}{\sinh\alpha_i-\alpha_i\cosh\alpha_i} \tag{7.68b}$$

Rotation constants of the member, under compressive and tensile axial loads, as derived above, are plotted for a wide range of values of φ. Table 7.1 also gives these values at closer intervals, which are useful in calculating the critical buckling load.

7.5.4 Transfer Function under Axial Compressive Load

Figure 7.22 shows unit translation at the jth end of the beam under axial compressive load.

By taking moments of all forces about the kth end of the member, we get:

$$P_a(1)-k_{pr}+k_{rr}\left(L_i\right)-k_{qr}=0 \tag{7.69}$$

$$k_{rr} = \frac{P_a(1)-\left(k_{pr}+k_{qr}\right)}{L_i} \tag{7.70}$$

$$k_{sr} = -k_{rr} \tag{7.71}$$

TABLE 7.1
Stability Functions (Negative Sign Indicates Tensile Axial Load)

Phi	r	c	t	Phi	r	c	t
−10	11.1864	0.1118	4.9678	**−6**	9.0436	0.1483	3.8512
−9.9	11.1382	0.1124	4.9429	**−5.9**	8.9830	0.1497	3.8192
−9.8	11.0897	0.1131	4.9179	**−5.8**	8.9220	0.1511	3.7869
−9.7	11.0410	0.1137	4.8928	**−5.7**	8.8605	0.1526	3.7544
−9.6	10.9921	0.1144	4.8675	**−5.6**	8.7986	0.1541	3.7216
−9.5	10.9430	0.1150	4.8421	**−5.5**	8.7362	0.1556	3.6885
−9.4	10.8936	0.1157	4.8166	**−5.4**	8.6733	0.1572	3.6551
−9.3	10.8440	0.1164	4.7909	**−5.3**	8.6100	0.1588	3.6215
−9.2	10.7941	0.1171	4.7652	**−5.2**	8.5461	0.1604	3.5875
−9.1	10.7440	0.1178	4.7392	**−5.1**	8.4818	0.1621	3.5532
−9	10.6937	0.1185	4.7131	**−5**	8.4169	0.1639	3.5187
−8.9	10.6431	0.1193	4.6869	**−4.9**	8.3515	0.1657	3.4838
−8.8	10.5922	0.1200	4.6606	**−4.8**	8.2855	0.1676	3.4485
−8.7	10.5411	0.1208	4.6341	**−4.7**	8.2190	0.1695	3.4129
−8.6	10.4897	0.1215	4.6074	**−4.6**	8.1520	0.1715	3.3770
−8.5	10.4380	0.1223	4.5806	**−4.5**	8.0843	0.1735	3.3407
−8.4	10.3860	0.1231	4.5536	**−4.4**	8.0161	0.1757	3.3040
−8.3	10.3338	0.1239	4.5265	**−4.3**	7.9472	0.1778	3.2669
−8.2	10.2813	0.1248	4.4992	**−4.2**	7.8777	0.1801	3.2295
−8.1	10.2285	0.1256	4.4717	**−4.1**	7.8075	0.1824	3.1916
−8	10.1754	0.1265	4.4441	**−4**	7.7367	0.1848	3.1533
−7.9	10.1220	0.1274	4.4163	**−3.9**	7.6652	0.1873	3.1146
−7.8	10.0683	0.1283	4.3884	**−3.8**	7.5930	0.1899	3.0755
−7.7	10.0143	0.1292	4.3602	**−3.7**	7.5201	0.1926	3.0359
−7.6	9.9600	0.1301	4.3319	**−3.6**	7.4465	0.1954	2.9958
−7.5	9.9054	0.1311	4.3034	**−3.5**	7.3721	0.1983	2.9552
−7.4	9.8504	0.1321	4.2747	**−3.4**	7.2969	0.2013	2.9141
−7.3	9.7951	0.1331	4.2458	**−3.3**	7.2209	0.2044	2.8725
−7.2	9.7395	0.1341	4.2167	**−3.2**	7.1441	0.2076	2.8304
−7.1	9.6835	0.1351	4.1875	**−3.1**	7.0664	0.2110	2.7877
−7	9.6272	0.1362	4.1580	**−3**	6.9878	0.2145	2.7444
−6.9	9.5706	0.1373	4.1283	**−2.9**	6.9084	0.2182	2.7005
−6.8	9.5136	0.1384	4.0984	**−2.8**	6.8280	0.2220	2.6560
−6.7	9.4562	0.1395	4.0683	**−2.7**	6.7466	0.2260	2.6108
−6.6	9.3984	0.1407	4.0380	**−2.6**	6.6642	0.2302	2.5650
−6.5	9.3403	0.1419	4.0074	**−2.5**	6.5808	0.2346	2.5185
−6.4	9.2817	0.1431	3.9766	**−2.4**	6.4963	0.2392	2.4712
−6.3	9.2228	0.1444	3.9456	**−2.3**	6.4107	0.2440	2.4232
−6.2	9.1635	0.1457	3.9144	**−2.2**	6.3239	0.2491	2.3744
−6.1	9.1037	0.1470	3.8829	**−2.1**	6.2360	0.2544	2.3248

(Continued)

TABLE 7.1 *(Continued)*
Stability Functions (Negative Sign Indicates Tensile Axial Load)

Phi	r	c	t	Phi	r	c	t
−2	6.1468	0.2600	2.2743	**0.2**	3.7297	0.5550	0.8298
−1.9	6.0564	0.2659	2.2230	**0.21**	3.7158	0.5581	0.8210
−1.8	5.9645	0.2721	2.1706	**0.22**	3.7019	0.5612	0.8122
−1.7	5.8714	0.2787	2.1174	**0.23**	3.6879	0.5644	0.8033
−1.6	5.7767	0.2857	2.0631	**0.24**	3.6739	0.5676	0.7943
−1.5	5.6806	0.2931	2.0077	**0.25**	3.6598	0.5708	0.7854
−1.4	5.5828	0.3010	1.9512	**0.26**	3.6457	0.5741	0.7764
−1.3	5.4835	0.3094	1.8935	**0.27**	3.6315	0.5774	0.7674
−1.2	5.3824	0.3183	1.8346	**0.28**	3.6174	0.5807	0.7584
−1.1	5.2795	0.3279	1.7743	**0.29**	3.6031	0.5841	0.7493
−1	5.1748	0.3381	1.7127	**0.3**	3.5889	0.5875	0.7402
−0.9	5.0681	0.3490	1.6496	**0.31**	3.5746	0.5910	0.7310
−0.8	4.9593	0.3608	1.5850	**0.32**	3.5602	0.5945	0.7218
−0.7	4.8483	0.3735	1.5187	**0.33**	3.5458	0.5981	0.7126
−0.6	4.7351	0.3872	1.4508	**0.34**	3.5314	0.6017	0.7034
−0.5	4.6194	0.4021	1.3809	**0.35**	3.5169	0.6053	0.6941
−0.4	4.5013	0.4183	1.3092	**0.36**	3.5024	0.6090	0.6848
−0.3	4.3804	0.4360	1.2354	**0.37**	3.4878	0.6127	0.6754
−0.2	4.2567	0.4553	1.1593	**0.38**	3.4732	0.6165	0.6660
−0.1	4.1299	0.4765	1.0809	**0.39**	3.4586	0.6203	0.6566
0	4.0000	0.5000	1.0000	**0.4**	3.4439	0.6242	0.6471
0.01	3.9868	0.5025	0.9918	**0.41**	3.4292	0.6281	0.6376
0.02	3.9736	0.5050	0.9835	**0.42**	3.4144	0.6321	0.6281
0.03	3.9604	0.5075	0.9752	**0.43**	3.3995	0.6361	0.6185
0.04	3.9471	0.5101	0.9669	**0.44**	3.3847	0.6402	0.6089
0.05	3.9338	0.5127	0.9585	**0.45**	3.3698	0.6443	0.5992
0.06	3.9204	0.5153	0.9502	**0.46**	3.3548	0.6485	0.5895
0.07	3.9070	0.5179	0.9418	**0.47**	3.3398	0.6528	0.5798
0.08	3.8936	0.5206	0.9333	**0.48**	3.3247	0.6571	0.5701
0.09	3.8802	0.5233	0.9249	**0.49**	3.3096	0.6614	0.5603
0.1	3.8667	0.5260	0.9164	**0.5**	3.2945	0.6659	0.5504
0.11	3.8531	0.5288	0.9078	**0.51**	3.2793	0.6703	0.5405
0.12	3.8396	0.5316	0.8993	**0.52**	3.2640	0.6749	0.5306
0.13	3.8260	0.5344	0.8907	**0.53**	3.2487	0.6795	0.5206
0.14	3.8123	0.5372	0.8821	**0.54**	3.2334	0.6841	0.5106
0.15	3.7987	0.5401	0.8735	**0.55**	3.2180	0.6889	0.5006
0.16	3.7849	0.5430	0.8648	**0.56**	3.2025	0.6937	0.4905
0.17	3.7712	0.5460	0.8561	**0.57**	3.1870	0.6985	0.4804
0.18	3.7574	0.5490	0.8474	**0.58**	3.1715	0.7035	0.4702
0.19	3.7436	0.5520	0.8386	**0.59**	3.1559	0.7085	0.4600

(Continued)

TABLE 7.1 *(Continued)*
Stability Functions (Negative Sign Indicates Tensile Axial Load)

Phi	r	c	t	Phi	r	c	t
0.6	3.1403	0.7136	0.4498	**1**	2.4674	1.0000	0.0000
0.61	3.1246	0.7187	0.4395	**1.01**	2.4493	1.0101	−0.0124
0.62	3.1088	0.7239	0.4291	**1.02**	2.4311	1.0204	−0.0248
0.63	3.0930	0.7292	0.4187	**1.03**	2.4128	1.0309	−0.0373
0.64	3.0771	0.7346	0.4083	**1.04**	2.3944	1.0416	−0.0498
0.65	3.0612	0.7401	0.3978	**1.05**	2.3760	1.0526	−0.0625
0.66	3.0453	0.7456	0.3873	**1.06**	2.3575	1.0638	−0.0752
0.67	3.0293	0.7513	0.3768	**1.07**	2.3389	1.0752	−0.0879
0.68	3.0132	0.7570	0.3661	**1.08**	2.3202	1.0868	−0.1007
0.69	2.9971	0.7628	0.3555	**1.09**	2.3015	1.0987	−0.1136
0.7	2.9809	0.7687	0.3448	**1.1**	2.2827	1.1109	−0.1266
0.71	2.9646	0.7746	0.3340	**1.11**	2.2638	1.1233	−0.1396
0.72	2.9484	0.7807	0.3233	**1.12**	2.2448	1.1360	−0.1527
0.73	2.9320	0.7869	0.3124	**1.13**	2.2258	1.1490	−0.1658
0.74	2.9156	0.7932	0.3015	**1.14**	2.2066	1.1623	−0.1790
0.75	2.8991	0.7995	0.2906	**1.15**	2.1874	1.1759	−0.1923
0.76	2.8826	0.8060	0.2796	**1.16**	2.1681	1.1898	−0.2057
0.77	2.8660	0.8126	0.2686	**1.17**	2.1487	1.2040	−0.2192
0.78	2.8494	0.8193	0.2575	**1.18**	2.1293	1.2185	−0.2327
0.79	2.8327	0.8261	0.2463	**1.19**	2.1097	1.2335	−0.2463
0.8	2.8159	0.8330	0.2351	**1.2**	2.0901	1.2487	−0.2599
0.81	2.7991	0.8400	0.2239	**1.21**	2.0704	1.2644	−0.2737
0.82	2.7822	0.8472	0.2126	**1.22**	2.0506	1.2804	−0.2875
0.83	2.7653	0.8544	0.2013	**1.23**	2.0307	1.2968	−0.3014
0.84	2.7483	0.8618	0.1899	**1.24**	2.0107	1.3137	−0.3153
0.85	2.7312	0.8693	0.1784	**1.25**	1.9906	1.3309	−0.3294
0.86	2.7141	0.8770	0.1669	**1.26**	1.9705	1.3487	−0.3435
0.87	2.6969	0.8848	0.1554	**1.27**	1.9502	1.3669	−0.3577
0.88	2.6797	0.8927	0.1438	**1.28**	1.9299	1.3855	−0.3720
0.89	2.6623	0.9008	0.1321	**1.29**	1.9094	1.4047	−0.3864
0.9	2.6450	0.9090	0.1204	**1.3**	1.8889	1.4244	−0.4009
0.91	2.6275	0.9173	0.1086	**1.31**	1.8683	1.4447	−0.4154
0.92	2.6100	0.9258	0.0968	**1.32**	1.8476	1.4655	−0.4300
0.93	2.5924	0.9345	0.0849	**1.33**	1.8267	1.4869	−0.4447
0.94	2.5748	0.9433	0.0729	**1.34**	1.8058	1.5089	−0.4595
0.95	2.5570	0.9523	0.0609	**1.35**	1.7848	1.5316	−0.4744
0.96	2.5392	0.9615	0.0489	**1.36**	1.7637	1.5549	−0.4894
0.97	2.5214	0.9709	0.0367	**1.37**	1.7425	1.5790	−0.5044
0.98	2.5035	0.9804	0.0246	**1.38**	1.7212	1.6038	−0.5196
0.99	2.4855	0.9901	0.0123	**1.39**	1.6997	1.6293	−0.5348

(Continued)

TABLE 7.1 *(Continued)*
Stability Functions (Negative Sign Indicates Tensile Axial Load)

Phi	r	c	t	Phi	r	c	t
1.4	1.6782	1.6557	−0.5502	**1.8**	0.7170	4.4969	−1.2537
1.41	1.6566	1.6828	−0.5656	**1.81**	0.6900	4.6924	−1.2739
1.42	1.6348	1.7109	−0.5811	**1.82**	0.6629	4.9051	−1.2944
1.43	1.6130	1.7399	−0.5967	**1.83**	0.6356	5.1377	−1.3149
1.44	1.5910	1.7699	−0.6125	**1.84**	0.6081	5.3929	−1.3357
1.45	1.5690	1.8009	−0.6283	**1.85**	0.5804	5.6742	−1.3566
1.46	1.5468	1.8329	−0.6442	**1.86**	0.5526	5.9859	−1.3776
1.47	1.5245	1.8661	−0.6602	**1.87**	0.5246	6.3331	−1.3989
1.48	1.5021	1.9005	−0.6763	**1.88**	0.4964	6.7223	−1.4202
1.49	1.4796	1.9361	−0.6925	**1.89**	0.4680	7.1616	−1.4418
1.5	1.4570	1.9731	−0.7089	**1.9**	0.4394	7.6612	−1.4635
1.51	1.4342	2.0114	−0.7253	**1.91**	0.4107	8.2345	−1.4855
1.52	1.4114	2.0512	−0.7418	**1.92**	0.3817	8.8990	−1.5076
1.53	1.3884	2.0926	−0.7585	**1.93**	0.3526	9.6785	−1.5298
1.54	1.3653	2.1356	−0.7752	**1.94**	0.3232	10.6056	−1.5523
1.55	1.3420	2.1804	−0.7921	**1.95**	0.2937	11.7264	−1.5749
1.56	1.3187	2.2271	−0.8090	**1.96**	0.2639	13.1087	−1.5978
1.57	1.2952	2.2757	−0.8261	**1.97**	0.2339	14.8562	−1.6208
1.58	1.2716	2.3264	−0.8433	**1.98**	0.2038	17.1355	−1.6440
1.59	1.2479	2.3794	−0.8606	**1.99**	0.1734	20.2327	−1.6674
1.6	1.2240	2.4348	−0.8781	**2**	0.1428	24.6841	−1.6910
1.61	1.2000	2.4927	−0.8956	**2.01**	0.1120	31.6264	−1.7148
1.62	1.1759	2.5534	−0.9133	**2.02**	0.0809	43.9616	−1.7388
1.63	1.1516	2.6170	−0.9311	**2.03**	0.0497	71.9627	−1.7631
1.64	1.1272	2.6838	−0.9490	**2.04**	0.0182	197.3863	−1.7875
1.65	1.1027	2.7540	−0.9670	**2.05**	−0.0135	−267.2161	−1.8122
1.66	1.0780	2.8278	−0.9852	**2.06**	−0.0455	−79.8138	−1.8371
1.67	1.0532	2.9056	−1.0035	**2.07**	−0.0777	−46.9612	−1.8622
1.68	1.0282	2.9877	−1.0219	**2.08**	−0.1101	−33.2921	−1.8875
1.69	1.0031	3.0744	−1.0405	**2.09**	−0.1428	−25.8013	−1.9130
1.7	0.9779	3.1662	−1.0592	**2.1**	−0.1757	−21.0722	−1.9388
1.71	0.9525	3.2635	−1.0780	**2.11**	−0.2089	−17.8154	−1.9648
1.72	0.9270	3.3667	−1.0969	**2.12**	−0.2423	−15.4361	−1.9911
1.73	0.9013	3.4766	−1.1160	**2.13**	−0.2760	−13.6217	−2.0176
1.74	0.8754	3.5936	−1.1353	**2.14**	−0.3099	−12.1925	−2.0444
1.75	0.8494	3.7187	−1.1546	**2.15**	−0.3441	−11.0376	−2.0714
1.76	0.8233	3.8524	−1.1741	**2.16**	−0.3786	−10.0850	−2.0986
1.77	0.7969	3.9960	−1.1938	**2.17**	−0.4134	−9.2858	−2.1261
1.78	0.7705	4.1504	−1.2136	**2.18**	−0.4485	−8.6059	−2.1539
1.79	0.7438	4.3169	−1.2336	**2.19**	−0.4838	−8.0203	−2.1820

(Continued)

TABLE 7.1 *(Continued)*
Stability Functions (Negative Sign Indicates Tensile Axial Load)

Phi	r	c	t	Phi	r	c	t
2.2	−0.5194	−7.5107	−2.2103	**2.6**	−2.2490	−2.2312	−3.6335
2.21	−0.5553	−7.0632	−2.2389	**2.61**	−2.3020	−2.1959	−3.6785
2.22	−0.5916	−6.6673	−2.2678	**2.62**	−2.3557	−2.1618	−3.7241
2.23	−0.6281	−6.3143	−2.2970	**2.63**	−2.4100	−2.1289	−3.7704
2.24	−0.6649	−5.9978	−2.3264	**2.64**	−2.4650	−2.0971	−3.8172
2.25	−0.7020	−5.7124	−2.3562	**2.65**	−2.5206	−2.0665	−3.8647
2.26	−0.7395	−5.4537	−2.3863	**2.66**	−2.5769	−2.0369	−3.9128
2.27	−0.7773	−5.2181	−2.4166	**2.67**	−2.6339	−2.0082	−3.9616
2.28	−0.8154	−5.0027	−2.4473	**2.68**	−2.6915	−1.9805	−4.0111
2.29	−0.8538	−4.8050	−2.4783	**2.69**	−2.7499	−1.9538	−4.0613
2.3	−0.8926	−4.6230	−2.5096	**2.7**	−2.8091	−1.9278	−4.1122
2.31	−0.9318	−4.4547	−2.5413	**2.71**	−2.8690	−1.9027	−4.1639
2.32	−0.9713	−4.2988	−2.5733	**2.72**	−2.9296	−1.8784	−4.2162
2.33	−1.0111	−4.1540	−2.6056	**2.73**	−2.9911	−1.8548	−4.2694
2.34	−1.0513	−4.0190	−2.6383	**2.74**	−3.0533	−1.8319	−4.3233
2.35	−1.0919	−3.8930	−2.6713	**2.75**	−3.1164	−1.8097	−4.3781
2.36	−1.1328	−3.7750	−2.7047	**2.76**	−3.1803	−1.7882	−4.4336
2.37	−1.1742	−3.6644	−2.7384	**2.77**	−3.2451	−1.7673	−4.4900
2.38	−1.2159	−3.5604	−2.7725	**2.78**	−3.3108	−1.7470	−4.5473
2.39	−1.2580	−3.4626	−2.8070	**2.79**	−3.3774	−1.7273	−4.6055
2.4	−1.3006	−3.3703	−2.8419	**2.8**	−3.4449	−1.7081	−4.6645
2.41	−1.3435	−3.2831	−2.8772	**2.81**	−3.5133	−1.6895	−4.7245
2.42	−1.3869	−3.2006	−2.9129	**2.82**	−3.5828	−1.6714	−4.7854
2.43	−1.4307	−3.1225	−2.9490	**2.83**	−3.6532	−1.6538	−4.8474
2.44	−1.4749	−3.0484	−2.9855	**2.84**	−3.7246	−1.6366	−4.9103
2.45	−1.5196	−2.9780	−3.0224	**2.85**	−3.7972	−1.6200	−4.9742
2.46	−1.5647	−2.9111	−3.0598	**2.86**	−3.8707	−1.6038	−5.0392
2.47	−1.6103	−2.8473	−3.0976	**2.87**	−3.9454	−1.5880	−5.1053
2.48	−1.6563	−2.7865	−3.1359	**2.88**	−4.0213	−1.5726	−5.1725
2.49	−1.7028	−2.7286	−3.1746	**2.89**	−4.0983	−1.5576	−5.2409
2.5	−1.7499	−2.6732	−3.2138	**2.9**	−4.1765	−1.5430	−5.3104
2.51	−1.7974	−2.6202	−3.2534	**2.91**	−4.2559	−1.5288	−5.3811
2.52	−1.8454	−2.5695	−3.2936	**2.92**	−4.3366	−1.5149	−5.4531
2.53	−1.8939	−2.5210	−3.3342	**2.93**	−4.4186	−1.5014	−5.5263
2.54	−1.9430	−2.4744	−3.3754	**2.94**	−4.5019	−1.4882	−5.6008
2.55	−1.9926	−2.4298	−3.4170	**2.95**	−4.5866	−1.4754	−5.6767
2.56	−2.0427	−2.3869	−3.4592	**2.96**	−4.6727	−1.4628	−5.7540
2.57	−2.0934	−2.3457	−3.5020	**2.97**	−4.7602	−1.4506	−5.8327
2.58	−2.1447	−2.3061	−3.5452	**2.98**	−4.8492	−1.4387	−5.9129
2.59	−2.1965	−2.2680	−3.5891	**2.99**	−4.9398	−1.4270	−5.9946

(Continued)

TABLE 7.1 *(Continued)*
Stability Functions (Negative Sign Indicates Tensile Axial Load)

Phi	r	c	t	Phi	r	c	t
3	−5.0320	−1.4157	−6.0778	**3.4**	−10.9082	−1.1223	−11.5754
3.01	−5.1258	−1.4046	−6.1627	**3.41**	−11.1485	−1.1179	−11.8055
3.02	−5.2212	−1.3937	−6.2491	**3.42**	−11.3965	−1.1135	−12.0435
3.03	−5.3184	−1.3832	−6.3373	**3.43**	−11.6528	−1.1093	−12.2897
3.04	−5.4174	−1.3728	−6.4273	**3.44**	−11.9178	−1.1052	−12.5445
3.05	−5.5182	−1.3628	−6.5191	**3.45**	−12.1919	−1.1011	−12.8084
3.06	−5.6209	−1.3529	−6.6127	**3.46**	−12.4757	−1.0972	−13.0820
3.07	−5.7256	−1.3433	−6.7083	**3.47**	−12.7697	−1.0933	−13.3657
3.08	−5.8323	−1.3339	−6.8058	**3.48**	−13.0745	−1.0896	−13.6602
3.09	−5.9410	−1.3247	−6.9055	**3.49**	−13.3907	−1.0859	−13.9661
3.1	−6.0519	−1.3157	−7.0072	**3.5**	−13.7190	−1.0824	−14.2840
3.11	−6.1651	−1.3069	−7.1111	**3.51**	−14.0601	−1.0789	−14.6148
3.12	−6.2805	−1.2983	−7.2174	**3.52**	−14.4149	−1.0755	−14.9591
3.13	−6.3984	−1.2899	−7.3259	**3.53**	−14.7842	−1.0722	−15.3180
3.14	−6.5186	−1.2817	−7.4369	**3.54**	−15.1689	−1.0690	−15.6922
3.15	−6.6415	−1.2737	−7.5505	**3.55**	−15.5702	−1.0659	−16.0829
3.16	−6.7669	−1.2659	−7.6666	**3.56**	−15.9890	−1.0628	−16.4912
3.17	−6.8951	−1.2582	−7.7854	**3.57**	−16.4267	−1.0599	−16.9183
3.18	−7.0262	−1.2508	−7.9071	**3.58**	−16.8845	−1.0570	−17.3655
3.19	−7.1601	−1.2434	−8.0316	**3.59**	−17.3640	−1.0542	−17.8343
3.2	−7.2971	−1.2363	−8.1592	**3.6**	−17.8668	−1.0514	−18.3264
3.21	−7.4373	−1.2293	−8.2899	**3.61**	−18.3946	−1.0488	−18.8435
3.22	−7.5807	−1.2224	−8.4238	**3.62**	−18.9494	−1.0462	−19.3875
3.23	−7.7276	−1.2157	−8.5611	**3.63**	−19.5335	−1.0438	−19.9608
3.24	−7.8779	−1.2092	−8.7019	**3.64**	−20.1492	−1.0413	−20.5657
3.25	−8.0320	−1.2028	−8.8464	**3.65**	−20.7993	−1.0390	−21.2049
3.26	−8.1899	−1.1965	−8.9947	**3.66**	−21.4868	−1.0367	−21.8814
3.27	−8.3518	−1.1904	−9.1469	**3.67**	−22.2150	−1.0345	−22.5987
3.28	−8.5178	−1.1844	−9.3032	**3.68**	−22.9879	−1.0324	−23.3606
3.29	−8.6881	−1.1786	−9.4639	**3.69**	−23.8096	−1.0304	−24.1713
3.3	−8.8629	−1.1729	−9.6290	**3.7**	−24.6852	−1.0284	−25.0358
3.31	−9.0425	−1.1673	−9.7988	**3.71**	−25.6201	−1.0265	−25.9597
3.32	−9.2269	−1.1618	−9.9734	**3.72**	−26.6208	−1.0247	−26.9492
3.33	−9.4165	−1.1565	−10.1532	**3.73**	−27.6945	−1.0229	−28.0117
3.34	−9.6114	−1.1512	−10.3382	**3.74**	−28.8496	−1.0212	−29.1556
3.35	−9.8119	−1.1461	−10.5289	**3.75**	−30.0960	−1.0196	−30.3907
3.36	−10.0183	−1.1412	−10.7253	**3.76**	−31.4449	−1.0180	−31.7284
3.37	−10.2308	−1.1363	−10.9279	**3.77**	−32.9098	−1.0165	−33.1820
3.38	−10.4497	−1.1315	−11.1369	**3.78**	−34.5066	−1.0151	−34.7673
3.39	−10.6755	−1.1269	−11.3526	**3.79**	−36.2539	−1.0138	−36.5033

(Continued)

TABLE 7.1 *(Continued)*
Stability Functions (Negative Sign Indicates Tensile Axial Load)

Phi	r	c	t
3.8	−38.1745	−1.0125	−38.4124
3.81	−40.2956	−1.0112	−40.5220
3.82	−42.6506	−1.0101	−42.8655
3.83	−45.2809	−1.0090	−45.4842
3.84	−48.2381	−1.0079	−48.4298
3.85	−51.5874	−1.0070	−51.7674
3.86	−55.4130	−1.0061	−55.5813
3.87	−59.8247	−1.0052	−59.9813
3.88	−64.9691	−1.0045	−65.1139
3.89	−71.0459	−1.0037	−71.1789
3.9	−78.3349	−1.0031	−78.4560
3.91	−87.2400	−1.0025	−87.3492
3.92	−98.3675	−1.0020	−98.4647
3.93	−112.6696	−1.0015	−112.7548
3.94	−131.7337	−1.0011	−131.8069
3.95	−158.4168	−1.0008	−158.4779
3.96	−198.4334	−1.0005	−198.4823
3.97	−265.1166	−1.0003	−265.1534
3.98	−398.4666	−1.0001	−398.4912
3.99	−798.4833	−1.0000	−798.4956
4	65535.0000	−1.0000	INF

With reference to the figure, the following expressions can be written. Due to the end rotations, we get the end moments as below:

$$k_{pr} = r_i \left(\frac{EI}{L_i}\right)\left(\frac{1}{L_i}\right) + r_i c_i \left(\frac{EI}{L_i}\right)\left(\frac{1}{L_i}\right) \tag{7.72a}$$

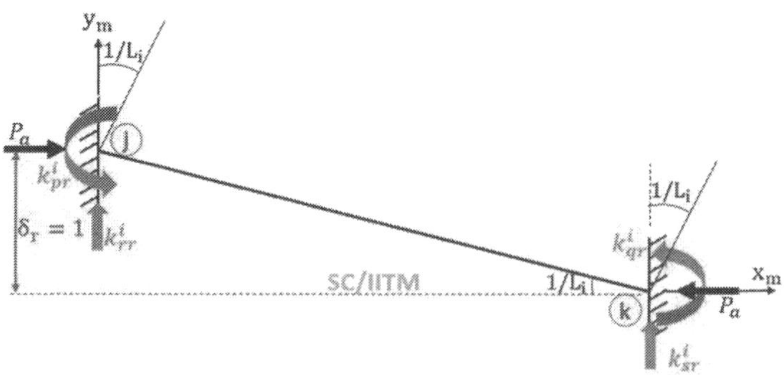

FIGURE 7.22 Unit translation at the jth end of the fixed beam under axial load.

$$k_{qr} = r_i c_i \left(\frac{EI}{L_i}\right)\left(\frac{1}{L_i}\right) + r_i \left(\frac{EI}{L_i}\right)\left(\frac{1}{L_i}\right) \tag{7.72b}$$

$$k_{qr} = k_{pr} = r_i (1 + c_i)\left(\frac{EI}{L_i^2}\right) \tag{7.72c}$$

Substituting Eq. (7.72) in Eq. (7.70), we get the expressions for end shear as below:

$$k_{rr} = -k_{sr} = 2r_i (1 + c_i)\left(\frac{EI}{L_i^3}\right) - \frac{P_a}{L_i} \tag{7.73}$$

For zero axial load, the equation for the end shear reduces to the following form:

$$k_{rr} = -k_{sr} = 2r_i (1 + c_i)\left(\frac{EI}{L_i^3}\right) \tag{7.74}$$

Hence, to make a general form, both in the presence and absence of axial load, the following function holds good to express end shear:

$$k_{rr} = -k_{sr} = t_i \left[2r_i (1 + c_i)\left(\frac{EI}{L_i^3}\right)\right] \tag{7.75}$$

Where t_i is known as the translation function. To obtain the translation function, let us now equate the axial load function as discussed below:
we know the following relationship (Eq. 7.45):

$$P_a = \frac{\pi^2 \varphi_i EI}{L_i^2} \quad for\, n = 1$$

Now, equating Eqs. (7.75) and (7.73), we get:

$$t_i \left[2r_i (1 + c_i)\left(\frac{EI}{L_i^3}\right)\right] = 2r_i (1 + c_i)\left(\frac{EI}{L_i^3}\right) - \frac{P_a}{L_i} \tag{7.76}$$

$$t_i = 1 - \left\{\left[\frac{P_a}{2r_i (1 + c_i)}\right]\left[\frac{L_i^2}{EI}\right]\right\} \tag{7.77}$$

Substituting for P_a, we get:

$$t_i = 1 - \frac{\pi^2 \varphi_i}{2r_i (1 + c_i)} \tag{7.78}$$

Similarly, by inducing unit rotation at the kth end of the beam, other coefficients (k_{ps}, k_{qs}, k_{rs}, k_{ss}) will be generated, as given below:

$$k_{ps} = k_{rs} = - r_i\left(1+c_i\right)\left(\frac{EI}{L_i^2}\right) \tag{7.79a}$$

$$k_{rs} = -k_{ss} = -2t_i r_i\left(1+c_i\right)\left(\frac{EI}{L_i^3}\right) \tag{7.79b}$$

By neglecting the axial deformation, member stiffness matrix of a fixed beam under axial load (either compressive or tensile) is given by the following form:

$$\left[K_i\right] = EI\begin{bmatrix} \frac{r_i}{L_i} & \frac{c_i r_i}{L_i} & \frac{r_i(1+c_i)}{L_i^2} & -\frac{r_i(1+c_i)}{L_i^2} \\ \frac{c_i r_i}{L_i} & \frac{r_i}{L_i} & \frac{r_i(1+c_i)}{L_i^2} & -\frac{r_i(1+c_i)}{L_i^2} \\ \frac{r_i(1+c_i)}{L_i^2} & \frac{r_i(1+c_i)}{L_i^2} & \frac{2t_i r_i(1+c_i)}{L_i^3} & -\frac{2t_i r_i(1+c_i)}{L_i^3} \\ -\frac{r_i(1+c_i)}{L_i^2} & -\frac{r_i(1+c_i)}{L_i^2} & -\frac{2t_i r_i(1+c_i)}{L_i^3} & \frac{2t_i r_i(1+c_i)}{L_i^3} \end{bmatrix} \tag{7.80}$$

7.6 LATERAL LOAD FUNCTIONS UNDER UNIFORMLY DISTRIBUTED LOAD

In this section, we shall derive the lateral load's functions for the fixed beam loaded under uniformly distributed load in the presence of the axial load. Consider a fixed beam, loaded as shown in Figure 7.23. Here, M_p does not refer to the plastic moment of resistance but indicates moment at the jth end under the rotation of θ_p. Alternatively, M_p also refers to the fixed end moment (FeM_p).

Consider the free-body diagram, as shown in Figure 7.24.

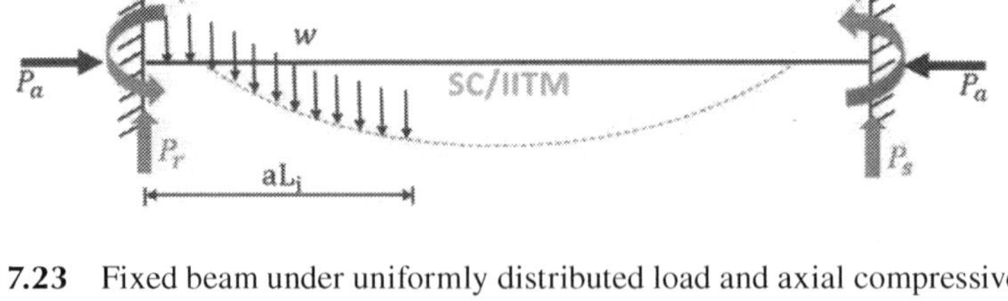

FIGURE 7.23 Fixed beam under uniformly distributed load and axial compressive load.

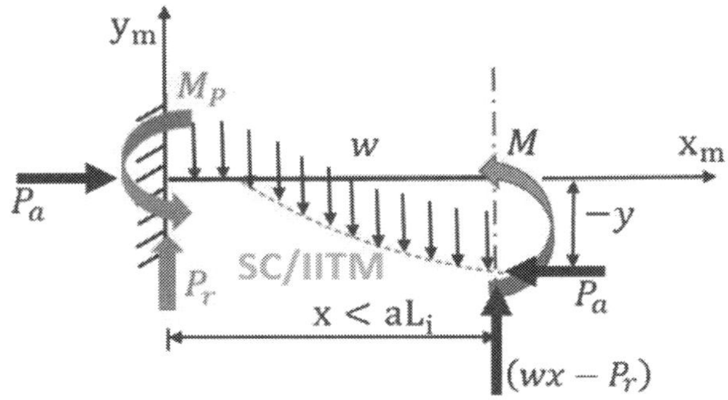

FIGURE 7.24 Free-body diagram ($x < aL_i$).

Under the fixed beam subjected to uniformly distributed load, end rotation and shear are developed as shown in the above figures. The following equation holds good:

$$EI\frac{d^2y}{dx^2} = M\ (for\ x < aL_i) \tag{7.81}$$

$$= -\left[M_p - P_a(-y) - P_r(x) + (-w)\frac{x^2}{2}\right] \tag{7.82}$$

The general solution to Eq. (7.82) is given by:

$$y = A\sin\left(\frac{\alpha_i x}{L_i}\right) + B\cos\left(\frac{\alpha_i x}{L_i}\right) - \frac{L_i^2}{\alpha_i^2 EI}\left[M_p + w\left(\frac{L_i^2}{\alpha_i^2}\right) - \frac{wx}{2}\left(\frac{2P_r}{w} + x\right)\right] \tag{7.83}$$

Differentiating once, we get:

$$\frac{dy}{dx} = \frac{\alpha_i}{L_i}A\cos\left(\frac{\alpha_i x}{L_i}\right) - \frac{\alpha_i}{L_i}B\sin\left(\frac{\alpha_i x}{L_i}\right) + \frac{L_i^2}{\alpha_i^2 EI}(P_r + wx) \tag{7.84}$$

Applying the boundary conditions:

$$@\,x = 0, \begin{cases} y = 0 \\ \dfrac{dy}{dx} = 0 \end{cases} \tag{7.85}$$

$$B = \frac{L_i^2}{\alpha_i^2 EI}\left[M_p + w\left(\frac{L_i^2}{\alpha_i^2}\right)\right] \tag{7.86a}$$

$$A = -\left[\frac{L_i^3}{\alpha_i^3 EI}P_r\right] \tag{7.86b}$$

Substituting Eq. (7.86) in Eq. (7.83), we get:

$$\left(\frac{\alpha_i^2 EI}{L_i^2}\right)y = -\left[\frac{L_i}{\alpha_i}P_r\right]\sin\left(\frac{\alpha_i x}{L_i}\right) + \left[M_p + w\left(\frac{L_i^2}{\alpha_i^2}\right)\right]\cos\left(\frac{\alpha_i x}{L_i}\right)$$
$$-M_p - w\left(\frac{L_i^2}{\alpha_i^2}\right) + \frac{wx}{2}\left(\frac{2P_r}{w} + x\right)\text{for } 0 \le x \le aL_i \quad (7.87a)$$

Substituting Eq. (7.86) in Eq. (7.84), we get:

$$\left(\frac{\alpha_i^2 EI}{L_i^2}\right)\frac{dy}{dx} = -\left(\frac{\alpha_i}{L_i}\right)P_r\cos\left(\frac{\alpha_i x}{L_i}\right) - \frac{\alpha_i}{L_i}\left[M_p + w\left(\frac{L_i^2}{\alpha_i^2}\right)\right]\sin\left(\frac{\alpha_i x}{L_i}\right) + \left(P_r + wx\right)$$
$$\text{for } 0 \le x \le aL_i \quad (7.87b)$$

For the position of the uniformly distributed load ($x > aL_i$), Eqs. (7.82 and 7.83) reduce to the following form (Figure 7.25):

$$EI\frac{d^2y}{dx^2} = M = -\left[M_p - P_a(-y) - P_r(x) + (-w\,aL_i)\left(x - \frac{aL_i}{2}\right)\right] \quad (7.88)$$

The general solution to Eq. (7.88) is given by:

$$y = A\sin\left(\frac{\alpha_i x}{L_i}\right) + B\cos\left(\frac{\alpha_i x}{L_i}\right) - \frac{L_i^2}{\alpha_i^2 EI}\left[M_p + \frac{wa^2L_i^2}{2} - \left(P_r + w\,aL_i\right)x\right] \quad (7.89)$$

Differentiating once, we get the following form:

$$\frac{dy}{dx} = \frac{\alpha_i}{L_i}A\cos\left(\frac{\alpha_i x}{L_i}\right) - \frac{\alpha_i}{L_i}B\sin\left(\frac{\alpha_i x}{L_i}\right) + \frac{L_i^2}{\alpha_i^2 EI}\left(P_r + w\,aL_i\right) \quad (7.90)$$

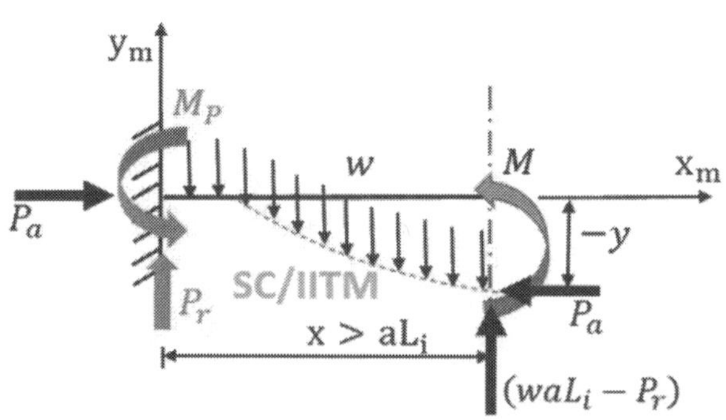

FIGURE 7.25 Free-body diagram ($x > aL_i$).

Applying the boundary conditions:

$$@x = L_i, \begin{cases} y = 0 \\ \dfrac{dy}{dx} = 0 \end{cases} \tag{7.91}$$

$$A = \frac{L_i^2}{\alpha_i^2 EI}\left[\left(M_p + \frac{wa^2L_i^2}{2}\right)\sin(\alpha_i) - L_i\left(P_r + waL_i\right)\left(\sin(\alpha_i) + \frac{\cos(\alpha_i)}{\alpha_i}\right)\right] \tag{7.92a}$$

$$B = \frac{L_i^2}{\alpha_i^2 EI}\left[\left(M_p + \frac{wa^2L_i^2}{2}\right)\cos(\alpha_i) - L_i\left(P_r + waL_i\right)\left(\cos(\alpha_i) - \frac{\sin(\alpha_i)}{\alpha_i}\right)\right] \tag{7.92b}$$

Substituting the above constants in Eqs. (7.89 and 7.90), we get:

$$\begin{aligned}\frac{\alpha_i^2 EI}{L_i^2}y = &\left[\left(M_p + \frac{wa^2L_i^2}{2}\right)\sin(\alpha_i) - L_i\left(P_r + waL_i\right)\left(\sin(\alpha_i) + \frac{\cos(\alpha_i)}{\alpha_i}\right)\right]\sin\left(\frac{\alpha_i x}{L_i}\right)\\ &+\left[\left(M_p + \frac{wa^2L_i^2}{2}\right)\cos(\alpha_i) - L_i\left(P_r + waL_i\right)\left(\cos(\alpha_i) - \frac{\sin(\alpha_i)}{\alpha_i}\right)\right]\cos\left(\frac{\alpha_i x}{L_i}\right)\\ &-\left[M_p + \frac{wa^2L_i^2}{2} - \left(P_r + waL_i\right)x\right] \text{for } aL_i \le x \le L_i\end{aligned} \tag{7.93a}$$

$$\begin{aligned}\frac{\alpha_i^2 EI}{L_i^2}\frac{dy}{dx} = &\frac{\alpha_i}{L_i}\left[\left(M_p + \frac{wa^2L_i^2}{2}\right)\sin(\alpha_i) - L_i\left(P_r + waL_i\right)\left(\sin(\alpha_i) + \frac{\cos(\alpha_i)}{\alpha_i}\right)\right]\\ &\cos\left(\frac{\alpha_i x}{L_i}\right) - \frac{\alpha_i}{L_i}\left[\left(M_p + \frac{wa^2L_i^2}{2}\right)\cos(\alpha_i) - L_i\left(P_r + waL_i\right)\left(\cos(\alpha_i) - \frac{\sin(\alpha_i)}{\alpha_i}\right)\right]\\ &\sin\left(\frac{\alpha_i x}{L_i}\right) + \left(P_r + waL_i\right) \text{for } aL_i \le x \le L_i\end{aligned} \tag{7.93b}$$

Now, for equating the displacement function (y) of both the segment lengths, one needs to equate Eqs. (7.87a and 7.93a). For equating the slope function (dy/dx) of both the segment lengths, one needs to equate Eqs. (7.87b and 7.93b). Equating as above and simplifying, we get the following:

$$P_r = waL_i\left(\frac{a}{2} - 1\right) + \frac{M_p + M_q}{L_i} \tag{7.94a}$$

$$M_p = -wL\left\{\left[\frac{\frac{a^2}{2}(1-\cos(\alpha_i)) + \left(\frac{\sin(a\alpha_i)}{\alpha_i} - a\right)\left(\frac{\sin(\alpha_i)}{\alpha_i} - \cos(\alpha_i)\right)}{2(1-cos(\alpha_i)) - \alpha_i\sin(\alpha_i)}\right] + \left[\frac{\frac{1-cos(a\alpha_i)}{\alpha_i}\left[\frac{1-cos(\alpha_i)}{\alpha_i} - \sin(\alpha_i)\right]}{2(1-cos(\alpha_i)) - \alpha_i\sin(\alpha_i)}\right]\right\} \tag{7.94b}$$

$$M_q = wL_i^2\left\{\left[\frac{a\left(1-\frac{a}{2}\right)(1-\cos(\alpha_i)) + \left(\frac{\sin(a\alpha_i)}{\alpha_i} - a\right)\left(\frac{\sin(\alpha_i)}{\alpha_i} - \cos(\alpha_i)\right)}{2(1-cos(\alpha_i)) - \alpha_i\sin(\alpha_i)}\right] + \left[\frac{\frac{1-cos(\alpha_i)}{\alpha_i}\left[\frac{1-cos(a\alpha_i)}{\alpha_i} - \sin(a\alpha_i)\right]}{2(1-cos(\alpha_i)) - \alpha_i\sin(\alpha_i)}\right]\right\} \tag{7.94c}$$

For the case of *zero axial loads*, one should apply L'Hospital rule to Eqs. (7.94a to 7.94c) to obtain the following results:

$$M_p = -wa^2L_i^2\left(\frac{1}{2} - \frac{2}{3}a + \frac{1}{4}a^2\right) \tag{7.95a}$$

$$M_q = wa^2L_i^2\left(\frac{1}{3} - \frac{1}{4}a\right) \tag{7.95b}$$

For the uniformly distributed load applied to the complete span of the beam, end moments shall take the following form:

$$M_p = -M_q = \frac{wL_i^2}{\alpha_i^2}\left[1 - \frac{\alpha_i\sin(\alpha_i)}{2(1-cos(\alpha_i))}\right] \tag{7.96}$$

Alternatively, they can also be expressed as:

$$M_p = -M_q = m_i\frac{wL_i^2}{\alpha_i^2} \tag{7.97a}$$

$$m_i = \frac{12}{\alpha_i^2}\left[1 - \frac{\alpha_i\sin(\alpha_i)}{2(1-cos(\alpha_i))}\right] \tag{7.97b}$$

Where m_i is termed as load function, which is to be multiplied by the end moments of a fixed beam under uniformly distributed load acting over the entire length of the member to account for the effect of the axial load. In the absence of axial compressive load ($\varphi_i = 0$), L'Hospital rule is applied to Eq. (7.96). This reduces to the following form, which is a standard expression for a fixed beam under uniformly distributed load acting upon its entire length.

$$\left(M_p\right)_{@\ \varphi_i=0} = -\left(M_q\right)_{@\ \varphi_i=0} = -\frac{wL_i^2}{12} \tag{7.98}$$

7.7 BEAM ELEMENT UNDER AXIAL TENSILE LOAD

Under tensile axial loads, the load function and end moments will take a different form. For a special case where (φ_i) is negative, Eqs. (7.94b and 7.94c) will reduce to the following form.

$$M_p = -wL_i^2\left\{\left[\frac{\frac{a^2}{2}\left(1-\cos h\left(\alpha_i\right)+\left(\frac{\sinh\left(a\alpha_i\right)}{\alpha_i}-a\right)\left(\frac{\sinh\left(\alpha_i\right)}{\alpha_i}-\cosh\left(\alpha_i\right)\right)\right.}{2\left(1-cosh\left(\alpha_i\right)\right)-\alpha_i\sinh\left(\alpha_i\right)}\right] - \left[\frac{\frac{1-cosh\left(a\alpha_i\right)}{\alpha_i}\left[\frac{1-cosh\left(\alpha_i\right)}{\alpha_i}-\sinh\left(\alpha_i\right)\right]}{2\left(1-cosh\left(\alpha_i\right)\right)-\alpha_i\sin h\left(\alpha_i\right)}\right]\right\} \tag{7.99a}$$

$$M_q = wL_i^2\left\{\left[\frac{a\left(1-\frac{a}{2}\right)\left(1-\cosh\left(\alpha_i\right)+\left(\frac{\sin h\left(a\alpha_i\right)}{\alpha_i}-a\right)\left(\frac{\sin h\left(\alpha_i\right)}{\alpha_i}-\cosh\left(\alpha_i\right)\right)\right.}{2\left(1-cosh\left(\alpha_i\right)\right)-\alpha_i\sinh\left(\alpha_i\right)}\right] - \left[\frac{\frac{1-cosh\left(\alpha_i\right)}{\alpha_i}\left[\frac{1-cosh\left(a\alpha_i\right)}{\alpha_i}-\sinh\left(a\alpha_i\right)\right]}{2\left(1-cosh\left(\alpha_i\right)\right)-\alpha_i\sin h\left(\alpha_i\right)}\right]\right\} \tag{7.99b}$$

For the uniformly distributed load applied to the complete span of the beam and in the presence of tensile axial load, end moments shall take the following form:

$$M_p = -M_q = \frac{wL_i^2}{\alpha_i^2}\left[1+\frac{\alpha_i\sinh\left(\alpha_i\right)}{2\left(1-cosh\left(\alpha_i\right)\right)}\right] \tag{7.100}$$

Alternatively, they can also be expressed as:

$$M_p = -M_q = m_i \frac{wL_i^2}{\alpha_i^2} \tag{7.101a}$$

$$m_i = -\frac{12}{\alpha_i^2}\left[1 - \frac{\alpha_i \sinh(\alpha_i)}{2(1 - cosh(\alpha_i))}\right] \tag{7.101b}$$

7.8 LATERAL LOAD FUNCTIONS UNDER CONCENTRATED LOAD

In this section, we shall derive the lateral load's functions for the fixed beam loaded under concentrated load in the presence of the axial load. Consider a fixed beam, loaded as shown in Figure 7.26.

Consider the free-body diagram, where ($x < aL_i$), as shown in Figure 7.27.

Under the fixed beam subjected to concentrated load, end rotations and shear are developed, as shown in the above figures. The following equation holds good:

$$EI \frac{d^2y}{dx^2} = M \ (for\, x < aL_i) \tag{7.102}$$

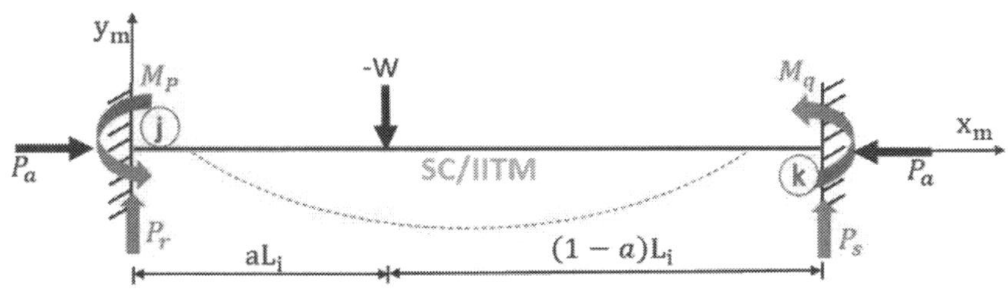

FIGURE 7.26 Fixed beam under concentrated load and axial compressive load.

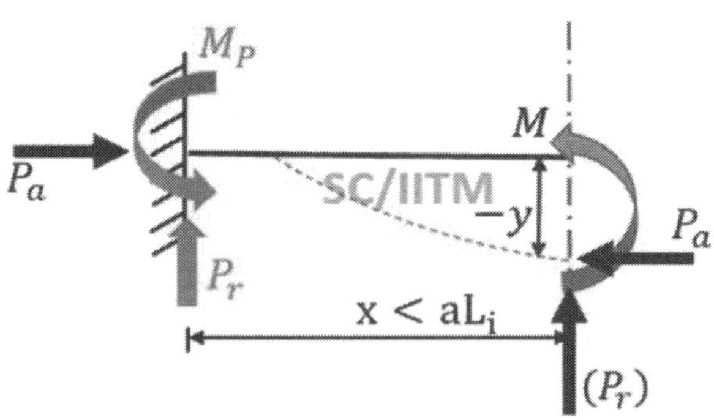

FIGURE 7.27 Free-body diagram with axial load and shear ($x < aL_i$).

$$=-\left[M_p-P_a(-y)-P_r(x)\right] \tag{7.103}$$

The general solution to Eq. (7.103) is given by:

$$y=A\sin\left(\frac{\alpha_i x}{L_i}\right)+B\cos\left(\frac{\alpha_i x}{L_i}\right)-\frac{L_i^2}{\alpha_i^2 EI}\left[M_p-P_r(x)\right] \tag{7.104}$$

Differentiating once, we get:

$$\frac{dy}{dx}=\frac{\alpha_i}{L_i}A\cos\left(\frac{\alpha_i x}{L_i}\right)-\frac{\alpha_i}{L_i}B\sin\left(\frac{\alpha_i x}{L_i}\right)+\frac{L_i^2}{\alpha_i^2 EI}(P_r) \tag{7.105}$$

Applying the boundary conditions:

$$@\,x=0,\begin{cases} y=0 \\ \dfrac{dy}{dx}=0 \end{cases} \tag{7.106}$$

$$B=\frac{L_i^2}{\alpha_i^2 EI}\left[M_p\right] \tag{7.107a}$$

$$A=-\left[\frac{L_i^3}{\alpha_i^3 EI}P_r\right] \tag{7.107b}$$

Substituting Eq. (7.107) in Eq. (7.104), we get:

$$\left(\frac{\alpha_i^2 EI}{L_i^2}\right)y=-\left[\frac{L_i}{\alpha_i}P_r\right]\sin\left(\frac{\alpha_i x}{L_i}\right)+M_p\cos\left(\frac{\alpha_i x}{L_i}\right)-\left[M_p-P_r(x)\right]\text{for } 0\le x\le aL_i \tag{7.108a}$$

Substituting Eq. (7.107) in Eq. (7.105), we get:

$$\left(\frac{\alpha_i^2 EI}{L_i^2}\right)\frac{dy}{dx}=-\left(\frac{\alpha_i}{L_i}\right)P_r\cos\left(\frac{\alpha_i x}{L_i}\right)-\frac{\alpha_i}{L_i}\left[M_p\right]\sin\left(\frac{\alpha_i x}{L_i}\right)+P_r\text{ for } 0\le x\le aL_i \tag{7.108b}$$

For the position of the concentrated load ($x > aL_i$), Eq. (7.102) reduces to the following form (Figure 7.28):

$$EI\frac{d^2y}{dx^2}=M=-\left[M_p-P_a(-y)-P_r(x)+(-W)(x-aL_i)\right] \tag{7.109}$$

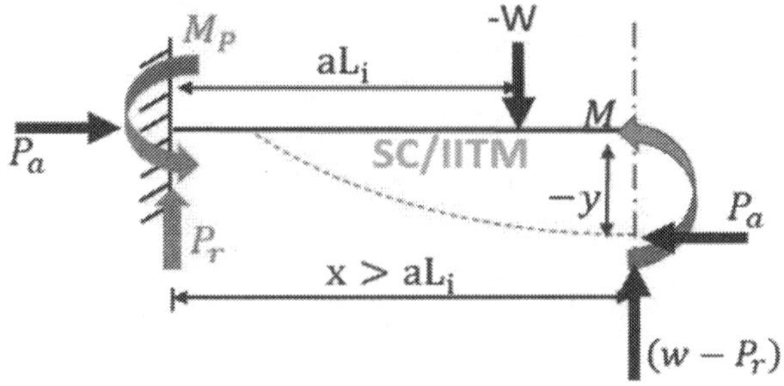

FIGURE 7.28 Free-body diagram with axial load and shear ($x > aL_i$).

The general solution to Eq. (7.109) is given by:

$$y = A\sin\left(\frac{\alpha_i x}{L_i}\right) + B\cos\left(\frac{\alpha_i x}{L_i}\right) - \frac{L_i^2}{\alpha_i^2 EI}\left[M_p + WaL_i - \left(P_r + W\right)x\right] \quad (7.110a)$$

Differentiating once, we get the following form:

$$\frac{dy}{dx} = \frac{\alpha_i}{L_i} A\cos\left(\frac{\alpha_i x}{L_i}\right) - \frac{\alpha_i}{L_i} B\sin\left(\frac{\alpha_i x}{L_i}\right) + \frac{L_i^2}{\alpha_i^2 EI}\left(P_r + W\right) \quad (7.110b)$$

Equating the expressions for deflection and slope for two segments (aL_i) and (1-a) L_i, we can evaluate the constants. Now, equating Eqs. (7.108a and 7.110a) for the deflection, and Eq. (7.108b) and Eq. (7.110b) for the slope, respectively, we get the following:

$$A = -\frac{L_i^3}{\alpha_i^3 EI}\left[P_r + W\cos\left(a\alpha_i\right)\right] \quad (7.111a)$$

$$B = \frac{L_i^2}{\alpha_i^2 EI}\left[M_p + \frac{L_i W}{\alpha_I} sin\left(a\alpha_i\right)\right] \quad (7.111b)$$

Substituting the above constants in Eqs. (7.110a and 7.110b), we get:

$$\frac{\alpha_i^2 EI}{L_i^2} y = -\frac{L_i}{\alpha_i}\left[P_r + W\cos\left(a\alpha_i\right)\right]\sin\left(\frac{\alpha_i x}{L_i}\right) + \left[M_p + \frac{L_i W}{\alpha_I} sin\left(a\alpha_i\right)\right]\cos\left(\frac{\alpha_i x}{L_i}\right)$$
$$-\left[M_p + WaL_i - \left(P_r + W\right)x\right] \text{ for } aL_i \le x \le L_i$$

(7.112a)

$$\frac{\alpha_i^2 EI}{L_i^2}\frac{dy}{dx} = -\left[P_r + W\cos(a\alpha_i)\right]\cos\left(\frac{\alpha_i x}{L_i}\right) - \frac{\alpha_i}{L_i}\left[M_p + \frac{L_i W}{\alpha_I} sin(a\alpha_i)\right]\sin\left(\frac{\alpha_i x}{L_i}\right)$$
$$+\left[P_r + W\right] \text{for } aL_i \le x \le L_i \tag{7.112b}$$

Applying the following boundary conditions

$$@\, x = L,\begin{cases} y = 0 \\ \dfrac{dy}{dx} = 0 \end{cases} \tag{7.113}$$

Eliminating P_r from Eq. (7.112a) and Eq. (7.112b) after applying the above boundary conditions, end moment assumes the following form:

$$M_p = -W\,L_i\left\{\left[\frac{\left[a\left(1-\cos(\alpha_i)\right)\right] - \left[\left(1-\cos(a\alpha_i)\right)\sin\left(\frac{\alpha_i x}{L_i}\right)\right] - \cos(\alpha_i)}{2\left(1-\cos(\alpha_i)\right) - \alpha_i\sin(\alpha_i)}\right]\right.$$
$$\left. + \left[\frac{\left[\dfrac{1-\cos(\alpha_i)}{\alpha_i} - \sin(\alpha_i)\right]\sin(a\alpha_i)}{2\left(1-\cos(\alpha_i)\right) - \alpha_i\sin(\alpha_i)}\right]\right\} \tag{7.114}$$

For the case of zero axial loads, one can estimate the end moment by applying L'Hospital rule.

$$\mathrm{M_p = WaL_i\left(a^2 - 1\right)} \tag{7.115}$$

Under tensile axial loads, which is a special case where (φ_i) is negative, Eq. (7.115) will reduce to the following form:

$$M_p = -WL_i\left\{\left[\frac{a\ (1-\cosh(\alpha_i) - (1-coh(a\alpha_i)\left(\dfrac{\sinh(\alpha_i)}{\alpha_i} - \cosh(\alpha_i)\right)}{2\left(1-cosh(\alpha_i)\right) + \alpha_i\sinh(\alpha_i)}\right]\right.$$
$$\left. + \left[\frac{\left(\dfrac{1-cosh(\alpha_i)}{\alpha_i} + \sinh(\alpha_i)\right)\sinh(a\alpha_i)}{2\left(1-cosh(\alpha_i)\right) + \alpha_i\sinh(\alpha_i)}\right]\right\} \tag{7.116}$$

7.9 EXAMPLE PROBLEMS WITH COMPUTER PROGRAMS

Exercise problem 1:

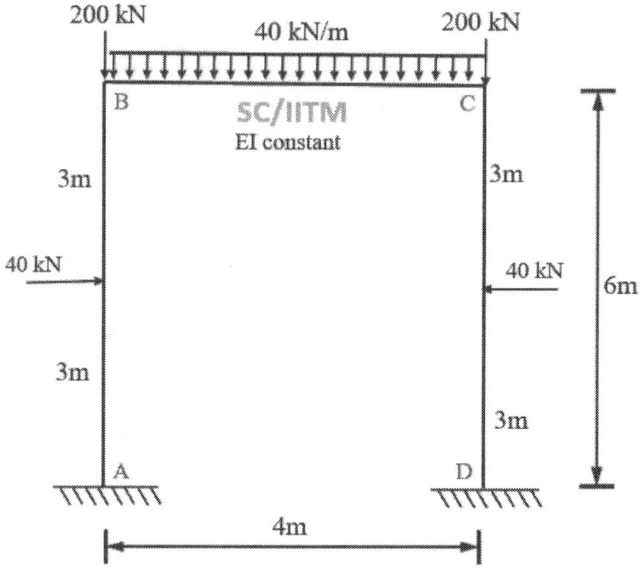

Take $I = 8.333e-6\,m^4, E = 2.1e11\,N/m^2$

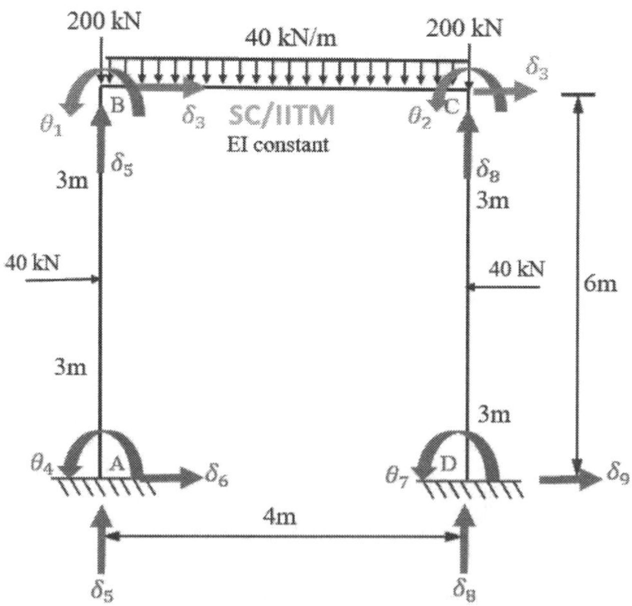

MATLAB program (without axial load):

```
%% This MATLAB code is for analysis of structures without considering
axial load
% Re-type the following code in MATLAB new script and run the file to get
the output.
%% Input
clc;
clear;
n = 3; % number of members
I = [1 1 1]*8.333e-6; %Moment of inertia in m4
E = [1 1 1]*2.1e11; % Young's modulus
L = [6 4 6]; % length in m
uu = 3; % Number of unrestrained degrees of freedom
ur = 6; % Number of restrained degrees of freedom
uul = [1 2 3]; % global labels of unrestained dof
url = [4 5 6 7 8 9]; % global labels of restained dof
l1 = [1 4 3 6]; % Global labels for member 1
l2 = [1 2 5 8]; % Global labels for member 2
l3 = [2 7 3 9]; % Global labels for member 3
l= [l1; l2; l3];
dof = uu+ur;
Ktotal = zeros (dof);
fem1= [-30 30 -20 -20]; % Local Fixed end moments of member 1
fem2= [53.333 -53.333 280 280]; % Local Fixed end moments of member 2
fem3= [30 -30 20 20]; % Local Fixed end moments of member 3

%% rotation coefficients for each member
rc1 = 4.*E.*I./L;
rc2 = 2.*E.*I./L;

%% stiffness matrix 4 by 4 (axial deformation neglected)
for i = 1:n
        Knew = zeros (dof);
        k1 = [rc1(i); rc2(i); (rc1(i)+rc2(i))/L(i); (-
(rc1(i)+rc2(i))/L(i))];
        k2 = [rc2(i); rc1(i); (rc1(i)+rc2(i))/L(i); (-
(rc1(i)+rc2(i))/L(i))];
        k3 = [(rc1(i)+rc2(i))/L(i); (rc1(i)+rc2(i))/L(i);
(2*(rc1(i)+rc2(i))/(L(i)^2)); (-2*(rc1(i)+rc2(i))/(L(i)^2))];
        k4 = -k3;
        K = [k1 k2 k3 k4];
        fprintf ('Member Number =');
        disp (i);
        fprintf ('Local Stiffness matrix of member, [K] = \n');
        disp (K);
        for p = 1:4
            for q = 1:4
                Knew((l(i,p)),(l(i,q))) =K(p,q);
            end
        end
```

```
        Ktotal = Ktotal + Knew;
        if i == 1
            Kg1=K;
        elseif i == 2
            Kg2 = K;
        else
            Kg3=K;
        end
end
fprintf ('Stiffness Matrix of complete structure, [Ktotal] = \n');
disp (Ktotal);
Kunr = zeros(uu);
for x=1:uu
    for y=1:uu
        Kunr(x,y)= Ktotal(x,y);
    end
end
fprintf ('Unrestrained Stiffness sub-matrix, [Kuu] = \n');
disp (Kunr);
KuuInv= inv(Kunr);
fprintf ('Inverse of Unrestrained Stiffness sub-matrix, [KuuInverse] =
\n');
disp (KuuInv);

%% Creation of joint load vector
jl= [-23.333; 23.333; 0; -30; -280; 20; 30; -280; -20]; % values
given in kN or kNm
jlu = jl(1:uu,1); % load vector in unrestrained dof
delu = KuuInv*jlu;
fprintf ('Joint Load vector, [Jl] = \n');
disp (jl');
fprintf ('Unrestrained displacements, [DelU] = \n');
disp (delu');
delr = zeros (ur,1);
del = [delu; delr];
deli= zeros (4,1);
for i = 1:n
    for p = 1:4
        deli(p,1) = del((l(i,p)),1) ;
    end
    if i == 1
            delbar1 = deli;
            mbar1= (Kg1 * delbar1)+fem1';
            fprintf ('Member Number =');
            disp (i);
            fprintf ('Global displacement matrix [DeltaBar] = \n');
            disp (delbar1');
            fprintf ('Global End moment matrix [MBar] = \n');
            disp (mbar1');
        elseif i == 2
```

```
            delbar2 = deli;
            mbar2= (Kg2 * delbar2)+fem2';
            fprintf ('Member Number =');
            disp (i);
            fprintf ('Global displacement matrix [DeltaBar] = \n');
            disp (delbar2');
            fprintf ('Global End moment matrix [MBar] = \n');
            disp (mbar2');
        else
            delbar3 = deli;
            mbar3= (Kg3 * delbar3)+fem3';
            fprintf ('Member Number =');
            disp (i);
            fprintf ('Global displacement matrix [DeltaBar] = \n');
            disp (delbar3');
            fprintf ('Global End moment matrix [MBar] = \n');
            disp (mbar3');
     end
end
%% check
mbar = [mbar1'; mbar2'; mbar3'];
jf = zeros(dof,1);
for a=1:n
    for b=1:4 % size of k matrix
        d = l(a,b);
        jfnew = zeros(dof,1);
        jfnew(d,1)=mbar(a,b);
        jf=jf+jfnew;
    end
end
fprintf ('Joint forces = \n');
disp (jf');
```

MATLAB output:

```
Member Number =     1
Local Stiffness matrix of member, [K] =
   1.0e+06 *
    1.1666    0.5833    0.2917   -0.2917
    0.5833    1.1666    0.2917   -0.2917
    0.2917    0.2917    0.0972   -0.0972
   -0.2917   -0.2917   -0.0972    0.0972
Member Number =     2
Local Stiffness matrix of member, [K] =
   1.0e+06 *
    1.7499    0.8750    0.6562   -0.6562
    0.8750    1.7499    0.6562   -0.6562
    0.6562    0.6562    0.3281   -0.3281
   -0.6562   -0.6562   -0.3281    0.3281
```

```
Member Number =      3
Local Stiffness matrix of member, [K] =
   1.0e+06 *
    1.1666     0.5833     0.2917   -0.2917
    0.5833     1.1666     0.2917   -0.2917
    0.2917     0.2917     0.0972   -0.0972
   -0.2917    -0.2917    -0.0972    0.0972
Stiffness Matrix of complete structure, [Ktotal] =
   1.0e+06 *
    2.9166    0.8750    0.2917    0.5833    0.6562   -0.2917         0   -0.6562         0
    0.8750    2.9166    0.2917         0    0.6562         0    0.5833   -0.6562   -0.2917
    0.2917    0.2917    0.1944    0.2917         0   -0.0972    0.2917         0   -0.0972
    0.5833         0    0.2917    1.1666         0   -0.2917         0         0         0
    0.6562    0.6562         0         0    0.3281         0         0   -0.3281         0
   -0.2917         0   -0.0972   -0.2917         0    0.0972         0         0         0
         0    0.5833    0.2917         0         0         0    1.1666         0   -0.2917
   -0.6562   -0.6562         0         0   -0.3281         0         0    0.3281         0
         0   -0.2917   -0.0972         0         0         0   -0.2917         0    0.0972
Unrestrained Stiffness sub-matrix, [Kuu] =
   1.0e+06 *
    2.9166     0.8750     0.2917
    0.8750     2.9166     0.2917
    0.2917     0.2917     0.1944
Inverse of Unrestrained Stiffness sub-matrix, [KuuInverse] =
   1.0e-05 *
    0.0416    -0.0073    -0.0514
   -0.0073     0.0416    -0.0514
   -0.0514    -0.0514     0.6686
Joint Load vector, [Jl] =
  -23.3330 23.3330        0 -30.0000 -280.0000  20.0000  30.0000
-280.0000  -20.0000
Unrestrained displacements, [DelU] =
   1.0e-04 *
   -0.1143     0.1143    -0.0000
Member Number =      1
Global displacement matrix [DeltaBar] =
   1.0e-04 *
   -0.1143          0    -0.0000          0
Global End moment matrix [MBar] =
  -43.3331    23.3334   -23.3333   -16.6667
Member Number =      2
Global displacement matrix [DeltaBar] =
   1.0e-04 *
   -0.1143     0.1143          0          0
Global End moment matrix [MBar] =
   43.3331   -43.3331   280.0000   280.0000
Member Number =      3
Global displacement matrix [DeltaBar] =
   1.0e-04 *
    0.1143          0    -0.0000          0
Global End moment matrix [MBar] =
   43.3331   -23.3334    23.3333    16.6667
```

```
Joint forces =
         0   -0.0000         0   23.3334  280.0000  -16.6667
-23.3334  280.0000   16.6667
>>
```

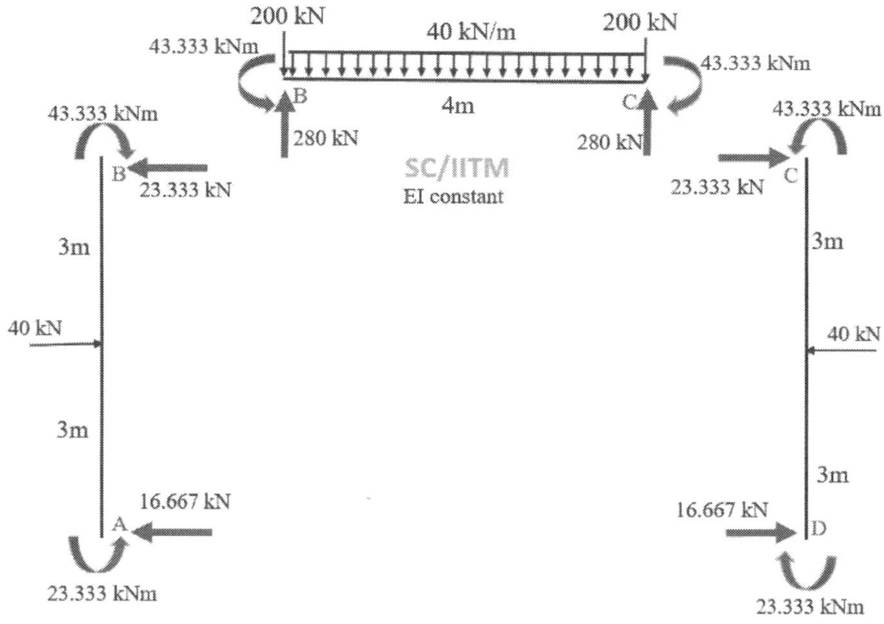

MATLAB Program (with axial load):

```
%% This MATLAB code is for stability analysis of structures, which
includes the effect of axial load on the end moments and shear
% Re-type the following code in MATLAB new script and run the file
to get the output.
%% Input
clc;
clear;
n = 3; % number of members
I = [1 1 1]*8.333e-6; %Moment of inertia in m4
E = [1 1 1]*2.1e11; % Young's modulus in N/m^2
L = [6 4 6]; % length in m
uu = 3; % Number of unrestrained degrees of freedom
ur = 6; % Number of restrained degrees of freedom
uul = [1 2 3]; % global labels of unrestrained dof
url = [4 5 6 7 8 9]; % global labels of restrained dof
l1 = [1 4 3 6]; % Global labels for member 1
l2 = [1 2 5 8]; % Global labels for member 2
```

```
l3 = [2 7 3 9]; % Global labels for member 3
l= [l1; l2; l3];
dof = uu+ur;
Ktotal = zeros (dof);
fem1= [-30 30 -20 -20]; % Local Fixed end moments of member 1
fem2= [53.333 -53.333 280 280]; % Local Fixed end moments of member 2
fem3= [30 -30 20 20]; % Local Fixed end moments of member 3
pa = [280 23.333 280]*1000; %Axial load in N
load = [1 1 1]; % 0-zero load, 1-compression, 2-tension

%% Load and angle calculation
pe = pi*pi.*E.*I./(L.*L); % Euler's load in N
phi = pa./pe;
alrad = pi.*sqrt(phi);
al = radtodeg (alrad);
r = zeros(1,n);
c = zeros(1,n);
t = zeros(1,n);
for i = 1:n
    if load(i)==1
        % rotation coefficients only for compression loads
        r(i) = (alrad(i).*((sind(al(i))-
(alrad(i).*cosd(al(i)))))./((2.*(1-cosd(al(i))))-
(alrad(i).*(sind(al(i))))); % rotation function
        c(i) = (alrad(i)-sind(al(i)))./(sind(al(i))-
(alrad(i).*cosd(al(i)))); % rotation function
        t(i) = 1-((pi*pi*phi(i))./(2.*r(i).*(1+c(i)))); % Translation
function
    elseif load(i)==2
        % rotation coefficients only for tension loads
        r(i)= (alrad(i)*((alrad(i)*cosh(alrad(i)))-
sinh(alrad(i))))/((2*(1-cosh(alrad(i))))+(alrad(i)*sinh(alra
d(i))));
        c(i) = (alrad(i)-sinh(alrad(i)))/
(sinh(alrad(i))-(alrad(i)*cosh(alrad(i))));
        t(i) = 1-((pi*pi*(-phi(i)))./(2.*r(i).*(1+c(i))));
    else
        r(i) = 4;
        c(i) = 0.5;
        t(i) = 1;
    end
end

%% rotation coefficients for each member
rc1 = r.*E.*I./L;
rc2 = c.*E.*I.*r./L;

%% stiffness matrix 4 by 4 (axial deformation neglected)
for i = 1:n
        Knew = zeros (dof);
```

```
        k1 = [rc1(i); rc2(i); (rc1(i)+rc2(i))/L(i); (-
(rc1(i)+rc2(i))/L(i))];
        k2 = [rc2(i); rc1(i); (rc1(i)+rc2(i))/L(i); (-
(rc1(i)+rc2(i))/L(i))];
        k3 = [(rc1(i)+rc2(i))/L(i); (rc1(i)+rc2(i))/L(i);
(2*t(i)*(rc1(i)+rc2(i))/(L(i)^2)); (-2*t(i)*(rc1(i)+rc2(i))/
(L(i)^2))];
        k4 = -k3;
        K = [k1 k2 k3 k4];
        fprintf ('Member Number =');
        disp (i);
        fprintf ('Local Stiffness matrix of member, [K] = \n');
        disp (K);
        for p = 1:4
            for q = 1:4
                Knew((l(i,p)),(l(i,q))) =K(p,q);
            end
        end
        Ktotal = Ktotal + Knew;
        if i == 1
            Kg1=K;
        elseif i == 2
            Kg2 = K;
        else
            Kg3=K;
        end
end
fprintf ('Stiffness Matrix of complete structure, [Ktotal] = \n');
disp (Ktotal);
Kunr = zeros(uu);
for x=1:uu
    for y=1:uu
        Kunr(x,y)= Ktotal(x,y);
    end
end
fprintf ('Unrestrained Stiffness sub-matrix, [Kuu] = \n');
disp (Kunr);
KuuInv= inv(Kunr);
fprintf ('Inverse of Unrestrained Stiffness sub-matrix, [KuuInverse] =
\n');
disp (KuuInv);

%% Creation of joint load vector
jl= [-23.333; 23.333; 0; -30; -280; 20; 30; -280; -20]; %
values given in kN or kNm
jlu = jl(1:uu,1); % load vector in unrestrained dof
delu = KuuInv*jlu;
fprintf ('Joint Load vector, [Jl] = \n');
disp (jl');
fprintf ('Unrestrained displacements, [DelU] = \n');
```

```
disp (delu');
delr = zeros (ur,1);
del = [delu; delr];
deli= zeros (4,1);
for i = 1:n
    for p = 1:4
        deli(p,1) = del((l(i,p)),1) ;
    end
    if i == 1
            delbar1 = deli;
            mbar1= (Kg1 * delbar1)+fem1';
            fprintf ('Member Number =');
            disp (i);
            fprintf ('Global displacement matrix [DeltaBar] = \n');
            disp (delbar1');
            fprintf ('Global End moment matrix [MBar] = \n');
            disp (mbar1');
        elseif i == 2
            delbar2 = deli;
            mbar2= (Kg2 * delbar2)+fem2';
            fprintf ('Member Number =');
            disp (i);
            fprintf ('Global displacement matrix [DeltaBar] = \n');
            disp (delbar2');
            fprintf ('Global End moment matrix [MBar] = \n');
            disp (mbar2');
        else
            delbar3 = deli;
            mbar3= (Kg3 * delbar3)+fem3';
            fprintf ('Member Number =');
            disp (i);
            fprintf ('Global displacement matrix [DeltaBar] = \n');
            disp (delbar3');
            fprintf ('Global End moment matrix [MBar] = \n');
            disp (mbar3');
     end
end
%% check
mbar = [mbar1'; mbar2'; mbar3'];
jf = zeros(dof,1);
for a=1:n
    for b=1:4 % size of k matrix
        d = l(a,b);
        jfnew = zeros(dof,1);
        jfnew(d,1)=mbar(a,b);
        jf=jf+jfnew;
    end
end
fprintf ('Joint forces = \n');
disp (jf');
```

MATLAB output:

```
Member Number =      1
Local Stiffness matrix of member, [K] =
   1.0e+05 *
    9.2333    6.5121    2.6242   -2.6242
    6.5121    9.2333    2.6242   -2.6242
    2.6242    2.6242    0.4081   -0.4081
   -2.6242   -2.6242   -0.4081    0.4081
Member Number =      2
Local Stiffness matrix of member, [K] =
   1.0e+06 *
    1.7375    0.8781    0.6539   -0.6539
    0.8781    1.7375    0.6539   -0.6539
    0.6539    0.6539    0.3211   -0.3211
   -0.6539   -0.6539   -0.3211    0.3211
Member Number =      3
Local Stiffness matrix of member, [K] =
   1.0e+05 *
    9.2333    6.5121    2.6242   -2.6242
    6.5121    9.2333    2.6242   -2.6242
    2.6242    2.6242    0.4081   -0.4081
   -2.6242   -2.6242   -0.4081    0.4081
Stiffness Matrix of complete structure, [Ktotal] =
   1.0e+06 *
    2.6608    0.8781    0.2624    0.6512    0.6539   -0.2624         0   -0.6539         0
    0.8781    2.6608    0.2624         0    0.6539         0    0.6512   -0.6539    0.2624
    0.2624    0.2624    0.0816    0.2624         0   -0.0408    0.2624         0   -0.0408
    0.6512         0    0.2624    0.9233         0   -0.2624         0         0         0
    0.6539    0.6539         0         0    0.3211         0         0   -0.3211         0
   -0.2624         0   -0.0408   -0.2624         0    0.0408         0         0         0
         0    0.6512    0.2624         0         0         0    0.9233         0    0.2624
   -0.6539   -0.6539         0         0   -0.3211         0         0    0.3211         0
         0   -0.2624   -0.0408         0         0         0   -0.2624         0    0.0408
Unrestrained Stiffness sub-matrix, [Kuu] =
   1.0e+06 *
    2.6608    0.8781    0.2624
    0.8781    2.6608    0.2624
    0.2624    0.2624    0.0816
Inverse of Unrestrained Stiffness sub-matrix, [KuuInverse] =
   1.0e-04 *
    0.0055   -0.0001   -0.0174
   -0.0001    0.0055   -0.0174
   -0.0174   -0.0174    0.2342
Joint Load vector, [Jl] =
  -23.3330  23.3330       0 -30.0000 -280.0000  20.0000   30.0000
 -280.0000  -20.0000
Unrestrained displacements, [DelU] =
   1.0e-04 *
   -0.1309    0.1309         0
Member Number =      1
Global displacement matrix [DeltaBar] =
   1.0e-04 *
```

```
   -0.1309          0          0          0
Global End moment matrix [MBar] =
  -42.0852   21.4765  -23.4348  -16.5652
Member Number =     2
Global displacement matrix [DeltaBar] =
   1.0e-04 *
   -0.1309    0.1309          0          0
Global End moment matrix [MBar] =
   42.0852  -42.0852  280.0000  280.0000
Member Number =     3
Global displacement matrix [DeltaBar] =
   1.0e-04 *
    0.1309          0          0          0
Global End moment matrix [MBar] =
   42.0852  -21.4765   23.4348   16.5652
Joint forces =
    0.0000   -0.0000          0   21.4765  280.0000  -16.5652
-21.4765  280.0000   16.5652
```

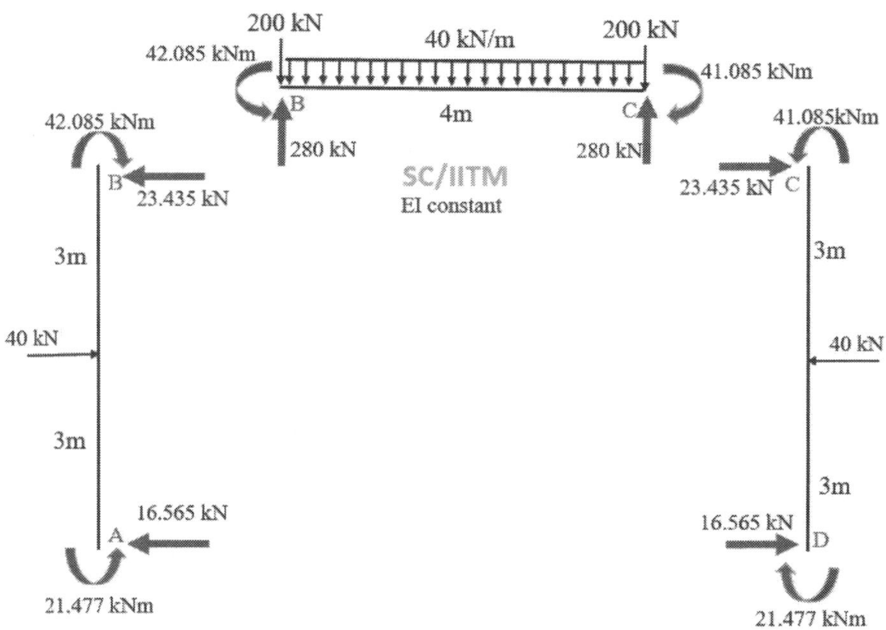

7.10 CRITICAL BUCKLING LOAD

One of the basic assumptions for obtaining the buckling load of a structural system with rigid joints is that the structure's deformation is sufficiently small compared to its initial condition. Therefore, one can apply the linear theory to estimate the critical buckling load. It is also important to note that structural system buckling loads are estimated because the structural system is transferring only axial forces. Members that encounter transverse loads develop additional moments, which significantly alter the members' relative stiffness. Under such cases, where there is a continuous interaction of the moment and the applied axial load (P-M interaction), critical buckling load is estimated only through the iterative scheme. However, for ideal members subjected to axial forces only, buckling loads can be estimated much simply. By neglecting axial deformations the members undergo, joint loads related to the unrestrained joint displacements are set to ZERO at all unrestrained joints.

$$\{J_{Lu}\} = \{0\} \tag{7.117}$$

However, from the fundamental equations of equilibrium, the following relationship holds good:

$$\{J_{Lu}\} = [K_{uu}]\{\Delta_u\} \tag{7.118}$$

where $[K_{uu}]$ is the sub-matrix of the stiffness matrix of the complete structural system, indicating only the unrestrained degrees of freedom, and Δ_u is the displacement vector of the unrestrained joints. By comparing the above two equations, it is obvious to note that Eq. (7.117) can be true only if Δ_u is a null vector. But this is a trivial solution, corresponding to no unrestrained joint displacements. Hence, to obtain the nontrivial solution to Eq. (7.117), the following conditions should be satisfied:

$$|K_{uu}| = 0 \tag{7.119}$$

The above is termed as the characteristic determinant, whose expansion will yield the buckling condition of the structural system. It is essential that, when the magnitude of the applied axial load is less than that of the critical buckling load, the displacements of the unrestrained joints would be zero. Hence, $|K_{uu}|$ will be positive, which corresponds to a stable condition. On the other hand, if the applied axial load exceeds the critical buckling load, then it will refer to an unstable condition; in such case, $|K_{uu}|$ will be negative.

In a structural system comprising more members, more than one member may be subjected to axial load. In such cases, it becomes necessary to establish a relationship (φ_i) between the axial and buckling loads, as explained in Eq. (7.44). This becomes more complicated when more than one member shares the applied axial load at any joint. The axial load shared by each member at that joint must be computed by a preliminary analysis. Then, the critical buckling load can be estimated using the following relationship:

$$P_{cr}^i = (\varphi_i)_{critical} \frac{\pi^2 EI}{L_i^2} \tag{7.120}$$

7.11 EXERCISE PROBLEMS WITH COMPUTER PROGRAMS

Example 7.1:

Estimate the critical buckling load of the column member using stability functions:

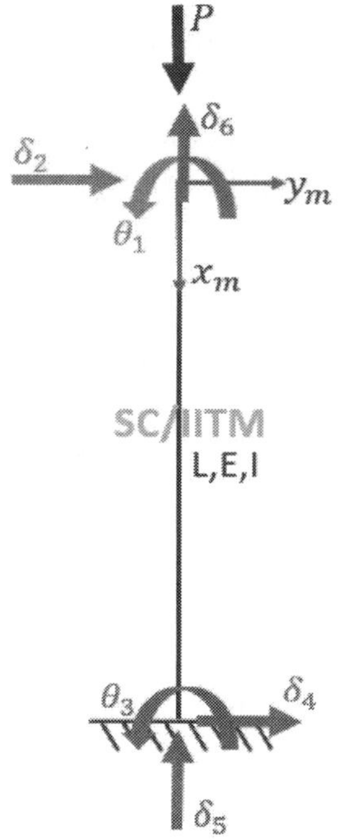

Concerning the above figure, both unrestrained and restrained, kinematic degrees of freedom, are marked. A detailed procedure for analyzing a member using the stiffness method is discussed in Chandrasekaran (2018). The member stiffness matrix is given below (the same as Eq. (7.80)). Numbers shown in circles are the labels of the degrees of freedom.

$$[K_i] = EI \begin{array}{c} \begin{array}{cccc} \quad 1 \quad & \quad 3 \quad & \quad 2 \quad & \quad 4 \quad \end{array} \\ \left[\begin{array}{cccc} \frac{r}{L} & \frac{cr}{L} & \frac{r(1+c)}{L^2} & -\frac{r(1+c)}{L^2} \\ \frac{cr}{L} & \frac{r}{L} & \frac{r(1+c)}{L^2} & -\frac{r(1+c)}{L^2} \\ \frac{r(1+c)}{L^2} & \frac{r(1+c)}{L^2} & \frac{2tr(1+c)}{L^3} & -\frac{2tr(1+c)}{L^3} \\ -\frac{r(1+c)}{L^2} & -\frac{r(1+c)}{L^2} & -\frac{2tr(1+c)}{L^3} & \frac{2tr(1+c)}{L^3} \end{array}\right] \end{array} \begin{array}{c} \\ 1 \\ 3 \\ 2 \\ 4 \end{array} \tag{7.121}$$

The above matrix can be partitioned to extract the unrestrained stiffness matrix, which will be of size 2×2, as this problem has two unrestrained degrees of freedom. The unrestrained stiffness matrix, as extracted from Eq. (7.121), is given by:

$$\begin{array}{c} \\ [K_{UU}] = EI \end{array}\begin{array}{c} \begin{array}{cc} 1 & 2 \end{array} \\ \begin{bmatrix} \dfrac{r}{L} & \dfrac{r(1+c)}{L^2} \\ \dfrac{r(1+c)}{L^2} & \dfrac{2tr(1+c)}{L^3} \end{bmatrix} \end{array}\begin{array}{c} \\ 1 \\ 2 \end{array} \tag{7.122}$$

As no joint load is applied in the unrestrained degrees of freedom, as per Eq. (7.119), let us set the determinant of Eq. (7.122) to zero. Expanding, we get as follows:

$$\left(\frac{r}{L}\right)\left[\frac{2tr(1+c)}{L^3}\right]-\left[\frac{r(1+c)}{L^2}\right]^2=0 \tag{7.123}$$

Simplifying the above expression, we get the following expression:

$$2t-c=1 \tag{7.124}$$

Eq. (7.124) is called as the Characteristic equation, defining the critical buckling condition in terms of the stability functions. Using Table 7.1, one can determine the value of φ_i, satisfying the Eq. (7.124). The MATLAB given below helps to scan the appropriate values of the stability functions and determines φ_i as 0.25. The corresponding values of stability functions are: r = 3.6598; t = 0.7854; c = 0.5708.

Using the Eq. (7.120), we get:

$$P_{cr}=\varphi\frac{\pi^2 EI}{L^2}=(0.25)\frac{\pi^2 EI}{L^2} \tag{7.125}$$

```
% getting phi value condition: 2t = 1+c
lhs = 2*t;
rhs = 1+c;
n = length(out);
for i=1:n
    if lhs(i) == rhs(i)
        phi_value = out(i,1);
    end
end
fprintf('Phi = %6.2f \n',phi_value);
```

MATLAB output:

```
Phi =   0.25
```

Example 7.2:

Estimate the critical buckling load of the structure using stability functions:

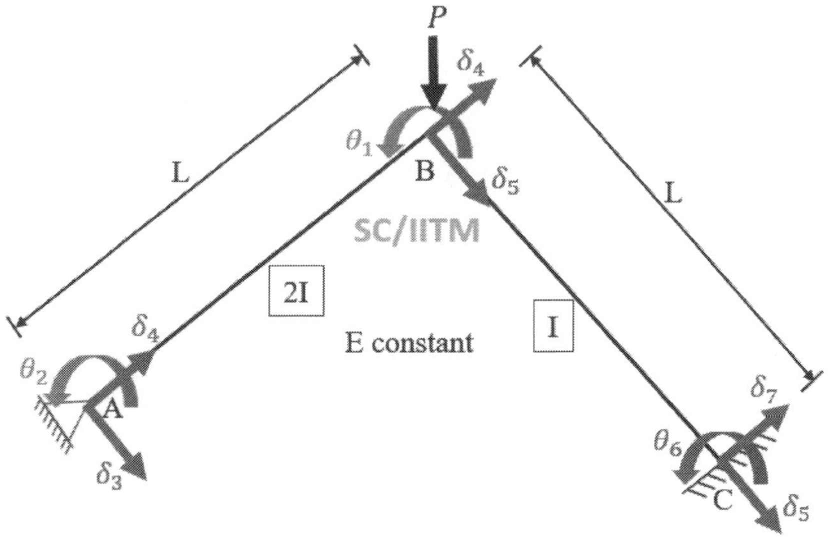

With reference to the above figure, kinematic degrees of freedom, both unrestrained and restrained, are marked. The member stiffness matrix for both members is given below.

Numbers shown in circles are the labels of the degrees of freedom of the respective member.

$$[K_1]=E(2I)\begin{array}{c}\begin{matrix}\quad 1\quad & \quad 2\quad & \quad 5\quad & \quad 3\quad\end{matrix}\\ \begin{bmatrix} \dfrac{r_1}{L_1} & \dfrac{c_1r_1}{L_1} & \dfrac{r_1(1+c_1)}{L_1^2} & -\dfrac{r_1(1+c_1)}{L_1^2} \\ \dfrac{c_1r_1}{L_1} & \dfrac{r_1}{L_1} & \dfrac{r_1(1+c_1)}{L_1^2} & -\dfrac{r_1(1+c_1)}{L_1^2} \\ \dfrac{r_1(1+c_1)}{L_1^2} & \dfrac{r_1(1+c_1)}{L_1^2} & \dfrac{2t_1r_1(1+c_1)}{L_1^3} & -\dfrac{2t_1r_1(1+c_1)}{L_1^3} \\ -\dfrac{r_1(1+c_1)}{L_1^2} & -\dfrac{r_1(1+c_1)}{L_1^2} & -\dfrac{2t_1r_1(1+c_1)}{L_1^3} & \dfrac{2t_1r_1(1+c_1)}{L_1^3} \end{bmatrix}\end{array}\begin{matrix}1\\[2ex]2\\[2ex]5\\[2ex]3\end{matrix} \tag{7.126}$$

$$
\begin{array}{cccccc}
 & 1 & 6 & 4 & 7 & \\
[K_2] = EI & \left[\begin{array}{cccc}
\frac{r_2}{L_2} & \frac{c_2 r_2}{L_2} & \frac{r_2(1+c_2)}{L_2^2} & -\frac{r_2(1+c_2)}{L_2^2} \\
\frac{c_2 r_2}{L_2} & \frac{r_2}{L_2} & \frac{r_2(1+c_2)}{L_2^2} & -\frac{r_2(1+c_2)}{L_2^2} \\
\frac{r_2(1+c_2)}{L_2^2} & \frac{r_2(1+c_2)}{L_2^2} & \frac{2t_2 r_2(1+c_2)}{L_2^3} & -\frac{2t_2 r_2(1+c_2)}{L_2^3} \\
-\frac{r_2(1+c_2)}{L_2^2} & -\frac{r_2(1+c_2)}{L_2^2} & -\frac{2t_2 r_2(1+c_2)}{L_2^3} & \frac{2t_2 r_2(1+c_2)}{L_2^3}
\end{array}\right] & \begin{array}{c} 1 \\ 6 \\ 4 \\ 6 \end{array}
\end{array}
\tag{7.127}
$$

The above matrices are assembled to obtain the unrestrained stiffness matrix, which will be of size 2×2, as this problem has two unrestrained degrees of freedom. In the above equations of the stiffness matrices, substitute $L_1 = L_2 = L$. Unrestrained stiffness matrix, as assembled by combining Eqs. (7.126 and 7.127), is given by:

$$
\begin{array}{ccc}
 & 1 \qquad 2 & \\
[K_{UU}] = EI & \left[\begin{array}{cc}
\frac{2r_1}{L} + \frac{r_2}{L} & \frac{2c_1 r_1}{L} \\
\frac{2c_1 r_1}{L} & \frac{2r_1}{L}
\end{array}\right] & \begin{array}{c} 1 \\ 2 \end{array}
\end{array}
\tag{7.128}
$$

As no joint load is applied in the unrestrained degrees of freedom, let us set the determinant of Eq. (7.128) to zero. Expanding, we get as follows:

$$\left(\frac{2r_1}{L} + \frac{r_2}{L}\right)\left(\frac{2r_1}{L}\right) - \left[\frac{2c_1 r_1}{L}\right]^2 = 0 \tag{7.129}$$

simplifying, we get the following expression:

$$2r_1\left(1 - c_1^2\right) + r_2 = 0 \tag{7.130}$$

Eq. (7.130) is the Characteristic equation, defining the critical buckling condition in terms of the stability functions. Also, we know the following relationship with reference to the problem loading:

$$\left(P_{cr}\right)_1 = \left(P_{cr}\right)_2 = P_{cr}(0.707) \tag{7.131}$$

$$\varphi_1 = \frac{P(0.707)}{\left[\frac{\pi^2 E(2I)}{L^2}\right]} \tag{7.132}$$

$$\varphi_2 = \frac{P(0.707)}{\left[\frac{\pi^2 EI}{L^2}\right]} \tag{7.133}$$

From the above equations, the following relationship can be deduced:

$$2\varphi_1 = \varphi_2 \tag{7.134}$$

Using Table 7.1, one can determine the value of φ_i, satisfying both Eq. (7.130) and Eq. (7.134). The MATLAB code given below helps to scan the appropriate values of the stability functions and determines φ_1 as 1.01 and φ_2 as 2.02. The corresponding values of stability functions are: $r_1 = 2.4493$; $t_1 = -0.0124$; $c_1 = 1.0101$ and $r_2 = 0.1120$; $t_2 = -1.7148$; $c_2 = 31.6264$.

Using Eq. (7.132), we get:

$$P_{cr} = \varphi_1 \frac{\pi^2 E(2I)}{(0.707)L^2} \tag{7.135}$$

```
% getting phi value condition: 2φ1 = φ2 and 2 r1 (1 - c1^2) + r2 = 0
lhs = 2*phi;
rhs = phi;
n = length (out);
phi1 = zeros(n,1);
phi2 = zeros(n,1);
u=1;
for  i=1:n
    for j=1:n
    if lhs(i) == rhs(j)
        r1 = r(i);
        c1 = c(i);
        r2 = r(j);
        if -0.00001 <= (2*r1*(1-(c1^2)))+r2 <=0.00001
            phi1(u) = phi(i);
            phi2(u) = phi(j);
            u=u+1;
        end
    end
    end
end
fprintf('Phi 1 = %6.2f \n',phi1(1));
fprintf('Phi 2 = %6.2f \n',phi2(1));
```

MATLAB output:

```
Phi 1 =   1.01
Phi 2 =   2.02
```

Example 7.3:

Estimate the critical buckling load of the structure using stability functions:

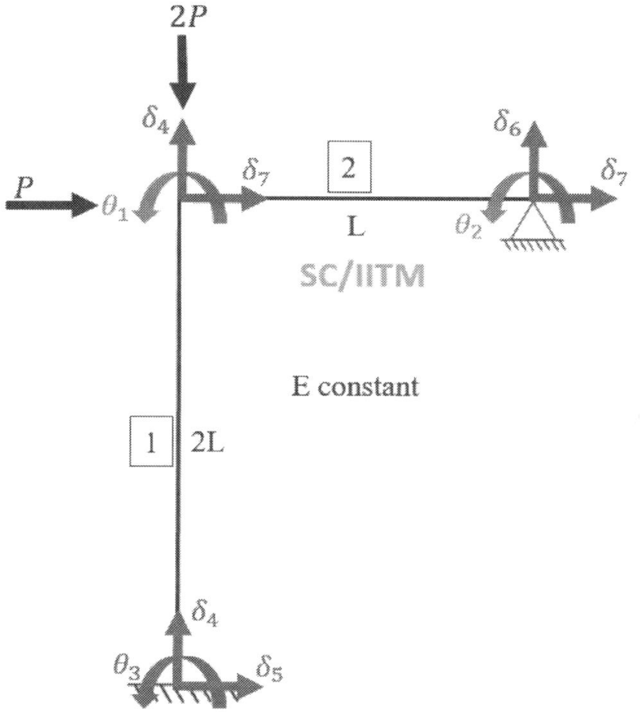

With reference to the above figure, kinematic degrees of freedom, both unrestrained and restrained, are marked. The member stiffness matrix for both members is given below. Numbers shown in circles are the labels of the degrees of freedom of the respective member.

$$[K_1] = EI \begin{array}{c} \begin{matrix} 1 & 3 & 7 & 5 \end{matrix} \\ \begin{bmatrix} \frac{r_1}{L_1} & \frac{c_1 r_1}{L_1} & \frac{r_1(1+c_1)}{L_1^2} & -\frac{r_1(1+c_1)}{L_1^2} \\ \frac{c_1 r_1}{L_1} & \frac{r_1}{L_1} & \frac{r_1(1+c_1)}{L_1^2} & -\frac{r_1(1+c_1)}{L_1^2} \\ \frac{r_1(1+c_1)}{L_1^2} & \frac{r_1(1+c_1)}{L_1^2} & \frac{2t_1 r_1(1+c_1)}{L_1^3} & -\frac{2t_1 r_1(1+c_1)}{L_1^3} \\ -\frac{r_1(1+c_1)}{L_1^2} & -\frac{r_1(1+c_1)}{L_1^2} & -\frac{2t_1 r_1(1+c_1)}{L_1^3} & \frac{2t_1 r_1(1+c_1)}{L_1^3} \end{bmatrix} \end{array} \begin{matrix} 1 \\ 3 \\ 7 \\ 5 \end{matrix} \quad (7.136)$$

$$[K_2] = EI \begin{bmatrix} \frac{r_2}{L_2} & \frac{c_2 r_2}{L_2} & \frac{r_2(1+c_2)}{L_2^2} & -\frac{r_2(1+c_2)}{L_2^2} \\ \frac{c_2 r_2}{L_2} & \frac{r_2}{L_2} & \frac{r_2(1+c_2)}{L_2^2} & -\frac{r_2(1+c_2)}{L_2^2} \\ \frac{r_2(1+c_2)}{L_2^2} & \frac{r_2(1+c_2)}{L_2^2} & \frac{2t_2 r_2(1+c_2)}{L_2^3} & -\frac{2t_2 r_2(1+c_2)}{L_2^3} \\ -\frac{r_2(1+c_2)}{L_2^2} & -\frac{r_2(1+c_2)}{L_2^2} & -\frac{2t_2 r_2(1+c_2)}{L_2^3} & \frac{2t_2 r_2(1+c_2)}{L_2^3} \end{bmatrix} \begin{matrix} 1 \\ 2 \\ 4 \\ 6 \end{matrix} \quad (\text{columns: } 1, 2, 4, 6) \tag{7.137}$$

The above matrices are assembled to obtain the unrestrained stiffness matrix, which will be of size 2×2, as this problem has two unrestrained degrees of freedom. In the above equations of the stiffness matrices, substitute $L_1 = 2L$ and $L_2 = L$. Unrestrained stiffness matrix, as assembled by combining Eqs. (7.136 and 7.137), is given by:

$$[K_{UU}] = EI \begin{bmatrix} \frac{r_1 + 2r_2}{2L} & \frac{c_2 r_2}{L} \\ \frac{c_2 r_2}{L} & \frac{r_2}{L} \end{bmatrix} \begin{matrix} 1 \\ 2 \end{matrix} \quad (\text{columns: } 1, 2) \tag{7.138}$$

As no joint load is applied in the unrestrained degrees of freedom, let us set the determinant of Eq. (7.138) to zero. Expanding, we get as follows:

$$\left(\frac{r_1 + 2r_2}{2L}\right)\left(\frac{r_2}{L}\right) - \left[\frac{c_2 r_2}{L}\right]^2 = 0 \tag{7.139}$$

simplifying, we get the following expression:

$$r_1 = 2r_2\left(c_2^2 - 1\right) \tag{7.140}$$

Eq. (7.140) is the Characteristic equation, defining the critical buckling condition in terms of the stability functions. Also, we know the following relationship with reference to the problem loading:

$$(P_{cr})_1 = 2P \tag{7.141}$$

$$(P_{cr})_2 = P \tag{7.142}$$

$$\varphi_1 = \frac{2P}{\left[\frac{\pi^2 EI}{4L^2}\right]} = \frac{8P}{\left[\frac{\pi^2 EI}{L^2}\right]} \tag{7.143}$$

$$\varphi_2 = \frac{P}{\left[\dfrac{\pi^2 EI}{L^2}\right]} \tag{7.144}$$

From the above equations, the following relationship can be deduced:

$$\varphi_1 = 8\varphi_2 \tag{7.145}$$

Using Table 7.1, one can determine the value of φ_i, satisfying both Eq. (7.140) and Eq. (7.145). MATLAB code given below helps to scan the appropriate values of the stability functions and determines φ_1 as 2.96 and φ_2 as 0.37. $r_1 = 4.6727$; $t_1 = -5.7540$; $c_1 = -1.4628$ and $r_2 = 3.4878$; $t_2 = 0.6754$; $c_2 = 0.6127$.

Using the Eq. (7.143), we get:

$$P_{cr} = \varphi_1 \frac{\pi^2 EI}{8L^2} \tag{7.146}$$

```
% getting phi value condition: φ1 = 8φ2 and r1 = 2r2 (c2^2 - 1)
lhs = phi;
rhs = 8*phi;
n = length (out);
phi1 = zeros(n,1);
phi2 = zeros(n,1);
u=1;
for  i=1:n
    for j=1:n
    if lhs(i) == rhs(j)
        r1 = r(i);
        c1 = c(i);
        r2 = r(j);
        c2 = c(j);
        if -0.00001 <= (r1-(2*r2*((c2^2)-1))) <=0.00001
            phi1(u) = phi(i);
            phi2(u) = phi(j);
            u=u+1;
        end
    end
    end
end
fprintf('Phi 1 = %6.2f \n',phi1(1));
fprintf('Phi 2 = %6.2f \n',phi2(1));
```

MATLAB output:

```
Phi 1 =   2.96
Phi 2 =   0.37
```

Example 7.4:

Estimate the critical buckling load of the structure using stability functions:

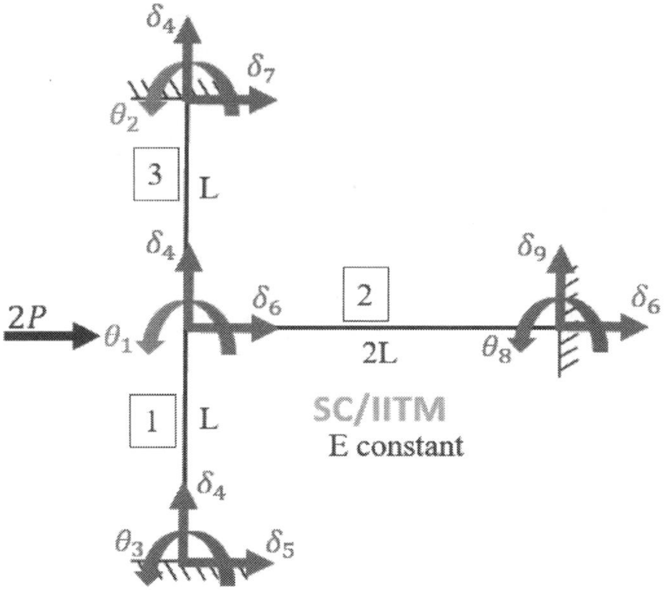

With reference to the above figure, kinematic degrees of freedom, both unrestrained and restrained, are marked. The member stiffness matrix for both members is given below. Numbers shown in circles are the labels of the degrees of freedom of the respective member.

$$[K_1] = EI \begin{array}{c} \begin{matrix} \quad 1 \quad & \quad 3 \quad & \quad 6 \quad & \quad 5 \quad \end{matrix} \\ \begin{bmatrix} \frac{r_1}{L_1} & \frac{c_1 r_1}{L_1} & \frac{r_1(1+c_1)}{L_1^2} & -\frac{r_1(1+c_1)}{L_1^2} \\ \frac{c_1 r_1}{L_1} & \frac{r_1}{L_1} & \frac{r_1(1+c_1)}{L_1^2} & -\frac{r_1(1+c_1)}{L_1^2} \\ \frac{r_1(1+c_1)}{L_1^2} & \frac{r_1(1+c_1)}{L_1^2} & \frac{2t_1 r_1(1+c_1)}{L_1^3} & -\frac{2t_1 r_1(1+c_1)}{L_1^3} \\ -\frac{r_1(1+c_1)}{L_1^2} & -\frac{r_1(1+c_1)}{L_1^2} & -\frac{2t_1 r_1(1+c_1)}{L_1^3} & \frac{2t_1 r_1(1+c_1)}{L_1^3} \end{bmatrix} \end{array} \begin{matrix} 1 \\ 3 \\ 6 \\ 5 \end{matrix} \qquad (7.147)$$

$$\begin{matrix} 2 & 1 & 7 & 6 \end{matrix}$$

$$[K_2] = EI \begin{bmatrix} \frac{r_2}{L_2} & \frac{c_2 r_2}{L_2} & \frac{r_2(1+c_2)}{L_2^2} & -\frac{r_2(1+c_2)}{L_2^2} \\ \frac{c_2 r_2}{L_2} & \frac{r_2}{L_2} & \frac{r_2(1+c_2)}{L_2^2} & -\frac{r_2(1+c_2)}{L_2^2} \\ \frac{r_2(1+c_2)}{L_2^2} & \frac{r_2(1+c_2)}{L_2^2} & \frac{2t_2 r_2(1+c_2)}{L_2^3} & -\frac{2t_2 r_2(1+c_2)}{L_2^3} \\ -\frac{r_2(1+c_2)}{L_2^2} & -\frac{r_2(1+c_2)}{L_2^2} & -\frac{2t_2 r_2(1+c_2)}{L_2^3} & \frac{2t_2 r_2(1+c_2)}{L_2^3} \end{bmatrix} \begin{matrix} 2 \\ 1 \\ 7 \\ 6 \end{matrix} \tag{7.148}$$

$$\begin{matrix} 1 & 8 & 4 & 9 \end{matrix}$$

$$[K_3] = EI \begin{bmatrix} \frac{r_3}{L_3} & \frac{c_3 r_3}{L_3} & \frac{r_3(1+c_3)}{L_3^2} & -\frac{r_3(1+c_3)}{L_3^2} \\ \frac{c_3 r_3}{L_3} & \frac{r_3}{L_3} & \frac{r_3(1+c_3)}{L_3^2} & -\frac{r_3(1+c_3)}{L_3^2} \\ \frac{r_3(1+c_3)}{L_3^2} & \frac{r_3(1+c_3)}{L_3^2} & \frac{2t_3 r_3(1+c_3)}{L_3^3} & -\frac{2t_3 r_3(1+c_3)}{L_3^3} \\ -\frac{r_3(1+c_3)}{L_3^2} & -\frac{r_3(1+c_3)}{L_3^2} & -\frac{2t_3 r_3(1+c_3)}{L_3^3} & \frac{2t_3 r_3(1+c_3)}{L_3^3} \end{bmatrix} \begin{matrix} 1 \\ 8 \\ 4 \\ 9 \end{matrix} \tag{7.149}$$

The above matrices are assembled to obtain the unrestrained stiffness matrix, which will be of size 2×2, as this problem has two unrestrained degrees of freedom. In the above equations of the stiffness matrices, substitute $L_1 = L_2 = L$; $L_3 = 2L$. After appropriate substitution, the unrestrained stiffness matrix, as assembled by combining Eqs. (7.147 to 7.149), is given by:

$$\begin{matrix} 1 & 2 \end{matrix}$$

$$[K_{UU}] = EI \begin{bmatrix} \frac{2r_1 + 2r_3}{2L} & \frac{c_2 r_2}{L} \\ \frac{c_2 r_2}{L} & \frac{r_2}{L} \end{bmatrix} \begin{matrix} 1 \\ 2 \end{matrix} \tag{7.150}$$

As no joint load is applied in the unrestrained degrees of freedom, let us set the determinant of Eq. (7.150) to zero. Expanding, we get as follows:

$$\left(\frac{2r_1 + r_3}{2L}\right)\left(\frac{r_2}{L}\right) - \left[\frac{c_2 r_2}{L}\right]^2 = 0 \tag{7.151}$$

simplifying, we get the following expression:

$$2r_1 + r_3 = 2r_2 c_2^2 \tag{7.152}$$

Eq. (7.152) is the Characteristic equation, defining the critical buckling condition in terms of the stability functions. Also, we know the following relationship with reference to the problem loading:

$$\varphi_1 = \varphi_2 = 0 \tag{7.153}$$

$$\left(P_{cr}\right)_3 = 2P \tag{7.154}$$

$$\varphi_3 = \frac{2P}{\left[\dfrac{\pi^2 EI}{4L^2}\right]} = \frac{8P}{\left[\dfrac{\pi^2 EI}{L^2}\right]} \tag{7.155}$$

Using Table 7.1, one can determine the value of φ_i, satisfying Eqs. (7.152, 7.153, and 4.155). MATLAB code given below helps to scan the appropriate values of the stability functions and determines φ_3 as 3.10; $r_3 = -6.0519$; $t_3 = -7.0072$; $c_3 = -1.3157$.

Using the Eq. (7.155), we get:

$$P_{cr} = \varphi_3 \frac{\pi^2 EI}{8L^2} \tag{7.156}$$

```
% getting phi value condition: phi1 = phi2 =0
r1 = 4;
r2 = 4;
c2 = 0.5;
r3 = (2*r2*c2*c2)-(2*r1);
n = length (out);
phi3 = zeros(n,1);
u=1;
for i = 1:n
    if r3 == round(r(i))
        if r(i)-r3 <=0.00001
            phi3(u) = phi(i);
            u=u+1;
        end
    end
end
fprintf('Phi 3 = %6.2f \n',phi3(1));
```

MATLAB output:

```
Phi 3 =   3.10
```

Example 7.5:

Estimate the critical buckling load of the structure using stability functions:

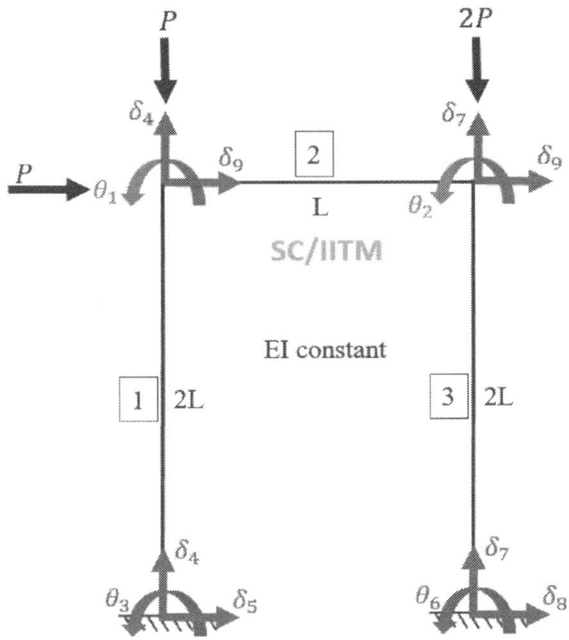

With reference to the above figure, kinematic degrees of freedom, both unrestrained and restrained, are marked. The member stiffness matrix for both members is given below. Numbers shown in circles are the labels of the degrees of freedom of the respective member.

$$[K_1] = EI \begin{bmatrix} \frac{r_1}{L_1} & \frac{c_1 r_1}{L_1} & \frac{r_1(1+c_1)}{L_1^2} & -\frac{r_1(1+c_1)}{L_1^2} \\ \frac{c_1 r_1}{L_1} & \frac{r_1}{L_1} & \frac{r_1(1+c_1)}{L_1^2} & -\frac{r_1(1+c_1)}{L_1^2} \\ \frac{r_1(1+c_1)}{L_1^2} & \frac{r_1(1+c_1)}{L_1^2} & \frac{2t_1 r_1(1+c_1)}{L_1^3} & -\frac{2t_1 r_1(1+c_1)}{L_1^3} \\ -\frac{r_1(1+c_1)}{L_1^2} & -\frac{r_1(1+c_1)}{L_1^2} & -\frac{2t_1 r_1(1+c_1)}{L_1^3} & \frac{2t_1 r_1(1+c_1)}{L_1^3} \end{bmatrix} \begin{matrix} 1 \\ 3 \\ 9 \\ 5 \end{matrix} \quad \text{(columns: 1, 3, 9, 5)} \tag{7.157}$$

$$[K_2]=EI\begin{array}{c} \\ \left[\begin{array}{cccc} 1 & 2 & 4 & 7 \\ \frac{r_2}{L_2} & \frac{c_2r_2}{L_2} & \frac{r_2(1+c_2)}{L_2^2} & -\frac{r_2(1+c_2)}{L_2^2} \\ \frac{c_2r_2}{L_2} & \frac{r_2}{L_2} & \frac{r_2(1+c_2)}{L_2^2} & -\frac{r_2(1+c_2)}{L_2^2} \\ \frac{r_2(1+c_2)}{L_2^2} & \frac{r_2(1+c_2)}{L_2^2} & \frac{2t_2r_2(1+c_2)}{L_2^3} & -\frac{2t_2r_2(1+c_2)}{L_2^3} \\ -\frac{r_2(1+c_2)}{L_2^2} & -\frac{r_2(1+c_2)}{L_2^2} & -\frac{2t_2r_2(1+c_2)}{L_2^3} & \frac{2t_2r_2(1+c_2)}{L_2^3} \end{array}\right]\end{array}\begin{array}{c} \\ 1 \\ 2 \\ 4 \\ 7 \end{array} \tag{7.158}$$

$$[K_3]=EI\begin{array}{c} \\ \left[\begin{array}{cccc} 2 & 6 & 9 & 8 \\ \frac{r_3}{L_3} & \frac{c_3r_3}{L_3} & \frac{r_3(1+c_3)}{L_3^2} & -\frac{r_3(1+c_3)}{L_3^2} \\ \frac{c_3r_3}{L_3} & \frac{r_3}{L_3} & \frac{r_3(1+c_3)}{L_3^2} & -\frac{r_3(1+c_3)}{L_3^2} \\ \frac{r_3(1+c_3)}{L_3^2} & \frac{r_3(1+c_3)}{L_3^2} & \frac{2t_3r_3(1+c_3)}{L_3^3} & -\frac{2t_3r_3(1+c_3)}{L_3^3} \\ -\frac{r_3(1+c_3)}{L_3^2} & -\frac{r_3(1+c_3)}{L_3^2} & -\frac{2t_3r_3(1+c_3)}{L_3^3} & \frac{2t_3r_3(1+c_3)}{L_3^3} \end{array}\right]\end{array}\begin{array}{c} \\ 2 \\ 6 \\ 9 \\ 8 \end{array} \tag{7.159}$$

The above matrices are assembled to obtain the unrestrained stiffness matrix, which will be of size 2×2, as there are two unrestrained degrees of freedom for this problem. In the above equations of the stiffness matrices, substitute $L_1 = L_3 = 2L$; $L_2 = L$. After appropriate substitution, the unrestrained stiffness matrix, as assembled by combining Eqs. (7.157 to 7.159), is given by:

$$[K_{UU}]=EI\begin{array}{c} \\ \left[\begin{array}{cc} 1 & 2 \\ \frac{r_1+2r_2}{2L} & \frac{c_2r_2}{L} \\ \frac{c_2r_2}{L} & \frac{2r_2+r_3}{2L} \end{array}\right]\end{array}\begin{array}{c} \\ 1 \\ 2 \end{array} \tag{7.160}$$

As no joint load is applied in the unrestrained degrees of freedom, let us set the determinant of Eq. (7.160) to zero. Expanding, we get as follows:

$$\left(\frac{r_1+2r_2}{2L}\right)\left(\frac{2r_2+r_3}{2L}\right)-\left[\frac{c_2r_2}{L}\right]^2=0 \tag{7.161}$$

simplifying, we get the following expression:

$$r_1(2r_2+r_3)+r_2(4r_2+2r_3)=4r_2^2c_2^2 \tag{7.162}$$

Equation (7.162) is the Characteristic equation, defining the critical buckling condition in terms of the stability functions. Also, we know the following relationship with reference to the problem loading:

$$\varphi_1 = \frac{P}{\left[\dfrac{\pi^2 EI}{4L^2}\right]} = \frac{4P}{\left[\dfrac{\pi^2 EI}{L^2}\right]} \tag{7.163}$$

$$\varphi_2 = \frac{P}{\left[\dfrac{\pi^2 EI}{L^2}\right]} \tag{7.164}$$

$$\varphi_3 = \frac{2P}{\left[\dfrac{\pi^2 EI}{4L^2}\right]} = \frac{8P}{\left[\dfrac{\pi^2 EI}{L^2}\right]} \tag{7.165}$$

$$\varphi_1 = 4\varphi_2 \tag{7.166}$$

$$\varphi_3 = 8\varphi_2 \tag{7.167}$$

Using Table 7.1, one can determine the value of φ_i, satisfying Eqs. (7.162, 7.166, and 7.167). The MATLAB code given below helps to scan the appropriate values of the stability functions and determines $\varphi_2 = 0.38$. The corresponding stability coefficients are $r_2 = 3.4732$; $t_2 = 0.6660$; $c_2 = 0.6165$.

Using the Eq. (7.164), we get:

$$P_{cr} = \varphi_2 \frac{\pi^2 EI}{L^2} \tag{7.168}$$

```
% getting phi value condition: phi1 = 4phi2
n = length (out);
lhs = phi;
rhs = 4*phi;
phi1 = zeros(n,1);
phi2 = zeros(n,1);
phi3 = zeros(n,1);
u=1;
for  i=1:n
    for j=1:n
    if lhs(i) == rhs(j)
        r1 = r(i);
        c1 = c(i);
        phi22 = phi(j);
        r2 = r(j);
        c2 = c(j);
        for k = 1:n
            if (8*phi22) == phi(k)
                r3 = r(k);
```

```
                  if -0.00001 <= ((r1*((2*r2)+r3))+(r2*((2*r3)+(4*r2)))-
(4*r2*r2*c2*c2)) <= 0.00001
                       phi1(u) = phi(i);
                       phi2(u) = phi(j);
                       phi3(u) = phi(k);
                       u=u+1;
                   end
               end
           end
       end
       end
end
fprintf('Phi 1 = %6.2f \n',phi1(1));
fprintf('Phi 2 = %6.2f \n',phi2(1));
fprintf('Phi 3 = %6.2f \n',phi3(1));
```

MATLAB output:

```
Phi 1 =   1.52
Phi 2 =   0.38
Phi 3 =   3.04
```

MATLAB code for preparing the stability chart

```
%% This MATLAB code is for plotting the stability indices
% Re-type the following code in MATLAB new script and run the file to
get the output.
clc;
clear;
phi=0.01:0.01:4;
alrad = pi.*sqrt(phi);
al = radtodeg (alrad);
% Compression
r = (alrad.*((sind(al))-(alrad.*cosd(al))))./
((2.*(1-cosd(al)))-(alrad.*(sind(al)))); % rotation function
c = (alrad-sind(al))./(sind(al)-(alrad.*cosd(al))); % rotation
function
t = 1-((pi*pi*phi)./(2.*r.*(1+c))); % Translation function
outcomp = [phi' r' c' t'];
% for zero
outzero = [0 4.0 0.5 1.0];
% Tension
phit =10:-0.1:0.1;
alrad = pi.*sqrt(phit);
al = radtodeg (alrad);
rt = (alrad.*((alrad.*cosh(alrad))-sinh(alrad)))./((2.*(1-
cosh(alrad)))+(alrad.*sinh(alrad)));
ct = (alrad-sinh(alrad))./(sinh(alrad)-(alrad.*cosh(alrad)));
tt = 1-((pi*pi*(-phit))./(2.*rt.*(1+ct)));
outten = [-phit' rt' ct' tt'];
% Stability chart
out = [outten; outzero; outcomp];
```

```
filename='StabilityChart.xlsx';
sheet=1;
xlswrite(filename,out,sheet);
phi = out(:,1);
r = out(:,2);
c = out(:,3);
t = out(:,4);
figure;
plot (phi,r,'b','linewidth',2)
xlabel('Phi');
ylabel('r');
grid on;
figure;
plot (phi,c,'b','linewidth',2)
xlabel('Phi');
ylabel('c');
grid on;
figure;
plot (phi,t, 'b','linewidth',2)
xlabel('Phi');
ylabel('t');
grid on;
```

MATLAB output:

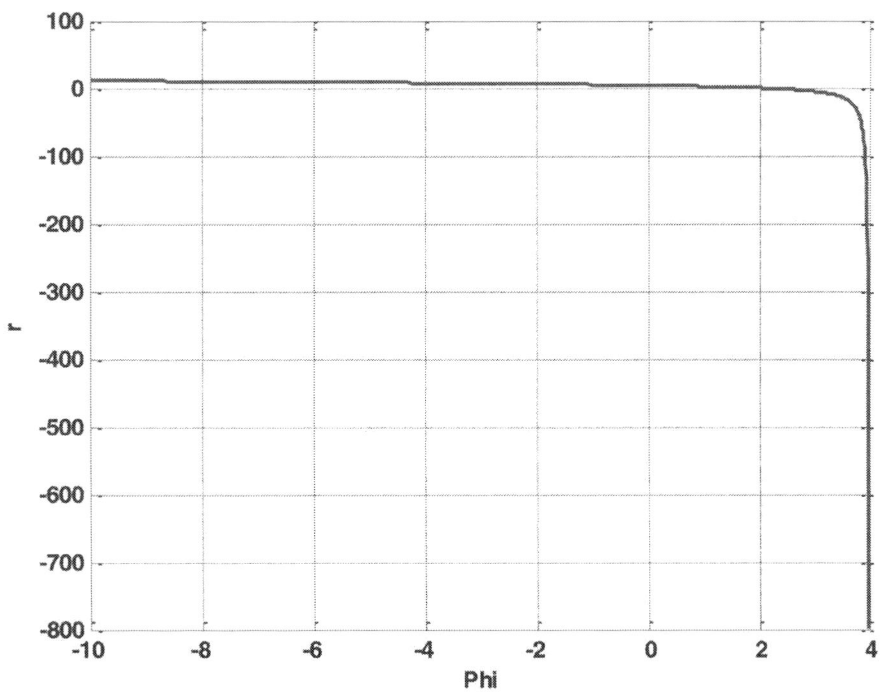

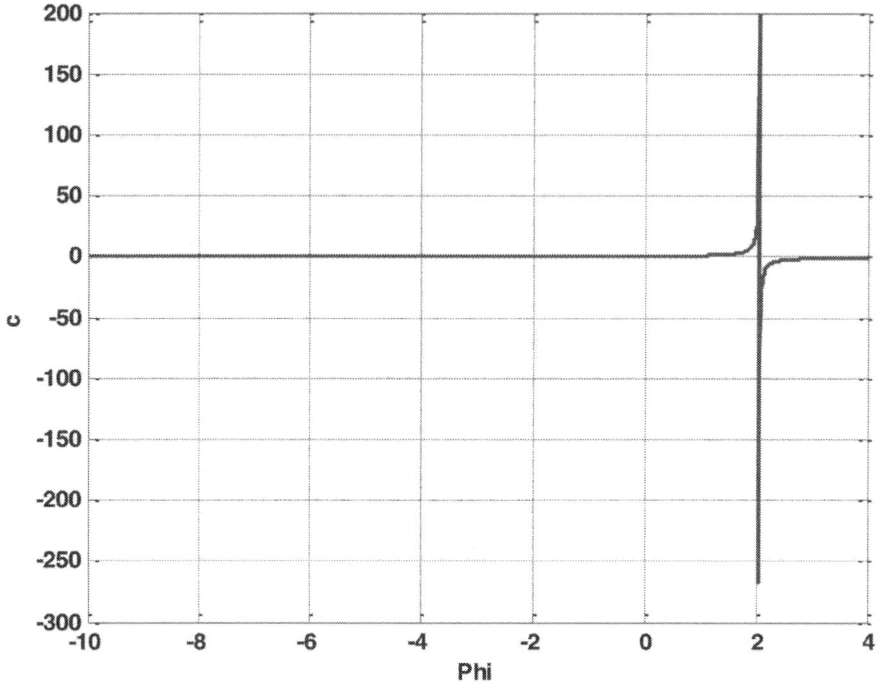
200
150
100
50
0
-50
-100
-150
-200
-250
-300
c
-10
-8
-6
-4
-2
0
2
4
Phi

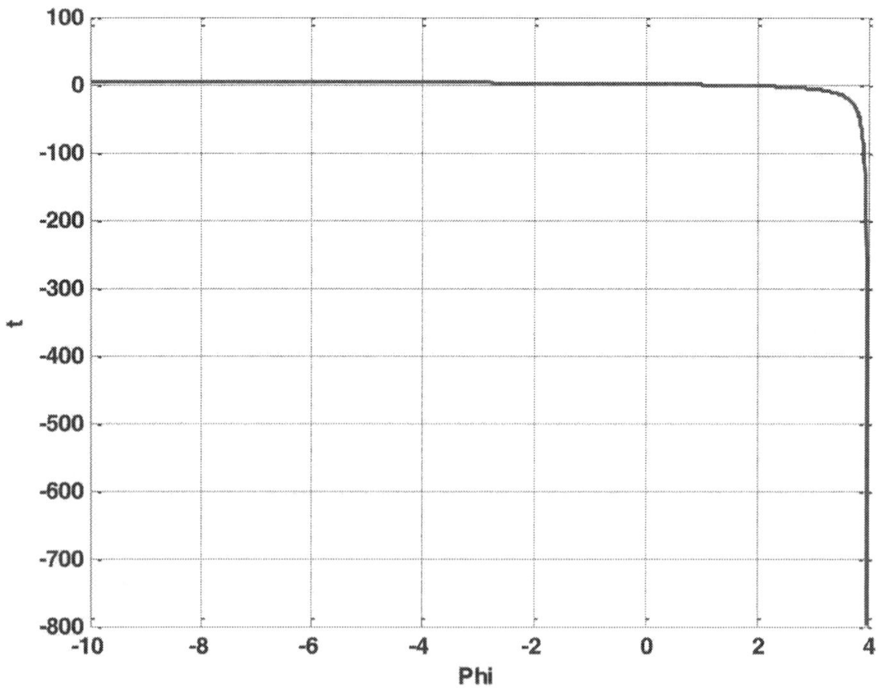
100
0
-100
-200
-300
-400
-500
-600
-700
-800
t
-10
-8
-6
-4
-2
0
2
4
Phi

References

Amin Ghali, A., Neville, A., and Brown, T.G. 2017. *Structural Analysis: A Unified Classical and Matrix Approach*. CRC Press, London.

Beaufait, F.W., Rowan, W.H., Jr., Hoadley, P.G., and Hackett, R.M. 1970. *Computer Methods of Structural Analysis*. Prentice-Hall, Englewood Cliffs, NJ.

Bhatt, P. 1986. *Programming the Matrix Analysis of Skeletal Structures*. Wiley, New York.

Chandrasekaran, S. 2013. Advanced Marine Structures, Video course, available at https://nptel.ac.in/courses/114106037.

Chandrasekaran, S. 2014. *Advanced Theory on Offshore Plant FEED Engineering*. Changwon National University Press, Republic of South Korea, pp. 237.

Chandrasekaran, S. 2015. *Advanced Marine Structures*. CRC Press, FL.

Chandrasekaran, S. 2015. *Dynamic Analysis and Design of Ocean Structures*, Springer, India.

Chandrasekaran, S. 2016. *Offshore Structural Engineering: Reliability and Risk Assessment*. CRC Press, FL.

Chandrasekaran, S. 2017. *Dynamic Analysis and Design of Ocean Structures*, 2nd ed. Springer, Singapore.

Chandrasekaran, S. 2018. *Advanced Structural Analysis with MATLAB*. CRC Press, FL.

Chandrasekaran, S. 2018. Structural Health Monitoring, Video Course on MOOC at https://nptel.ac.in/courses/114106046.

Chandrasekaran, S. 2019a. *Advanced Steel Design of Structures*. CRC Press, FL.

Chandrasekaran, S. 2019b. *Structural Health Monitoring with Application to Offshore Structures*. World Scientific Publishing Co, Singapore.

Chandrasekaran, S. 2020. *Offshore Semi-Submersible Platform Engineering*. CRC Press, FL, p. 240.

Chandrasekaran, S. 2024. Advanced Design of Steel structures, Re-run Course in MOOC/NPTEL Portal.

Chandrasekaran, S. 2024. Computer Methods of Structural Analysis of Offshore Structures, Re-run Course in MOOC at https://onlinecourses.nptel.ac.in/noc24_oe05/preview.

Chandrasekaran, S. 2024. Offshore Structures under Special Environmental Loads Including Fire Resistance, Re-run Course in MOOC/NPTEL Portal.

Chandrasekaran, S. and Ajesh Kumar, P.T. 2017. Damage Assessment in Concrete Marine Structures Using Damage Plasticity Model, Lisbon, Portugal, May 8–10.

Chandrasekaran, S. and Ajesh Kumar, P.T. 2019. Damage detection in reinforced concrete berthing jetty using a plasticity model approach. *Journal of Marine Science and Application*, 18:482–491.

Chandrasekaran, S. and Chandak, N.R. 2004. Seismic behavior of road bridge abutments, Concrete India. *Journal of IC-ACI*, 19(1):3–5.

Chandrasekaran, S. and Chinu, P. 2024. Dynamic analysis of offshore triceratops supporting wind turbine: Preliminary studies. *Maritime Technology and Research*, 6(1):265564, https://doi.org/10.33175/mtr.2024.265564

Chandrasekaran, S. and Chithambaram, T. 2016. Health Monitoring of Offshore Structures Using Wireless Sensor Network: Experimental Investigations. In: *Proceedings of SPIE 9804, Nondestructive Characterization and Monitoring of Advanced Materials, Aerospace, and Civil Infrastructure 2016*, Las Vegas, NV, March 20–24.

Chandrasekaran, S. and Gupta, V. 2007a. Nonlinear Seismic Analysis of Buildings. Kolkata, 28th Jan–03rd Feb 2007, CD-ROM.

Chandrasekaran, S. and Gupta, V. 2007b. Pushover Analysis of RC Framed Structures. Structural Engineering, India, Dec 2007, CD-ROM.

Chandrasekaran, S. and Harinder. 2010. Proposed Mechanism for Mechanical Wave Energy Converter, *Ship and Offshore Newsletter*, 4:101.

Chandrasekaran, S. and Harinder. 2011. Design and Efficiency Analysis of Mechanical Wave Energy Converter, OMAE 2011, Rotterdam, The Netherlands, June 19–24, OMAE 2011-49830.

Chandrasekaran, S. and Harinder. 2012. Power generation using mechanical wave energy converter, *The International Journal of Ocean and Climate Systems*, 3(1):57–70.

Chandrasekaran, S. and Jain, A.K. 2016. *Ocean Structures: Construction, Materials and Operations*. CRC Press, FL.

Chandrasekaran, S. and Kiran, P.A. 2018. Mathieu stability of buoyant leg storage and regasification platform, *Ocean Systems Engineering*, 8(3):345–360, https://doi.org/10.12989/ose.2018.8.3.345.

Chandrasekaran, S. and Kiran, P.A. 2018. Mathieu stability of offshore triceratops under postulated failure, *Ships and Offshore Structures*, 13(2):143–148.

Chandrasekaran, S. and Koshti, Y. 2013. Dynamic analysis of a tension leg platform under extreme waves, *Journal of Naval Architecture and Marine Engineering*, 10:59–68, https://doi.org/10.3329/jname.v10i1.14518.

Chandrasekaran, S. and Kumar, M. 2023. Dynamic Analysis of Drillship under Critical Environmental Loads, *Proceedings of the Thirty-third (2023) International Ocean and Polar Engineering Conference*, Ottawa, Canada, June 19–23, 2023, pp. 2366–2373. www.isope.org

Chandrasekaran, S. and Lognath, R.S. 2015. Dynamic analyses of buoyant leg storage regasification platform (BLSRP) under regular waves: Experimental investigations, *Ships and Offshore Structures*, 12(2):227–232.

Chandrasekaran, S. and Lognath, R.S. 2017. Dynamic analyses of buoyant leg storage and regasification platforms: Numerical studies, *Marine Systems & Ocean Technology*, 12(2):39–48.

Chandrasekaran, S. and Madhavi N. 2014a. Retrofitting of offshore structural member using perforated cylinders, *SFA Newsletter*, 13:10–11.

Chandrasekaran, S. and Madhavi N. 2014d. Variations of Water Particle Kinematics of Offshore TLPs with Perforated Members: Numerical Investigations. In: *Proceedings of International Conference in Structural Engineering*, IIT Delhi, India, Dec 22–24.

Chandrasekaran, S. and Madhavi N. 2015a. Variations of Water Particle Kinematics of Offshore TLPS with Perforated Members: Numerical Investigations. In: V. Matsagar (ed). *Advances in Structural Engineering* (pp. 629–645). Springer. https://doi.org/10.1007/978-81-322-2190-6_51

Chandrasekaran, S. and Madhavi N. 2015e. Variation of the flow field around twin cylinders with and without outer perforated cylinder: Numerical studies, *China Ocean Engineering*, 30(5):763–771.

Chandrasekaran, S. and Madhavi N. 2015f. Flow field around an outer perforated circular cylinder under regular waves: Numerical study. *Marine Systems & Ocean Technology*, 10:91–100, https://doi.org/10.1007/s40868-015-0008-1.

Chandrasekaran, S. and Madhavi, N. 2014b. *Variation of Water Particle Kinematics with Perforated Cylinder under Regular Waves*. Busan, South Korea.

Chandrasekaran, S. and Madhavi, N. 2014c. Hydrodynamic performance of retrofitted structural member under regular waves. *International Journal of Forensic Engineering*, 2(2):100–121.

Chandrasekaran, S. and Madhavi, N. 2015b. Estimation of Force Reduction on Ocean Structures with Perforated Members. In: *Proceedings of the ASME 2015 34th International Conference on Ocean, Offshore and Arctic Engineering*, St. Johns, Newfoundland, Canada, May 31–June 5, https://doi.org/10.1115/OMAE2015-41153.

Chandrasekaran, S. and Madhavi, N. 2015c. Design aids for offshore structures with perforated members. *Ship and Offshore Structures*, 10(2):183–203.

Chandrasekaran, S. and Madhavi, N. 2015d. Retrofitting of offshore cylindrical structures with different geometrical configurations of perforated outer cover. *International Journal of Shipbuilding*, 62(1–2):43–56, https://doi.org/3233/ISP-150115.

Chandrasekaran, S. and Madhavi, N. 2016. Numerical study on geometrical configurations of perforated cylindrical structures. *Journal of Performance of Constructional Facilities*, 30(1):04014185. https://doi.org/10.1061/(ASCE)CF.1943-5509.0000687

Chandrasekaran, S. and Madhuri, S. 2015. Dynamic response of offshore triceratops: Numerical and experimental investigations. *Ocean Engineering*, 109(15):401–409.

Chandrasekaran, S. and Mayank, S. 2017. Dynamic analyses of stiffened triceratops under regular waves: Experimental investigations. *Ships and Offshore Structures*, 12(5):697–705.

Chandrasekaran, S. and Nagavinothini, R. 2018. Tether analyses of offshore triceratops under wind, wave, and current. *Marine Systems & Ocean Technology*, 13:34–42. https://doi.org/10.1007/s40868-018-0043-9

Chandrasekaran, S. and Nagavinothini, R. 2020. Dynamic response of offshore triceratops with elliptical buoyant legs. *International Journal of Innovative Infrastructure Solutions*, 5(47):1–14.

Chandrasekaran, S. and Nagavinothini, R. 2020a. Parametric studies on the impact response of offshore triceratops in ultra-deep waters. *Structure and Infrastructure Engineering*, 16(7):1002–1018.

Chandrasekaran, S. and Nagavinothini, R. 2020b. Offshore triceratops under impact forces in ultra-deep arctic waters. *International Journal of Steel Structures*, 20(2): 464–479.

Chandrasekaran, S. and Nagavinothini, R. 2020c. Behavior of stiffened deck plates under hydrocarbon fire. *Marine Systems & Ocean Technology*, 15:95–109.

Chandrasekaran, S. and Nannaware, M. 2013. Response analyses of the offshore triceratops to seismic activities. *Ship and Offshore Structures*, 9(6):633–642.

Chandrasekaran, S. and Nassery, J. 2017. Nonlinear response of stiffened triceratops under impact and non-impact waves. *Ocean Systems Engineering*, 7(3):179–193. https://doi.org/10.12989/ose.2017.7.3.179

Chandrasekaran, S. and Nassery, J. 2017. Ringing response of offshore triceratops. *Journal of Innovative Infrastructure Solutions*, 2(23):1–6.

Chandrasekaran, S. and Pachaiappan, S. 2023. Displacement-controlled non-linear analysis of offshore platform topside under accidental loads. *Arabian Journal for Science and Engineering*, 48:5619–5635. https://doi.org/10.1007/s13369-022-07509-7

Chandrasekaran, S. and Pradeepa, M. 2023. Structural Response of Offshore Topside Subjected to Hydrocarbon Fire. In: *Proceedings of International Conference on Petroleum, Hydrogen and Decarbonization*, IIT Guwahati, Assam, India, Nov 3–5.

Chandrasekaran, S. and Raghavi, B. 2015. Design, Development, and Experimentation of Deep Ocean Wave Energy Converter System. In: *Proceedings of International Conference on Alternate Energy in Developing Countries and Emerging Economics*, Bangkok, Thailand, May 28–29.

Chandrasekaran, S. and Rao, G.S. 2024. Postulated tendon failure of offshore triceratops. *Journal of Marine Science and Application*, 23:182–200. https://doi.org/10.1007/s11804-024-00397-3

Chandrasekaran, S. and Raphel D.C. 2013. Conceptual Design and Experimental Investigations on Deep Ocean Wave Energy Converter. In: *International Conference on Ocean Engineering*, Chennai, India, Dec 2–5.

Chandrasekaran, S. and Roy, A. 2006. Seismic evaluation of multi-storey RC frame using modal pushover analysis. *Nonlinear Dynamics*, 43(4):329–342.

Chandrasekaran, S. and Seeram, M. 2012. Stability studies on offshore triceratops. *International Journal of Research & Development*, 1(10):398–404.

Chandrasekaran, S. and Shyba, A. 2024. Dynamic Response Analysis of Buoyant Leg Storage and Regasification Platform under Random Waves: Numerical studies. In: *Proceedings of International Conference on Oceanography and Marine Biology*, Barcelona, Spain, Sep 9–10.

Chandrasekaran, S. and Sricharan, V.V.S. 2018. Design and Development of Near-shore Mechanical Wave Energy Converter. *International Conference on Marine Technology (MARTEC-2018)*, Kuala Lumpur, Malaysia, Aug 13–14.

Chandrasekaran, S. and Sricharan, V.V.S. 2019. Improved efficiency of a floating wave energy converter under different wave-approach angles: Numerical and experimental investigations. *Journal of Ocean Engineering and Marine Energy*, 5(1):41–50. https://doi.org/10.1007/s40722-019-00128-9

Chandrasekaran, S. and SriCharan, V.V.S. 2020. Analysis of a Floating Wave Energy Converter with Hydraulic Mechanical Power Take-Off Using WEC-Sim and Simscape. *International Conference on Wave Energy* (Virtual Conf.) Fort Lauderdale, FL, Aug 3–7.

Chandrasekaran, S. and Sricharan, V.V.S. 2020. Numerical analysis of a new multi-body mechanical wave energy converter with a linear power-take off the system. *Renewable Energy*, 159:250–271.

Chandrasekaran, S. and Sricharan, V.V.S. 2021. Numerical study of bean-float wave energy converter with float number parametrization using WEC-Sim in regular waves with the Levelized Cost of Electricity assessment for Indian sea states. *Ocean Engineering*, 237:109591.

Chandrasekaran, S. and Sricharan, V.V.S. 2021. Time-domain analysis of a bean-shaped multi-body floating wave energy converter with a hydraulic power take-off using WEC-Sim. *Energy*, 223:119985.

Chandrasekaran, S. and Srivastava, G. 2017. *Design Aids for Offshore Structures under Special Environmental Loads Including Fire Resistance*. Springer, Singapore.

Chandrasekaran, S. and Srivastava, G. 2022. Fire Resistant Design of Structures. CRC Press, p. 212.

Chandrasekaran, S. and Srivastava, S. 2007. Nonlinear Dynamic Analysis of Asymmetric Buildings under Seismic Forces. *International Conference on Structural Engineering*, Kolkata, 28th Jan–3rd Feb 2007, CD-ROM.

Chandrasekaran, S. and Suja, T.P. 2024. Response control of TLP with single TMD under wind, wave, and current. *Maritime Technology and Research*, 7(1):272515. https://doi.org/10.33175/mtr.2025.272515

Chandrasekaran, S. and Thailammai, C.T. 2018. Health monitoring of tension leg platform using wireless sensor networking: Experimental investigations. *Journal of Marine Science and Technology*, 24:60–72. https://doi.org/10.1007/s00773-018-0531-9

Chandrasekaran, S. and Thailammai, C.T. 2018. Structural health monitoring of offshore buoyant leg storage and re-gasification platform: Experimental investigations. *Journal of Marine Science and Application*, 17:87–100.

Chandrasekaran, S. and Thennavan, M. 2023. Numerical Analysis of Coped Beam with X52 Steel and FGM Used in Offshore Topside. In: *Proceedings of the Thirty-third (2023) International Ocean and Polar Engineering Conference*, Ottawa, Canada, June 19–23, 2023, pp. 1067–1074. www.isope.org

Chandrasekaran, S. and Thennavan, M. 2024. Steel coped beam with functionally graded material for offshore topside: Numerical investigations. *International Journal of Steel Structures*, 24(3):506–514. https://doi.org/10.1007/s13296-024-00832-9

Chandrasekaran, S. and Tripati, U.K. 2004. Seismic Vulnerability of Irregular Buildings. *International Conference on Earthquake Engineering*, Singapore, Dec 1–3.

Chandrasekaran, S. and Uddin, S. 2020. Postulated failure analyses of spread-moored semi-submersible. *Innovative Infrastructure Solutions*, 5:36. https://doi.org/10.1007/s41062-020-0284-2.

Chandrasekaran, S. and Uddin, S.A. 2021. Dynamic analysis of semi-submersible under the postulated failure of restraining system with buoy. *International Journal of Steel Structures*, 21(1):118–131.

Chandrasekaran, S., Amarkarthik, A., Sivakumar, K., and Sinhmar, H. 2012. Laboratory experiments using the non-floating body to generate electrical energy from water waves. *Frontiers in Energy*, 6(4):361–365. https://doi.org/10.1007/s11708-012-0210-1

Chandrasekaran, S., Amarkathik, A., Sivakumar, K., Selvamuthukumaran, D., and Sidney, S. 2013. Experimental investigation and ANN modeling on improved performance of an innovative method of using heave response of a non-floating object for ocean wave energy conversion. *Frontiers in Energy*, 7(3):279–287. https://doi.org/10.1007/s11708-013-0268-4

Chandrasekaran, S., Dubey, A.K., and Tripathi, U.K. 2006. Seismic behavior of SMRF under various structural irregularities. School of Civil Engineering, SASTRA, Deemed University, Tanjavur, Tamil Nadu, Feb 25–26.

Chandrasekaran, S., Hari, M., and Rao, V.R. 2019. Numerical analysis on triceratops restraining system: Failure conditions of tethers. *International Journal of Environmental and Ecological Engineering*, 13(9):588–592.

Chandrasekaran, S., Hari, M., and Rao, V.R. 2019. Numerical Analysis on Triceratops Restraining System: Failure Conditions of Tethers. *International Conference on Ocean Systems*, Paris, France, Sep 19–20.

Chandrasekaran, S., Jain, A.K., and Gupta, V. 2006. Performance-based Design for RC Framed Buildings. *International Conference on Structural Engineering*, CD-ROM, IIT, Guwahati, Dec 8–10.

Chandrasekaran, S., Jain, A.K., and Madhuri, S. 2013. Aerodynamic response of offshore triceratops. *Ship and Offshore Structures*, 8(2):123–140.

Chandrasekaran, S., Jain, A.K., Serino, G., Spizzuoco, M., Srivastava, S., and Gupta, V. 2007. Risk Assessment of Seismic Vulnerabilities of RC Framed Buildings Due to Asymmetricity. *International Conference on Earthquake Engineering*, Singapore, Dec 5–7.

Chandrasekaran, S., Jain, A.K., Shafiq, N., Mubarak, M., and Wahab, A. 2021. *Design Aids for Offshore Platforms under Special Loads*. CRC Press, FL, p. 280.

Chandrasekaran, S., Khan, F., and Abbassi, R. 2022. *Wave Energy Devices: Design, Development and Experimental Studies*. CRC Press, FL, p. 271.

Chandrasekaran, S., Kumar, D., and Ramanathan, R. 2013. Dynamic response of tension leg platform with tuned mass dampers. *Journal of Naval Architecture and Marine Engineering*, 10(2):149–156.

Chandrasekaran, S., Kumar, D., and Ramanathan, R. 2014. Response Control of TLP Using Tuned Mass Dampers. *International Conference on Offshore Mechanics and Arctic Engineering, OMAE 2014*, San Francisco, CA, June 8–13.

Chandrasekaran, S., Kumar, S., and Madhuri, S. (Eds). 2021. *Recent Advances in Structural Engineering*. Springer.

Chandrasekaran, S., Madhavi, N., and Natarajan, C. 2014. Variations of hydrodynamic characteristics with the perforated cylinder. *International Conference on Offshore Mechanics and Arctic Engineering, OMAE 2014*, San Francisco, CA, (paper accepted - 23455), June 8–13.

Chandrasekaran, S., Madhavi, N., and Sampath, S. 2015. Force Reduction on Ocean Structures with Perforated Members. In: V. Matsagar (ed). *Advances in Structural Engineering* (pp. 647–661). Springer. https://doi.org/10.1007/978-81-322-2190-6_52

Chandrasekaran, S., Nunziante, L., Gupta, V., and Carannante, F. 2008. Nonlinear seismic analyses of high-rise reinforced concrete buildings. *ICFAI Journal of Structural Design*, 1(1):7–24.

Chandrasekaran, S., Nunziante, L., Seriino, G., and Caranannate, F. 2009. *Seismic Design Aids for Nonlinear Analysis of Reinforced Concrete Structures*. CRC Press, FL.

Chandrasekaran, S., Nunziante, L., Serino, G., and Carannante, F. 2011. Curvature ductility of RC sections based on Euro Code: Analytical procedure. *KSCE Journal of Civil Engineering*, 15(1):131–144.

Chandrasekaran, S., Patil, B.I., and Prasad, M. 2024. Power Take-off Mechanism to Improve Efficiency of TSUSUCA DOLPHIN. In: *Proceedings of International Conference on Ships and Offshore Structures (ICSOS 2024)*, IIT Madras, Chennai, India, Sep 8–12.

Chandrasekaran, S., Phoemsapthawee, S., Krishna, S., and Hari, S. 2024. *Fundamentals of Offshore Engineering*. CRC Press, FL, p. 280.

Chandrasekaran, S., Raphel, D.C., and Saishri, P. 2014. Deep ocean wave energy systems: Experimental investigations. *Journal of Naval Architecture and Marine Engineering*, 11:139–147.

Chandrasekaran, S., Roy, A., and Chandak, N.R. 2005. Modal Pushover Analysis of Multistorey Moment Resisting Framed Buildings. *International Conference on Structural Engineering*, IIT-Kharagpur, India.

Chandrasekaran, S., Serino G.G., and Miranda, S. 2011. Springing and ringing response of triangular TLPs. *International Shipbuilding Progress*, 58(2–3):141–163.

Chandrasekaran, S., Serino, G.G., and Spizzuoco, M. 2024. *Earthquake Engineering and Structural Control: Theory and Applications*. CRC Press, FL, p. 432.

Chandrasekaran, S., Serino, GG., Chowhan, Y.J., Sanghvi, C., and Gohil, A. 2024. Feasibility study of offshore triceratops supporting floating offshore wind turbine. *Journal of Institution of Engineers (India): Series A*, 105:295–305. https://doi.org/10.1007/s40030-024-00800-w.

Chandrasekaran, S., Shah, B., and Chauhan, Y. 2022a. Fatigue assessment of offshore triceratops restraining system under hurricane-driven metocean conditions. *International Journal of Steel Structures*, 23:208–224. https://doi.org/10.1007/s13296-022-00689-w

Chandrasekaran, S., Shah, B., and Chauhan, Y. 2023. Tether response of offshore triceratops under hurricane conditions. *Structures*, 51:513–527.

Chandrasekaran, S., Shah, B., and Chouhan, Y. 2022b. Dynamic analyses of triceratops under hurricane-drive metocean conditions in Gulf of Mexico. *Ocean Engineering*, 256:111511.

Chandrasekaran, S., Shukla, S.K., Das, B.B., and Kolathayar, S (Eds). 2020. *Smart Technologies for Sustainable Development*. Springer.

Chandrasekaran, S., Thaillammai, C., and Khader, S.A. 2016. Structural health monitoring of offshore structures using wireless sensor networking under operational and environmental variability. *International Journal of Environmental and Ecological Engineering*, 10(1):33–39.

Chandrasekaran, S., Thennavan, M., and Phimpisan, P. 2024. Experimental and numerical investigations of steel coped beam using DIC method. *International Journal of Mechanical Science and Technology*, 38(8):4153–4160.

Chapman, S.J. 2015. *MATLAB Programming for Engineers*. Cengage Learning, UK.

Derucher, K.N., Putcha, C., and Kim, U. 2013. *Indeterminate Structural Analysis*. Edwin Mellen Press, New York.

Derucher, K.N., Putcha, C., Kim, U., Hota, V.S., and Rao, G. 2022. *Static Analysis of Determinate and Indeterminate Structures*. CRC Press, Boca Raton, FL.

Fleming, J.F. 1989. *Computer Analysis of Structural Systems*. McGraw-Hill, New York.

Forsythe, G. and Moler, C.B. 1969. Computer solution of linear algebraic systems. *The Mathematical Gazette*, 53(384):221–222. https://doi.org/10.2307/3614613

Hibbeler, R.C. 1984. *Structural Analysis*, 9th ed. Pearson Prentice Hall, New Jersey.
Holzer, S.H. 1985. *Computer Analysis of Structures*. Elsevier, New York.
Holzer, S.M. 1985. *Computer Analysis of Structures: Matrix Structural Analysis Structured Programming*. Elsevier, CA.
Householder, A.S. 2013. *The Theory of Matrices in Numerical Analysis*. Courier Corporation, North Chelmsford, NY.
Jarquio, R.V. 2007. *Structural Analysis: The Analytical Method*. CRC Press, Boca Raton, FL.
Kardestuncer, H. 1974. *Elementary Matrix Analysis of Structures*. McGraw-Hill, New York.
Karnovsky, Igor A. and Lebed. O. 2010. *Advanced Methods of Structural Analysis*. Springer, New York.
Kassimali, A. 2009. *Structural Analysis*. Cengage Learning Publishing, Boston, MA.
Leet, K., Uang, Chia-Ming, and Lanning, J. 2002. *Fundamentals of Structural Analysis*. McGraw-Hill, Chichester, New York.
Livesley, R.K. 2013. *Matrix Methods of Structural Analysis*. Pergamon International Library of Science, Elsevier, UK.
Madav, J.A., Chandrasekaran, S., and Singh, S. 2024. *Structures for Architects: Planning, Analysis and Design*. CRC Press, FL.
Mau, S.T. 2012. *Introduction to Structural Analysis: Displacement and Force Methods*. CRC Press, Boca Raton, FL.
McCormick, J.C. 2007. *Structural Analysis Using Classical and Matrix Methods*. Wiley, UK.
McGuire, W. and Gallagher, R.H. 1979. *Matrix Structural Analysis*. Wiley, New York.
McGuire, W., Gallagher, R.H., and Ziemian, R.D. 2000. *Matrix Structural Analysis*. Bucknell University Press, Lewisburg, PA.
McKenzie, William M.C. 2006. *Examples in Structural Analysis*. CRC Press, London.
Meek, J.L.1971. *Matrix Structural Analysis*. McGraw-Hill, New York.
Meyers, V.J. 1983. *Matrix Analysis of Structures*. Harper and Row, New York.
Patil, B.I., Chandrasekaran, S., Prasad, M., and Saengsupavanich, C. 2024a. Energy harvest on TSUSUCA DOLPHIN under irregular waves: Experimental studies. *Energy*, 299:131379.
Patil, B.I., Chandrasekaran, S., Prasad, M., and Saengsupavanich, C. 2024b. Dynamic Response Analysis of Bio-Inspired Tsusuca Dolphin. In: *Proceedings of 43rd International Conference on Ocean, Offshore and Arctic Engineering*, Paper No. OMAE2024-122083, Singapore EXPO, Singapore, June 9–14.
Pérez López, C. 2014. *MATLAB Programming for Numerical Analysis*. Apress Berkeley, CA.
Pradeepa, M. and Chandrasekaran, S. 2024. Analysis of Offshore Topside under Hydrocarbon Fire. In: *Proceeding of International Conference on Ships and Offshore Structures (ICSOS 2024)*, IIT Madras, Chennai, India, Sep 8–12.
Pradeepa, M. and Chandrasekaran, S. 2024. Modelling the influence of hydrocarbon fire on offshore topside. *Journal of Mechanical Science and Technology*, 38(11):5927–5959. https://doi.org/10.1007/s12206-024-1013-0
Przemieniecki, J.S. 1985. *Theory of Matrix Structural Analysis*. Courier Corporation, North Chelmsford, NY.
Reddy, C.S. 2011. *Basic Structural Analysis*. Tata Mc Graw-Hill, India.
Rubinstein, M.F. 1970. *Structural Systems-Statics, Dynamics, and Stability*. Prentice-Hall, Englewood Cliffs, NJ.
Sack, R.L. 1994. *Matrix Structural Analysis*. Waveland Press Inc., Sennett.
Siauw, T. and Bayen, A.M. 2014. *An Introduction to MATLAB Programming and Numerical Methods*. Elsevier.
Smith David, M. 2010. *Engineering Computation with MATLAB*, 2nd ed. Addison/Wesley, Boston, MA.

Unnikrishnan, A., Panindre, A., and Chandrasekaran, S. 2024. Stability Analysis of Offshore Triceratops under Tether Failure Condition in Adverse Sea States. In: *Proceedings of International Conference on Ships and Offshore Structures (ICSOS 2024)*, IIT Madras, Chennai, India, Sep 8–12.

Vanderbilt, M. D. 1974. *Matrix Structural Analysis.* Quantum, New York.

Wang, C.K. 1970. *Matrix Methods of Structural Analysis*, 2nd ed. Scranton, Pennsylvania, PA.

Wang, Chu-Kia. 1953. *Statically Indeterminate Structures.* McGraw Hill, New York.

Weaver, W. and James, M. 2012. *Matrix Analysis of Framed Structures.* Springer Science & Business Media, New York.

Wilson, Edward L. 2002. *Three-dimensional Static and Dynamic Analysis of Structures: A Physical Approach with Emphasis on Earthquake Engineering.* Computers and Structures Inc.

Index

Note: Page numbers in **bold** and *italics* refer to tables and figures, respectively.

A

Angle of rotation (θ), 90
Anticlockwise moment, 37, 254
Applied bending moment, 212
Arbitrary end displacements, 32
Artificial intelligence (AI) tools, 7
Axes system, 129, 131
Axial deformation, 30, 36, 50, 86, 250
  beams with, *87*
Axial displacements, 85
  degrees of freedom, 85
  modular beam with, *86*

B

Beam(s)
  with axial deformation, *87*
  elements, 160
  with rotation, **48**
    simply supported beam, **48**
  under unit moment, *36*, *38*, *41–42*
  with varying depth, *39*
  with varying moments of inertia, *40*
  with settlement of support, **48**
Beam element, 249–255
  under axial tensile load, 279–280
  rotational and translational displacements of, *31*
  standard, *30*, 30–36, *31*
Bending, 211
Bending axis, 39
Bending moment, 49–50, 248
Bending moment diagram (BMD), 46, 49–50
Bending stresses, 209, 212
Bio-inspired geometric forms and simulation, 3, 5–7, *6*
  design overview, 5
  energy efficiency, 6
  load distribution, 6
  resilience, 6
  sustainability, 6
Bridge, *9*
Bridges, structural analysis, 8
Buckling and instability, 243–247
Buildings, structural analysis, 8

C

Cantilever, 11
Cargo vessel with idealized fixed beams, *9*
Clockwise couple balances, 37
Cofactor matrix, 25
Compatibility equations, 19
Compatibility of deformations, 19
Compressive forces, 242
Compressive stress, 211
Conjugate beam, *38*
Constant flexural rigidity (EI), 30
Continuous beam, 11–12, *20*, 20–21
  with fixed support, *13*
Crane hooks, 234–241
Critical buckling load, 294–295
Critical load, 248
Critical structural systems, 3
Curved beams, 225
  large initial curvature, 229–233
  simplified equations to estimate stresses, 233–234
  small initial curvature, 226–229
Curved beam with small initial curvature, 226

D

Deflection, 211
  of curved beam with small initial curvature, 228
Deformation, 230
Degree of kinematic indeterminacy, 11
Degree of static indeterminacy, 10, 12
Degrees of freedom (DOF), 11, 31, 33, 36, 44, 51, 60, 86, 88, 90, 94, 251
  axial deformation neglected, *15*
  axial displacements, 85
  fixed beam under unit rotation, *37*
  kinematic degrees of freedom, 13–18
  and labeling, 129
  in local axes system, *161*
  modular fixed beam with, *44*
  of non-orthogonal frame, *88*
  planar truss, 129, *129*
  restrained, 131
  of single bay single-story frame, *45*
  static degrees of freedom, 11–13

unrestrained, 131
unrestrained and restrained, 60, *61*
Direction cosines, 166, 168
Displacement method, 19
Displacements, 66
Distortional buckling, 245
DOF, *see* Degrees of freedom
Dynamic loading, 8

E

End moments, 66–68, 98, 180
due to joint displacements, **48**
and shear, 54
End shear, 33
Environmental loads, 8
Equation of equilibrium, 19
Equations of static equilibrium, 7
Euler column, *247*
Euler critical load, 247–249
Euler's criterion, 242–243
Extrados, 229

F

FEM, *see* Fixed end moments
Finite element analysis (FEA) software, 19
Fixed beams, 8, 21, *21*, 30
bridges, 8–10
buildings, 8
with central clockwise moment, **47**
with central concentrated load, **47**
with eccentric anticlockwise moment, **47**
with eccentric concentrated load, **47**
frames and 3D space structures, 10
ships and offshore structures, 10
with triangular loading, **47**
with uniformly distributed load, **47**
with uniformly varying load, **47**
under unit displacement, *32–33*
under unit rotation, *31–32*, *37*
Fixed end moments (FEM), **47**, 52–53, 63–65, 93
Fixed support beam, 11, *11*
Flexibility approach, 20
Flexibility coefficients, 35–36
Flexibility matrix, 19, 37, 42
of beam element, 41
Flexibility method, 18–20
Flexural buckling (FB), 245
Flexural rigidity (EI), 39
Flexural-torsional buckling (FTB), 245
Floating offshore platforms, 2
Force-displacement relationships, 19, 93
Force method, 19
Force reduction, 5
Form-dominant approach, 2
Frame, *16*, 22, *22*
stepped frame, *17*
Free-body diagram of column member, 248, *248*
Functionally graded materials (FGM), 5

G

Geometric forms, architecture, 1–2
Geometric stability, 242
Global and local axis system, *86*, *88*
Global axes, 94, 129, *130*
Global stiffness matrix, 97, 135

H

Harvesting systems, 8
High-magnitude pretension, 243
Hinged support beam, 11, *11*
Hooke's Law, 7
Hydrodynamic stability, 242

I

Indeterminacy, 10
degrees of freedom (DOF), 11
kinematic degrees of freedom, 13–18
static degrees of freedom, 11–13
kinematic indeterminacy, 10–11
static indeterminacy, 10
Inertia, 209
Intrados, 229
Inverse of matrix, 24–25

J

Joint equilibrium of the planar structure, 160
Joint loads, 46
Joint load vectors, 53–54, 65, 96, 132, 137, 150–151
Joint translation, 129

K

Kinematic degrees of freedom, 13, **14**
axial deformation neglected, **18**
Kinematic indeterminacies, 10–11, 20

L

Lateral load functions under concentrated load, 280–283
Lateral load functions under uniformly distributed load, 274–279
Lateral-torsional buckling (LTB), 245
of I-section, *246*
Liapunov's condition, 243
Liapunov's dynamic criterion, 242

Linear equations, 23–24
Load-bearing elements, 8
Load-carrying capacity, 242, 244
Load distribution in trusses and frames, 8
Loads, 211
Load vector, 131–132
Local axes system, 94, 129, *130*, *176*
Local stiffness matrix, 95, 134
Local-torsional buckling, 245

M

Matix operations, 23
MATLAB, 26–27
- actions, 27
- arithmetic operations, 27
- capabilities, 27
- 3D analysis of space frame, 190
- non-orthogonal frame analysis, 100
- relational and logical operators, 27–28
- stiffness matrix, 68–72, 79–84, 100, 139
- syntax and operations, 27
- unsymmetrical bending, 216

Matrix algebra
- inverse of matrix, 24–25
- linear equations, 23–24
- matix operations, 23
- solution of linear equations, 26

Matrix stiffness matrix, 87
Maximum load capacity, 233
Member forces, 137, 151
Member index, 250
Member rotation, *167*
Member rotation matrix, 167–168
Modern and contemporary forms, architecture, 2–3
- outrigger structures, 4–5
- tube-in-tube structures, 3–4

Modification factor, 246
Modular beam with axial displacement, *86*
Modular fixed beam with degrees of freedom, *44*
Moment of inertia, 39, 212
Moments, 211

N

Neutral axis, 212, 217, 226, 229
Non-orthogonal frame, 86
Non-orthogonality, 85
Nontrivial equilibrium, 242
Nonuniform bending, *see* Unsymmetric bending

O

Offshore platform, *9*
Orthogonal frame, 85
Outrigger structures, 3, *4*, 4–5
- applications, 5
- design overview, 4
- lateral stiffness, 4–5
- reduction of drift, 5

P

Planar non-orthogonal structure, 85–89
- load vector and end moments, 93–94
- MATLAB, example problems using, 94–128
- transformation matrix, 89–90
- transformation of end moments, reactions, and displacements, 90–93

Planar orthogonal frame, example problem, 60–72
Planar orthogonal structures, 43–46
- beam element with varying flexural rigidity, 39–43
- computer-aided analysis, 50
- continuous beams
  - example problem, 50–56
  - shear force and bending moment diagrams, 56–59
- frame, example problem, 60–72
- rotation coefficients, estimation, 36
- shear force and bending moment, 46–49
  - bending moment diagram (BMD), 49–50
  - steps to draw shear force diagram (SFD), 49–50
- standard beam element, 30–36
- step frame example problem, 73–84
- stiffness matrix of beam element, 36–39

Planar truss structures, 129
- analysis using MATLAB, 132
- computer-aided analysis, framework, 132
- degree of freedom and labeling, 129
- element stiffness matrix, 131
- example problems with a computer program, 133–159
- global stiffness matrix, 131
  - axes systems, 131
- load vector, 131–132
- transformation matrix, 129–130

Potential energy stability criterion, 242–243
Ψ angle, 168, 174–176

R

Reference axes system, 165, *176*
Restrained degrees-of-freedom, 131, 133
Rotational coefficients, 34–36, 43, 254–259
Rotation coefficients, 39
Rotation functions
- under axial compression load, 259–262
- under axial tensile load, 264–265
- under no axial load, 262–264

S

Shear, 98
Shear center, 245
Shear force, 49, 66–68
Shear force diagram (SFD), 46, 49–50
Simply supported beam, 11, *12*, 21, *21*
Small initial curvature, 226
Solution of linear equations, 26
Stability analysis, 247
Stability analysis element, 249
Stability calculations of ships, 243
Stability functions, 259
 rotation functions under axial compression load, 259–262
 rotation functions under axial tensile load, 264–265
 rotation functions under no axial load, 262–264
 transfer function under axial compressive load, 265–274
Stability of structural systems
 beam element, 249–255
  under axial tensile load, 279–280
 buckling and instability, 243–247
 conditions for stability, 242–243
 critical buckling load, 294–295
 Euler critical load, 247–249
 example problems with computer programs, 284–294
 exercise problems with computer programs, 296–312
 lateral load functions under concentrated load, 280–283
 lateral load functions under uniformly distributed load, 274–279
 rotational coefficients, 255–259
 stability functions, 259
  rotation functions under axial compression load, 259–262
  rotation functions under axial tensile load, 264–265
  rotation functions under no axial load, 262–264
  transfer function under axial compressive load, 265–274
Stable equilibrium, 248
Standard beam
 element, *250*
 rotational and translational moments, *250*
 rotation coefficients, *253*
Standard beam element, *30*, 30–36, *31*
Standard reference axes system (X-Y-Z), 166
Standard stiffness matrix, 131
Statically determinate model, 7
Statically indeterminate model, 7
Static buckling load, 247
Static degree of indeterminacy, 13
Static indeterminacy, 10
Steel structure, *9*
Step frame example problem, 73–78
Stepped frame, *17*
Stiffness approach, 20
Stiffness coefficients, 31–33, 35–37, 162
 for unit rotation, 34, *34*
Stiffness matrix, 19, 36, 39, 43, 51–52, 86, 162, 165, 249
 approach, 85
 formulation, 62
Stiffness matrix of 3D beam element, *164*
Stiffness method, 18–19
Stress correction factors, 233–234
Stress distribution, 212
Stresses, 211, 230, 248
Stress variation, 229
Structural analysis and design, 7
 fixed beams, 8–10
 importance of, 7–8
  buildings and bridges, 8
  ships and offshore structures, 8
 methods of analysis, 18–19
  flexibility method, 18–20
  stiffness method, 18–19
 statically determinate model, 7
 statically indeterminate model, 7
Structural forms and functions, 1
 evolution of, 2–3
 geometric forms, 1–2
 modern and contemporary forms, 2–3
  outrigger structures, 4–5
  tube-in-tube structures, 3–4
Structural health monitoring (SHM), 7
Structural stability, 243
Structural systems, 244
Symmetrical bending, 207

T

Tensile/compressive stress, 232
Tensile stress, 211
Three-dimensional analysis of structures, 160
 analysis of space frame, 176–179
 beam element, 161–162
 member rotation matrix, 167–168
 problems solved using MATLAB program, 179–206
 sign convention, 160–161
 space frame, framework of analysis of, 179
 stiffness matrix, 162–165
 transformation matrix, 165–167
 Y-Z-X transformation, 168–172
 Z-Y-X transformation, 172–174
 Ψ angle, 174–176
Torsional buckling (TB), 245

Torsional constant, 163–164
Total stiffness matrix, 136, 150, 180
Transfer function under axial compressive load, 265–274
Transformation matrix, 88–90, 96, 135
Transformation matrix coefficients, 94
Transformation of member geometry, *92*
Tube-in-tube structures, *3*, 3–4
  application, 4
  design overview, 4
  efficiency and cost-effectiveness, 4
  lateral load resistance, 4
Two axes systems, 129
Two-span continuous beam, *15*

U

Uniform bending, 207
Uniplanar bending, 211–212
Unit axial displacement, 87
Unrestrained degrees-of-freedom, 131, 133; *see also* Degrees of freedom
Unrestrained displacements, 54
Unstable structural system, 244
Unsymmetric bending, 207–211
  crane hooks, 234–241
  curved beams, 225
    large initial curvature, 229–233
    simplified equations to estimate stresses, 233–234
    small initial curvature, 226–229
  exercise problems, 212–225
  uniplanar bending, 211–212

W

Wave energy, 2
Winkler Bach equation, 232

Y

Y-Z-X transformation, 168–172, *175*

Z

Z-axis rotation, *170*
Zoning rules, 1
Z-Y-X transformation, 172–174, *175*